Plant Extracts Used to Control Microbial Growth: Efficacy, Stability and Safety Issues for Food Applications

Plant Extracts Used to Control Microbial Growth: Efficacy, Stability and Safety Issues for Food Applications

Guest Editors

Loris Pinto

Jesus Fernando Ayala-Zavala

MDPI

Basel • Beijing • Wuhan • Barcelona • Belgrade • Novi Sad • Cluj • Manchester

Guest Editors

Loris Pinto
Institute of Sciences of Food Production
National Research Council of Italy (CNR-ISPA)
Bari
Italy

Jesus Fernando Ayala-Zavala
Coordinación de Tecnología de Alimentos de Origen Vegetal
Centro de Investigación en Alimentación y Desarrollo (CIAD)
Hermosillo
Mexico

Editorial Office

MDPI AG
Grosspeteranlage 5
4052 Basel, Switzerland

This is a reprint of the Special Issue, published open access by the journal *Foods* (ISSN 2304-8158), freely accessible at: https://www.mdpi.com/journal/foods/special_issues/plant_extracts_control_microbial.

For citation purposes, cite each article independently as indicated on the article page online and as indicated below:

Lastname, A.A.; Lastname, B.B. Article Title. *Journal Name* **Year**, *Volume Number*, Page Range.

ISBN 978-3-7258-7767-6 (Hbk)
ISBN 978-3-7258-7768-3 (PDF)
https://doi.org/10.3390/books978-3-7258-7768-3

Contents

About the Editors

Loris Pinto

Loris Pinto graduated in Food Science and Technology and received his PhD in Food Microbiology from the University of Bari (Italy). Since 2020, he has been a Researcher at the National Research Council of Italy (CNR)—Institute of Sciences of Food Production (ISPA), Bari. His main research area is food microbiology with expertise in the characterization of pro-technological and spoilage bacteria, and the use of antimicrobial compounds against spoilage bacteria and foodborne pathogens. He is the co-author of over 60 scientific contributions (articles and scientific communications at national and international congresses).

Jesús Fernando Ayala-Zavala

Jesús Fernando Ayala-Zavala is a Professor and Researcher at the Centro de Investigación en Alimentación y Desarrollo (CIAD), Hermosillo, Mexico, within the Department of Plant-Based Food Technology. He holds a PhD in Food Science and has extensive experience in food chemistry, food biotechnology, food microbiology, and emerging technologies. His main research interests include bioactive compounds from plants (phenolics, flavonoids, and essential oils), their antioxidant, antimicrobial, and antibiofilm properties, and their application in food quality, safety, and shelf-life extension. He also integrates computational approaches into his research to explore mechanisms of action at the molecular level. He has authored or co-authored more than 227 scientific contributions, including journal articles, book chapters, and conference proceedings, and has participated in several national and international research collaborations.

foods MDPI

Editorial

Application of Plant Antimicrobials in the Food Sector: Where Do We Stand?

Loris Pinto [1,*] and Jesús Fernando Ayala-Zavala [2]

1 Institute of Sciences of Food Production, National Research Council of Italy, Via G. Amendola 122/O, 70126 Bari, Italy
2 Centro de Investigación en Alimentación y Desarrollo, A.C., Carretera Gustavo Enrique Astiazarán Rosas 46, Hermosillo 83304, Sonora, Mexico; jayala@ciad.mx
* Correspondence: loris.pinto@ispa.cnr.it

Abstract: The Special Issue "Plant Extracts Used to Control Microbial Growth: Efficacy, Stability and Safety Issues for Food Applications" explored the potential of plant-based extracts as natural antimicrobial agents in the food industry. Its purpose was to address the growing demand for natural, safe, and effective food preservation methods. The contributions highlighted various plant extracts' antimicrobial efficacy, including phenolic compounds, terpenes, and other bioactive substances. Research papers and one review were submitted from countries, including Spain, Portugal, Italy, Mexico, Turkey, India, USA, Romania, China, and Lithuania, showcasing a diverse international collaboration. Key topics covered in this issue included the chemical characterization of plant extracts, their stability under different processing and storage conditions, and their safety assessments. Advances were reported in using plant extracts to inhibit spoilage microorganisms and foodborne pathogens, enhance food safety, and extend shelf life. The published papers in the Special Issue studied various food types, including yogurt, catfish fillets, edible Mushrooms, red grapes, herring Fillets, and various food types covered in the review. This diversity demonstrates the broad applicability of plant extracts across different food products. Notable findings included the antimicrobial activities of fermented grapevine leaves, grapefruit seed extract, cinnamaldehyde, clove oil, and other plant-based compounds. In conclusion, this Special Issue demonstrated significant progress in applying plant extracts for food preservation, highlighting their potential to contribute to safer and more sustainable food systems worldwide.

1. Utilizing the Power of Plant Antimicrobials for Food Safety

The use of natural preservatives with antimicrobial properties has gotten significant interest in food and drug research due to the growing awareness of the negative impacts associated with synthetic preservatives. These impacts include potential health risks to consumers, the emergence of multidrug-resistant microorganisms, and the requirement for alternatives to traditional thermal treatments [1–3]. Plant antimicrobials offer a promising solution as they are rich sources of multiple bioactive compounds capable of reducing contamination levels of pathogenic bacteria and inhibiting the growth of spoilage microorganisms in various foods. The compounds found in plant antimicrobials are diverse and include polyphenols known for their antioxidant properties; polyphenols also exhibit strong antimicrobial activity [4,5]. They can disrupt microbial cell membranes and interfere with the functions of microbial enzymes and proteins [6,7]. Essential oils and their constituents, such as carvacrol, thymol, and eugenol, possess great antimicrobial properties [8–10]. Essential oils can penetrate microbial cell membranes, causing structural and functional damage [11,12]. Glucosinolate derivatives found in cruciferous vegetables have been shown to possess antimicrobial activity against a broad spectrum of microorganisms [13]. They act by releasing isothiocyanates upon hydrolysis, which are toxic to bacteria [14]. Alkaloids are nitrogen-containing compounds with antimicrobial properties that can inhibit the growth of bacteria, fungi, and viruses [15]. Alkaloids can interfere with DNA replication and protein synthesis in microorganisms and can attenuate bacterial pathogenesis [15,16]. Thiols are sulfur-containing compounds that exhibit strong antimicrobial activity [17]. They can disrupt microbial cell walls and membranes, leading to cell lysis and death.

The objective of the Special Issue entitled "Plant Extracts Used to Control Microbial Growth: Efficacy, Stability and Safety Issues for Food Applications" was to present the latest advances in the use of plant antimicrobials to reduce spoilage microorganisms and ensure food safety across different foodstuffs. The papers in this issue highlight the potential of plant extracts as natural preservatives that can be integrated into various food systems to enhance safety and extend shelf life.

2. Summary of Published Papers

The main studied sources of plant antimicrobials were the herbal plant *Potentilla kleiniana*, cranberry pomace and grape seeds, essential oils and their compounds, bog bilberry leaf extracts, spice extracts, grapefruit seed extract, and grapevine leaf extracts (Table 1). Among the compounds in the used raw extracts were identified phenolic acids, flavonoids, and terpenes from plant tissues.

Table 1. Plant antimicrobial compounds, sources, target food/microorganisms, and main results were obtained in the published papers in this special issue.

Plant Antimicrobial Compounds	Source	Target Food/Microorganisms	Main Results	Contribution
Oxymorphone and rutin	Methanol-phase extract from an edible herb *Potentilla kleiniana* Wight et Arn	More than 20 pathogenic bacteria	Inhibition rate of 68%, MIC values of 1.56–50 mg mL^{-1}, putative mechanism of action	[1]
Polyphenols and procyanidins	Cranberry pomace and grape seed extracts	Herring/*Listeria monocytogenes* and *Pseudomonas aeruginosa*	Film with grape seed extract showed bacteriostatic activity against *L. monocytogenes* and reduced the concentration of histamine and cadaverine	[2]
Carvacrol, thymol, geraniol, citral, L-menthol, menthone, anisaldehyde, linalool, citronellal, trans-2-hexenal, diallyl disulfide, trans-caryophyllene, piperone, eugenol, and anethole	RON Reagent Shanghai Yi En Chemical Technology Co., Ltd., Shanghai, China	Red grape fruit/*Alternaria alternata*	Significant reduction of the decay rate after carvacrol treatment	[3]
Phenolic acids, flavonols, flavanols	Leaf extracts of the bog bilberry	Gram-positive and Gram-negative bacteria, yeasts	MIC values of 8.9 or 17.8 mg mL^{-1}	[4]
Polyphenolic compounds	Sumac (*Rhus coriaria* L.), cumin (*Cuminum cyminum* L.), black pepper (*Piper nigrum*), and red pepper (*Capsicum annuum*) extracts	Gram-positive and Gram-negative bacteria	Sumac extract reduced the growth of foodborne pathogens and the production of biogenic amines	[5]

Table 1. *Cont.*

Plant Antimicrobial Compounds	Source	Target Food/Microorganisms	Main Results	Contribution
Flavonoids	Grapefruit seed extracts	Mushroom/Gram-positive and Gram-negative bacteria, yeast	MIC values from 162.5 μg mL^{-1} to 650 μg mL^{-1}, potential application to reduce yellowing on mushrooms	[6]
Cinnamaldheydhe and clove oil	Sigma-Aldrich (St. Louis, MO, USA) and Piping Rock Health Products LLC (Ronkonkoma, NY, USA)	Catfish fillet/*Shewanella baltica*, *Aeromonas hydrophila,* total bacteria	Reduction of 3 or 6 log cfu mL^{-1} of total bacteria on adsorbent pads	[7]
Polyphenols	Fermented grapevine leaves	Total yeasts and bacteria of yogurt	Fermented grapevine leaves showed a preserving effect equal to potassium sorbate	[8]

Tang et al. evaluated the antibacterial activity of the methanol phase extract from the edible herb *Potentilla kleiniana* against more than 20 pathogenic bacteria. MIC values of Fragment 1 ranged from 6.25 to 50 mg/mL against *Bacillus cereus, Shigella flexneri, Staphylococcus aureus,* and *Vibrio parahemolyticus* strains. Oxymorphone and rutin were identified in Fragment 1, and a putative mechanism of action involving the inhibition of energy supply and protein translation, the blocking of signal transduction, and the repression of ABC transporters was proposed [Contribution 1]. An alginate/pectin film containing grape seed extract showed bacteriostatic activity against *L. monocytogenes* on herring, reducing its load by 3 log cfu/g compared to unpacked fillets stored for 18 days at 4 °C. In addition, the accumulation of histamine, cadaverine, putrescine, and tyramine was significantly reduced starting from 12 days of storage in fillets packed using the active coating [Contribution 2].

Li et al. found that carvacrol displayed the lowest EC50 value against *A. alternata*, significantly reducing the decay rate when applied to contaminated red grapes. Grapes inoculated with *A. alternata* showed a decay rate higher than 60%, whereas contaminated fruit treated with carvacrol showed a decay rate lower than 15% after 12 days at room temperature [Contribution 3]. Bog bilberry leaf extracts obtained through ultrasound (UAE) extraction showed the lowest MIC values against *Candida parapsilosis* and *Salmonella enterica*; high-pressure (HPE) extracts showed the inhibition of *S. aureus* growth at sub-MIC levels [Contribution 4]. Kuley et al. demonstrated that, among four spice extracts, sumac (*Rhus coriaria* L.) extract reduced by 1–3 log cfu/mL the growth of *Enterococcus faecalis, Campylobacter jejuni,* and *Yersinia enterocolitica* in tyrosine decarboxylase broth and the production of histamine by *E. faecalis* and tyramine by *C. jejuni* [Contribution 5].

Grapefruit seed extract showed antibacterial action against *S. aureus* ATCC 6538, *P. aeruginosa* ATCC 9027, *P. fluorescens* wild type, *Escherichia coli* ATCC 8739, with MIC values ranging from 162.5 μg/mL to 650 μg/mL. Rutin, naringin, hesperidin, neohesperidin, and naringenin were identified in the extract. A preliminary trial on mushrooms showed that applying grapefruit seed extract can limit the development of yellowing on *Pleurotus eryngii* caused by *Pseudomonas* spp. [Contribution 6]. Cinnamaldehyde and clove oil showed in vitro antibacterial activity against *S. baltica* and *A. hydrophila*; their application on adsorbent pads in contact with catfish fillets reduced the total bacteria on pads by 3 to 6 log cfu/mL [Contribution 7].

Freitas et al. found that fermented grapevine leaves using *Saccharomyces cerevisiae*, in both solid and liquid media, preserved the microbiological quality of yogurt in the same manner as potassium sorbate without affecting the viability of lactic acid bacteria.

Further research is necessary to evaluate the antimicrobial effect of fermented grapevine leaves against yogurt spoilage microorganisms [Contribution 8]. Finally, Pinto et al. summarized recent findings on applying plant extracts and plant antimicrobial compounds against spoilage and pathogenic microorganisms in different foods. Interesting results were achieved by using combinations of plant antimicrobials, with synergistic or additive effects, and by integrating plant extracts with food technologies, ensuring an improved hurdle effect. The review highlighted the need for further research in fields such as the mode of action of plant antimicrobials, optimization of delivery systems, sensory properties of food including plant antimicrobial compounds, safety assessment of plant extracts, regulatory aspects, eco-friendly production methods, and consumer education [18].

3. Key Advances and Findings

The papers published in this issue highlighted several key advances and findings, demonstrating the potential of plant-based compounds to enhance food safety and quality. The methanol-phase extract from *Potentilla kleiniana* exhibited a 68% inhibition against over 20 pathogenic bacteria, with MIC values ranging from 1.56 to 50 mg/mL. This study identified oxymorphone and rutin as active components and proposed mechanisms involving inhibiting energy supply, protein translation, signal transduction, and repression of ABC transporters. Alginate/pectin films containing cranberry pomace and grape seed extracts showed bacteriostatic activity against *Listeria monocytogenes* on herring, significantly reducing the load of histamine and cadaverine during storage. Carvacrol was effective against *Alternaria alternata* in red grapes, reducing the decay rate to less than 15% compared to over 60% in untreated controls after 12 days at 25 °C. Bog bilberry leaf extracts obtained through ultrasound extraction showed the lowest MIC values against *Candida parapsilosis* and *Salmonella enterica*, with high-pressure extracts inhibiting *Staphylococcus aureus* at sub-MIC levels. Sumac extract significantly reduced the growth of several foodborne pathogens, including *Enterococcus faecalis*, *Campylobacter jejuni*, and *Yersinia enterocolitica*, and decreased the production of biogenic amines.

Grapefruit Seed Extract demonstrated broad-spectrum antibacterial activity against multiple strains, including *S. aureus*, *P. aeruginosa*, *P. fluorescens*, and *E. coli*, with MIC values between 162.5 µg/mL and 650 µg/mL. Additionally, it showed potential in preventing yellowing in mushrooms caused by *Pseudomonas* spp. Cinnamaldehyde and clove oil applied on absorbent pads in contact with catfish fillets reduced total bacterial counts by 3 to 6 log cfu/mL. Fermented grapevine leaves used in yogurt maintained microbial quality similar to potassium sorbate without compromising the viability of lactic acid bacteria.

Several studies focused on improving the stability of plant extracts under different processing and storage conditions. Techniques such as encapsulation [19–21], inclusion in biopolymers [22–24], and spray-drying [25–27] were explored to enhance the stability and effectiveness of plant antimicrobials. The research covered a wide range of food products, demonstrating the versatility of plant antimicrobials. These included yogurt, herring fillets, red grapes, catfish fillets, and mushrooms. This diversity illustrates the broad applicability of plant extracts across different food matrices, contributing to enhanced food safety and shelf life. Combining different plant extracts often resulted in synergistic or additive antimicrobial effects. This approach and integration into food technologies provided an improved hurdle effect, enhancing overall food preservation outcomes.

4. Future Research Directions

The Special Issue highlighted the need for further research in several key areas. Detailed studies are necessary to understand how plant antimicrobials exert their effects, focusing on their mode of action. Additionally, there is a need for the optimization of delivery systems to develop effective methods that maximize the efficacy of plant extracts. Investigating the impact of plant antimicrobials on the sensory attributes of food is crucial to ensure that these natural preservatives do not negatively affect taste, texture, or aroma. Comprehensive safety evaluations are required to ensure consumer health is not compromised. Addressing regulatory challenges will facilitate the commercial use of plant antimicrobials, ensuring they meet all necessary standards and guidelines. Developing sustainable production techniques for plant extracts is essential to promote eco-friendly methods that align with environmental goals. Lastly, increasing awareness and acceptance of natural food preservatives among consumers through effective education campaigns is vital for broader adoption and understanding of these innovations.

5. Conclusions

This Special Issue has significantly contributed to the field of food microbiology by advancing our knowledge of plant antimicrobials and their applications. The findings emphasize the potential of these natural compounds to revolutionize food preservation, paving the way for safer, more sustainable food systems that meet the growing demand for natural and health-friendly food additives.

Author Contributions: Writing—original draft preparation, L.P., J.F.A.-Z.; writing—review and editing, L.P., J.F.A.-Z. All authors have read and agreed to the published version of the manuscript.

Acknowledgments: The Editors would like to thank all contributing authors, the Editor-in-Chief, and the journal staff for their support.

Conflicts of Interest: The authors declare no conflicts of interest.

List of Contributions:

1. Tang, Y.; Yu, P.; Chen, L. Identification of Antibacterial Components and Modes in the Methanol-Phase Extract from a Herbal Plant Potentilla kleiniana Wight et Arn. *Foods* **2023**, *12*, 1640. https://doi.org/10.3390/foods12081640.
2. Urbonavičiūtė, G.; Dyglė, G.; Černauskas, D.; Šipailienė, A.; Venskutonis, P.R.; Leskauskaitė, D. Alginate/Pectin Film Containing Extracts Isolated from Cranberry Pomace and Grape Seeds for the Preservation of Herring. *Foods* **2023**, *12*, 1678. https://doi.org/10.3390/foods12081678.
3. Li, H.; Ding, J.; Liu, C.; Huang, P.; Yang, Y.; Jin, Z.; Qin, W. Carvacrol Treatment Reduces Decay and Maintains the Postharvest Quality of Red Grape Fruits (*Vitis vinifera* L.) Inoculated with *Alternaria alternata*. *Foods* **2023**, *12*, 4305. https://doi.org/10.3390/foods12234305.
4. Ştefănescu, B.E.; Socaci, S.A.; Fărcaș, A.C.; Nemeș, S.A.; Teleky, B.E.; Martău, G.A.; Călinoiu, L.F.; Mitrea, L.; Ranga, F.; Grigoroaea, D.; et al. Characterization of the Chemical Composition and Biological Activities of Bog Bilberry (*Vaccinium uliginosum* L.) Leaf Extracts Obtained via Various Extraction Techniques. *Foods* **2024**, *13*, 258. https://doi.org/10.3390/foods13020258.
5. Kuley, F.; Rathod, N.B.; Kuley, E.; Yilmaz, M.T.; Ozogul, F. Inhibition of Food-Borne Pathogen Growth and Biogenic Amine Synthesis by Spice Extracts. *Foods* **2024**, *13*, 364. https://doi.org/10.3390/foods13030364.
6. Murgia, M.; Pani, S.M.; Sanna, A.; Marras, L.; Manis, C.; Banchiero, A.; Coroneo, V. Antimicrobial Activity of Grapefruit Seed Extract on Edible Mushrooms Contaminations: Efficacy in Preventing *Pseudomonas* spp. in *Pleurotus eryngii*. *Foods* **2024**, *13*, 1161. https://doi.org/10.3390/foods13081161.
7. Ebirim, R.I.; Long, W., III. Evaluation of Antimicrobial and Preservative Effects of Cinnamaldehyde and Clove Oil in Catfish (*Ictalurus punctatus*) Fillets Stored at 4 °C. *Foods* **2024**, *13*, 1445. https://doi.org/10.3390/foods13101445.
8. Freitas, L.; Sousa-Dias, M.; Paula, V.B.; Dias, L.G.; Estevinho, L.M. Fermented Grapevine Leaves: Potential Preserving Agent in Yogurt. *Foods* **2024**, *13*, 2053. https://doi.org/10.3390/foods13132053.

References

1. Pisoschi, A.M.; Pop, A.; Georgescu, C.; Turcuş, V.; Olah, N.K.; Mathe, E. An overview of natural antimicrobials role in food. *Eur. J. Med. Chem.* **2018**, *143*, 922–935. [CrossRef] [PubMed]
2. Batiha, G.E.S.; Hussein, D.E.; Algammal, A.M.; George, T.T.; Jeandet, P.; Al-Snafi, A.E.; Tiwari, A.; Pagnossa, G.P.; Lima, C.M.; Thorat, N.D.; et al. Application of natural antimicrobials in food preservation: Recent views. *Food Control* **2021**, *126*, 108066. [CrossRef]
3. Quinto, E.J.; Caro, I.; Villalobos-Delgado, L.H.; Mateo, J.; De-Mateo-Silleras, B.; Redondo-Del-Río, M.P. Food Safety through Natural Antimicrobials. *Antibiotics* **2019**, *8*, 208. [CrossRef] [PubMed]

4. Bouarab Chibane, L.; Degraeve, P.; Ferhout, H.; Bouajila, J.; Oulahal, N. Plant antimicrobial polyphenols as potential natural food preservatives. *J. Sci. Food Agric.* **2019**, *99*, 1457–1474. [CrossRef] [PubMed]
5. Gerardi, C.; Pinto, L.; Baruzzi, F.; Giovinazzo, G. Comparison of Antibacterial and Antioxidant Properties of Red (cv. Negramaro) and White (cv. Fiano) Skin Pomace Extracts. *Molecules* **2021**, *26*, 5918. [CrossRef] [PubMed]
6. Oulahal, N.; Degraeve, P. Phenolic-Rich Plant Extracts with Antimicrobial Activity: An Alternative to Food Preservatives and Biocides? *Front. Microbiol.* **2022**, *12*, 3906. [CrossRef] [PubMed]
7. Bae, J.Y.; Seo, Y.H.; Oh, S.W. Antibacterial activities of polyphenols against foodborne pathogens and their application as antibacterial agents. *Food Sci. Biotechnol.* **2022**, *31*, 985–997. [CrossRef] [PubMed]
8. Al-Maqtari, Q.A.; Rehman, A.; Mahdi, A.A.; Al-Ansi, W.; Wei, M.; Yanyu, Z.; Phyo, H.M.; Galeboe, O.; Yao, W. Application of essential oils as preservatives in food systems: Challenges and future prospectives—A review. *Phytochem. Rev.* **2021**, *21*, 1209–1246. [CrossRef]
9. Pinto, L.; Cefola, M.; Bonifacio, M.A.; Cometa, S.; Bocchino, C.; Pace, B.; De Giglio, E.; Palumbo, M.; Sada, A.; Logrieco, A.F.; et al. Effect of red thyme oil (*Thymus vulgaris* L.) vapors on fungal decay, quality parameters, and shelf-life of oranges during cold storage. *Food Chem.* **2021**, *336*, 127590. [CrossRef]
10. Pinto, L.; Cervellieri, S.; Netti, T.; Lippolis, V.; Baruzzi, F. Antibacterial Activity of Oregano (*Origanum vulgare* L.) Essential Oil Vapors against Microbial Contaminants of Food-Contact Surfaces. *Antibiotics* **2024**, *13*, 371. [CrossRef]
11. Hou, T.; Sana, S.S.; Li, H.; Xing, Y.; Nanda, A.; Netala, V.R.; Zhang, Z. Essential oils and its antibacterial, antifungal and anti-oxidant activity applications: A review. *Food Biosci.* **2022**, *47*, 101716. [CrossRef]
12. Falleh, H.; Ben Jemaa, M.; Saada, M.; Ksouri, R. Essential oils: A promising eco-friendly food preservative. *Food Chem.* **2020**, *330*, 127268. [CrossRef]
13. Abdel-Massih, R.M.; Debs, E.; Othman, L.; Attieh, J.; Cabrerizo, F.M. Glucosinolates, a natural chemical arsenal: More to tell than the myrosinase story. *Front. Microbiol.* **2023**, *14*, 1130208. [CrossRef] [PubMed]
14. Andini, S.; Araya-Cloutier, C.; Lay, B.; Vreeke, G.; Hageman, J.; Vincken, J.P. QSAR-based physicochemical properties of isothiocyanate antimicrobials against gram-negative and gram-positive bacteria. *LWT-Food Sci. Technol.* **2021**, *144*, 111222. [CrossRef]
15. Yan, Y.; Li, X.; Zhang, C.; Lv, L.; Gao, B.; Li, M. Research progress on antibacterial activities and mechanisms of natural alkaloids: A review. *Antibiotics* **2021**, *10*, 318. [CrossRef]
16. Cushnie, T.P.T.; Cushnie, B.; Lamb, A.J. Alkaloids: An Overview of Their Antibacterial, Antibiotic-Enhancing and Antivirulence Activities. *Int. J. Antimicrob. Agents* **2014**, *44*, 377–386. [CrossRef]
17. Gutiérrez-del-Río, I.; Fernández, J.; Lombó, F. Plant nutraceuticals as antimicrobial agents in food preservation: Terpenoids, polyphenols and thiols. *Int. J. Antimicrob. Agents* **2018**, *52*, 309–315. [CrossRef] [PubMed]
18. Pinto, L.; Tapia-Rodríguez, M.R.; Baruzzi, F.; Ayala-Zavala, J.F. Plant Antimicrobials for Food Quality and Safety: Recent Views and Future Challenges. *Foods* **2023**, *12*, 2315. [CrossRef] [PubMed]
19. Homayonpour, P.; Jalali, H.; Shariatifar, N.; Amanlou, M. Effects of nano-chitosan coatings incorporating with free/nano-encapsulated cumin (*Cuminum cyminum* L.) essential oil on quality characteristics of sardine fillet. *Int. J. Food Microbiol.* **2021**, *341*, 109047. [CrossRef]
20. Oprea, I.; Fărcaș, A.C.; Leopold, L.F.; Diaconeasa, Z.; Coman, C.; Socaci, S.A. Nano-Encapsulation of Citrus Essential Oils: Methods and Applications of Interest for the Food Sector. *Polymers* **2022**, *14*, 4505. [CrossRef]
21. Plati, F.; Paraskevopoulou, A. Micro- and Nano-Encapsulation as Tools for Essential Oils Advantages' Exploitation in Food Applications: The Case of Oregano Essential Oil. *Food Bioprocess Technol.* **2022**, *15*, 949–977. [CrossRef]
22. Chen, L.; Wu, F.; Xiang, M.; Zhang, W.; Wu, Q.; Lu, Y.; Fu, J.; Chen, M.; Li, S.; Chen, Y.; et al. Encapsulation of tea polyphenols into high amylose corn starch composite nanofibrous film for active antimicrobial packaging. *Int. J. Biol. Macromol.* **2023**, *245*, 125245. [CrossRef] [PubMed]
23. Zhou, X.; Liu, X.; Wang, Q.; Lin, G.; Yang, H.; Yu, D.; Cui, S.W.; Xia, W. Antimicrobial and Antioxidant Films Formed by Bacterial Cellulose, Chitosan and Tea Polyphenol–Shelf Life Extension of Grass Carp. *Food Packag. Shelf Life* **2022**, *33*, 100866. [CrossRef]
24. Elshamy, S.; Khadizatul, K.; Uemura, K.; Nakajima, M.; Neves, M.A. Chitosan-based film incorporated with essential oil nanoemulsion foreseeing enhanced antimicrobial effect. *J. Food Sci. Technol.* **2021**, *58*, 3314–3327. [CrossRef] [PubMed]
25. Fernandes, M.R.V.; Dias, A.L.T.; Carvalho, R.R.; Souza, C.R.F.; Oliveira, W.P. Antioxidant and antimicrobial activities of *Psidium guajava* L. spray dried extracts. *Ind. Crops Prod.* **2014**, *60*, 39–44. [CrossRef]
26. do Valle Calomeni, A.; de Souza, V.B.; Tulini, F.L.; Thomazini, M.; Ostroschi, L.C.; de Alencar, S.M.; Massarioli, A.P.; de Carvalho Balieiro, J.C.; de Carvalho, R.A.; Favaro-Trindade, C.S. Characterization of antioxidant and antimicrobial properties of spray-dried extracts from peanut skins. *Food Bioprod. Process.* **2017**, *105*, 215–223. [CrossRef]
27. Radunz, M.; dos Santos Hackbart, H.C.; Camargo, T.M.; Nunes, C.F.P.; de Barros, F.A.P.; Dal Magro, J.; Filho, P.J.S.; Gandra, E.A.; Radünz, A.L.; da Rosa Zavareze, E. Antimicrobial potential of spray drying encapsulated thyme (*Thymus vulgaris*) essential oil on the conservation of hamburger-like meat products. *Int. J. Food Microbiol.* **2020**, *330*, 108696. [CrossRef]

foods

Review

Plant Antimicrobials for Food Quality and Safety: Recent Views and Future Challenges

Loris Pinto [1,*], Melvin R. Tapia-Rodríguez [2], Federico Baruzzi [1] and Jesús Fernando Ayala-Zavala [3]

1 Institute of Sciences of Food Production, National Research Council of Italy, Via G. Amendola 122/O, 70126 Bari, Italy; federico.baruzzi@ispa.cnr.it
2 Departamento de Biotecnología y Ciencias Alimentarias, Instituto Tecnológico de Sonora, 5 de Febrero 818 sur, Col. Centro, Ciudad Obregón, Obregón 85000, Sonora, Mexico; melvin.tapia14987@potros.itson.edu.mx
3 Centro de Investigación en Alimentación y Desarrollo, A.C, Carretera Gustavo Enrique Astiazarán Rosas 46, Hermosillo 83304, Sonora, Mexico; jayala@ciad.mx
* Correspondence: loris.pinto@ispa.cnr.it

Abstract: The increasing demand for natural, safe, and sustainable food preservation methods drove research towards the use of plant antimicrobials as an alternative to synthetic preservatives. This review article comprehensively discussed the potential applications of plant extracts, essential oils, and their compounds as antimicrobial agents in the food industry. The antimicrobial properties of several plant-derived substances against foodborne pathogens and spoilage microorganisms, along with their modes of action, factors affecting their efficacy, and potential negative sensory impacts, were presented. The review highlighted the synergistic or additive effects displayed by combinations of plant antimicrobials, as well as the successful integration of plant extracts with food technologies ensuring an improved hurdle effect, which can enhance food safety and shelf life. The review likewise emphasized the need for further research in fields such as mode of action, optimized formulations, sensory properties, safety assessment, regulatory aspects, eco-friendly production methods, and consumer education. By addressing these gaps, plant antimicrobials can pave the way for more effective, safe, and sustainable food preservation strategies in the future.

Keywords: antimicrobial activity; essential oils; food preservation; foodborne pathogens; polyphenols

1. Introduction

The use of natural antimicrobials in the food industry is gaining attention due to the consumers' demand for environmentally friendly production systems and products with clean labels, promoting the use of natural antimicrobial preservatives rather than synthetic ones [1,2]. Indeed, synthetic food preservatives such as nitrates, benzoates, sulfites, sorbates, and formaldehyde are known for allergic or carcinogenic effects [3]. Microbial food spoilage is responsible for about 25% of food losses [4]. According to the Food and Agriculture Organization (FAO), wasted food costs approximately 680$ billion in industrialized countries and 310$ billion in developing countries, with a high emission footprint for meat products [5]. Moreover, the growing consumption of fresh, minimally processed, and ready-to-eat foods increases the chance of microbial contamination by spoilage and pathogenic microorganisms [1]. Therefore, natural antimicrobials should be promoted to extend the shelf life of perishable foods, and to ensure the product's microbial food safety.

Plant antimicrobials represent the main group of natural preservatives, including secondary metabolites targeting microbial cells. Different parts of plants, such as seeds, fruit, peels, leaves, and roots are rich in plant antimicrobials such as phenolic compounds (e.g., simple phenols, phenolic acids, anthocyanins, flavonoids, quinones), tannins, essential oils and terpenoids, glucosinolates derivatives, alkaloids, and thiols [6,7]. Most of the plant extracts are generally recognized as safe (GRAS) and were granted the qualified presumption of safety (QPS) status in the USA and EU, respectively [8]. Plant extracts, such as moso bamboo (Takeguard™) with benzoquinone derivatives and tannin, or an antifungal blend (Biovia™ YM10) with green tea (*Camellia sinensis* L.) extract and mustard (*Brassica nigra* W.D.J. Koch) essential oil, are commercially available as alternatives to chemical preservatives [1]. Moreover, the European Food Safety Authority (EFSA) authorized rosemary (*Rosmarinus officinalis* L.) extract, endowed with antimicrobial activity, as a food additive (E 392) [9,10].

Plant antimicrobials were proposed to control the growth of microbial spoilage populations and foodborne pathogens. As regards the control of spoilage microorganisms, several applications were described in animal-based foods. In fish products, grape (*Vitis vinifera* L.) seed extract, tea polyphenols, thyme essential oil, and rosemary extract delayed the growth of lactic acid bacteria, *Enterobacteriaceae*, hydrogen sulfide-producing bacteria (HSPB), and psychrotrophic bacteria, well known to produce off-flavours [11]. In meat products, tannic acid or catechin showed good antimicrobial activity in camel sausages, whereas ethanolic extracts of rosemary and clove (*Syzygium aromaticum* L.) reduced spoilage bacterial counts in raw chicken meat. In beef sausages, the use of *Ziziphus* leaf extracts, rich in vanillic and ellagic acids, inhibited the growth of spoilage bacteria during cold storage [12]. Among essential oils, the application of *Ziziphora clinopodioides* Lam., rich in carvacrol, thymol, p-cymene, and γ-terpinene, showed the best antimicrobial activity against spoilage bacteria in beef patties [13]. In plant-based foods, as reviewed by Patrignani et al. [14], citral, hexanal, and 2-(E)-hexenal showed antimicrobial activity against yeasts responsible for spoilage of fresh-cut fruits, soft drinks, and fruit-based salads, whereas citral-based films or the application of oregano (*Origanum vulgare* L.) and thyme (*Thymus vulgaris* L.) oil during the washing step reduced spoilage bacterial populations on salad. The antimicrobial action of plant extracts against foodborne pathogens is well documented [1–3,7,15–17]. In particular, phenolic extracts and essential oils showed remarkable antibacterial action against Gram-positive and Gram-negative bacteria, including spore-forming bacteria. In addition to the effect against viable cells, plant antimicrobials inhibited the production of microbial toxins [18,19] and biofilm formation [20–22].

Despite the antimicrobial action of plant antimicrobials, their use in the food industry is hampered by chemical instability, limited dispersibility in food matrices, limited availability of ready-to-use commercial formulations, or unacceptable flavour profiles [6]. For these reasons, several stabilization techniques, such as nano-emulsions, encapsulation, and inclusion in active packaging, were proposed [6,23,24]. Moreover, these stabilization techniques ensure, in some cases, better antimicrobial activity of the bioactive compounds, and a controlled release during food storage.

However, some challenges remain, including potential negative sensory impacts, variations in antimicrobial effectiveness, and concerns about the possible development of microbial resistance. To address these issues, researchers explored synergistic combinations of plant antimicrobials and the application of hurdle technologies, which involve the simultaneous or successive use of multiple preservation techniques. Although plant extracts showed considerable potential in food preservation, limited information is available concerning their safety. In some instances, these extracts can be contaminated with various hazardous substances, such as heavy metals [25], mycotoxins [26], or crop protection residues [27]. The levels of contamination in plant extracts are affected by several factors, including the cultivation practices employed, the geographical location of the cultivation site, and the application of crop protection products. Further research is needed to establish proper guidelines and regulatory frameworks that can help minimize the risks associated with contaminants in plant extracts, ultimately ensuring the safe application of these natural preservatives in the food industry. Further research is also necessary to understand the modes of action of plant antimicrobials alone or in combination to optimize their formulation and the delivery of bioactive compounds. Addressing these gaps will help the acceptance of plant extracts as food preservatives and their use in different food industries.

This review aims to summarize the applications of plant antimicrobials in the food sector. After that, an overview of different classes of plant antimicrobials, antimicrobial activity against spoilage, and pathogenic microorganisms in different foods is described. Then, the stabilization techniques of plant extracts are presented followed by their use in different food matrices. This review also discusses the additive and synergistic effects of various combinations of plant antimicrobials, as well as the integration of plant extracts into different hurdle technologies, including mild or non-thermal treatments, to enhance food preservation. Finally, safety aspects and regulation related to the use of plant extracts are introduced. By providing a comprehensive overview of the current knowledge, this review aims to contribute to the ongoing development and optimization of food preservation techniques based on plant antimicrobials.

2. Classification and Antimicrobial Activity of Plant Antimicrobials

A great diversity of structures among plant secondary metabolites (PSMs) occurs in nature (e.g., more than 12,000 known alkaloids, more than 10,000 phenolic compounds, and over 25,000 different terpenoids) [1]. From a structural point of view, plant antimicrobials can be divided in two classes: PSMs with one or several nitrogen atoms into their structures, such as alkaloids, glucosinolates, and PSMs without nitrogen, such as terpenoids and phenolic substances. Alkaloids, glucosinolates, and phenolic substances are water-soluble compounds, whereas terpenoids are lipophilic PSMs [28]. The following sections summarize the different classes of plant antimicrobials and their antimicrobial action against main food-related microorganisms.

2.1. Polyphenols

Polyphenols are PSMs produced by higher plants, sharing a common chemical structure characterized by at least one aromatic ring with one or more hydroxyl groups [29]. Polyphenols can be classified as flavonoids and nonflavonoids. The latter includes the phenolic acids (e.g., derivatives of benzoic acid and cinnamic acid), stilbenes (e.g., resveratrol), tannins (e.g., proanthocyanidins, gallotannins, and ellagitannins), and lignins (e.g., secoisolariciresinol). Flavonoids can be divided into six subclasses: flavonols, flavones, flavanones, flavanols, anthocyanins, and isoflavones [30].

2.1.1. Phenolic Acids

Phenolic acids are divided into hydroxybenzoic acids (e.g., vanillic, gallic, salicylic, syringic, and protocatechuic acid) and hydroxycinnamic acids (ferulic, rosmarinic, p-coumaric, chlorogenic, cinnamic, and caffeic acid) [31]. The main phenolic acids showing antimicrobial action are gallic acid, ferulic acid, and p-coumaric acid [31]. A minimum inhibitory concentration (MIC) of 1000–2000 $\mu g\ mL^{-1}$ was found for gallic acid and ferulic acid against *Escherichia coli*, *Staphylococcus aureus*, and *Listeria monocytogenes* [32]. Ferulic acid and p-coumaric acid showed MIC values of 500–1000 $\mu g\ mL^{-1}$ against *Salmonella enteritidis* [33]. Li et al. [34] recently found that p-coumaric acid controlled the contamination of *Alicyclobacillus acidoterrestris* in apple juice. However, the antibacterial action of phenolic acids can be enhanced considering their derivatives, as demonstrated for alkyl ferulate and gallate esters against *L. monocytogenes* and *E. coli*, respectively [35,36]. Regarding the antifungal action, ferulic acid and p-coumaric acid showed antifungal activity against *Botrytis cinerea* and *Alternaria alternata* [37,38]. As reported for the antibacterial action, ester derivatives of phenolic acids showed enhanced antifungal action compared to phenolic acids. In particular, ethyl p-coumarate showed interesting antifungal activity against *Alt. alternata* [39].

2.1.2. Stilbenes, Tannins, and Lignins

Other polyphenols endowed with antimicrobial activity are stilbenoids, tannins, and lignins. Stilbenes such as resveratrol showed antibacterial action against foodborne pathogens, with MIC values of 100–200 $\mu g\ mL^{-1}$ for *S. aureus* and *Enterococcus faecalis*, and

>200 µg mL^{-1} for *E. coli* and *Sal. enterica* [40]. Cai et al. [41] found that pterostilbene had higher antifungal activity against ochratoxin A (OTA)-producing *Aspergillus carbonarius* than piceatannol and resveratrol. As regards tannins, they are classified into hydrolysable and condensed tannins. Hydrolysable tannins such as ellagitannins showed antibacterial action against *S. aureus* and *E. coli*. In particular, increased free galloyl groups enhanced antibacterial action against *S. aureus*, while large molecular size positively affected the antimicrobial effect against *E. coli* [42]. Condensed tannins such as proanthocyanidins from persimmon [43] or chokeberry [44] showed MIC values of 0.7–5 mg mL^{-1} against *S. aureus*. Regarding the antibacterial activity of lignin, different sources and extraction processes can result in different antibacterial performances. However, the ethanol fractionation of bamboo kraft lignin enhanced the antibacterial activity compared to non-fractionated lignin, and the ethanol fraction showed a MIC value of 2 mg mL^{-1} against *Bacillus subtilis* and *S. aureus* [45].

2.1.3. Flavonoids

Flavonoids are the main dietary polyphenols. They show a characteristic phenyl-benzopyrone structure and can be classified into anthocyanidins, flavan-3-ols, flavones, flavanones, flavonols, and isoflavonoids [29]. Among them, flavan-3-ols, flavonols, and flavanones showed the highest antibacterial activity against foodborne pathogens [31]. In particular, flavan-3-ols such as epigallocatechin-3-gallate showed antibiofilm activity against *L. monocytogenes* [46], and bactericidal effect against *E. coli* [47]. The main flavonol endowed with antibacterial activity is resveratrol. Resveratrol showed a MIC value lower than 10 mg mL^{-1} against *E. coli* O157:H7 and *Sal. enteritidis* [48] and 300–600 µg mL^{-1} against methicillin-resistant *S. aureus* [49]. However, the presence of rhamnose and additional hydroxyl groups in the flavonoids myricetin-3-O-rhamnoside and quercetin-3-O-rhamnoside resulted in reduced antibacterial activity compared to quercetin [31]. As regards the antifungal activity of flavonoids, quercetin at 0.25 mg mL^{-1} inhibited mycelial growth of *Penicillium expansum* [50] and showed a MIC value of 505 µg mL^{-1} against *Aspergillus flavus* [51]. Flavanones belong to a sub-class of flavonoids. The most interesting antibacterial activity was found for sophoraflavanone G against methicillin-resistant *S. aureus*, with MIC values ranging from 0.5 to 8 µg mL^{-1} [52]. Recently, other flavonoids, such as the mono-prenylated isoflavonoids showed high antifungal activity against *Zygosaccharomyces parabailii*, a spoilage yeast of acidic food products, with a minimum fungicidal concentration (MFC) of 12.5 µg mL^{-1} [53].

2.2. *Terpenes and Essential Oils*

Essential oils (EOs) are complex blends of aromatic metabolites extracted from different plant parts, including leaves, bark, flowers, and roots, using solvents, distillation, or microwaves [54]. Volatile compounds represent 90–95% of EOs, including monoterpenes, sesquiterpene hydrocarbons and their oxygenated derivatives, aldehydes, alcohols, and esters. The non-volatile portion (5–10% of the whole EO) comprises hydrocarbons, fatty acids, sterols, carotenoids, waxes, cumarines, and flavonoids. The main antimicrobial compounds present in EOs can be divided into different groups: terpenes (e.g., p-cymene, limonene), terpenoids (e.g., thymol, carvacrol), and phenylpropenes (e.g., eugenol, vanillin) [30].

Rosemary EO, rich in the monoterpenes α-pinene, 1,8-cineol, and camphor, showed antibacterial action against *E. coli* and *S. aureus* [55,56]. A recent study [57] showed that the geographic origin of rosemary EOs affected their composition and antimicrobial activity. EOs extracted from *Salvia officinalis* L., *Lavandula dentata* L., and *Laurus nobilis* L., rich in 1,8-cineol, inhibited the growth rate of *A. carbonarius* and the OTA production [58].

EOs with terpenoids such as thymol and carvacrol as main compounds paid great attention due to their broad spectrum of antimicrobial activity and potential application through direct contact and vapour phase. Oregano and thyme EOs showed antibacterial activity by direct contact against drug-resistant Gram-positive pathogens such as *S. aureus* and *Enterococcus faecium*, and Gram-negative pathogens such as *E. coli* and

Sal. thyphimurium [59,60]. These EOs showed antimicrobial activity in vapour phase, with MIC values of 0.16–4.00 μg mL^{-1} of air against *E. coli* and *Penicillium expansum* [61]. Moreover, oregano and thyme EOs vapours showed antifungal activity against different species of the genera *Aspergillus*, with MIC values of 15.6–62.5 μL L^{-1} of air [62]. Regarding p-cymene, this monoterpene has low antibacterial activity, high MIC values, and no antifungal action against *Rhizopus oryzae* and *A. niger* [63]. Similarly, in *B. cinerea*, *P. italicum*, and *Alt. alternata*, p-cymene showed higher MIC values than other monoterpenes such as thymol and γ-terpinene [64].

Phenylpropanoids such as eugenol and isoeugenol, both present in clove EO, showed antibacterial action against *E. coli* and *L. monocytogenes* with MIC values in the range 312.5–625 μg mL^{-1} [65]. Clove oil, with eugenol as the main compound, inhibited *P. italicum* growth on citrus fruit when applied at concentrations ranging from 0.05% to 0.8% (v/v) [66]. Other phenylpropanoids, such as vanillin, showed a bacteriostatic effect against foodborne pathogens, but MIC values were higher than that of pure compounds belonging to terpenes or terpenoids [67].

Other bioactive compounds occurring in EOs are the aldehydes citral and cinnamaldehyde, found in lemongrass (*Cymbopogon citratus* Stapf) EO and cinnamon (*Cinnamomum verum* Presl) bark EO, respectively. Free citral showed a MIC value of 0.8 mg mL^{-1} against *B. cereus* and 2 mg mL^{-1} against *E. coli* and *S. aureus* [68]. However, the main application of citral is its use as an antifungal agent, as demonstrated against different fungal strains [69–71]. As regards cinnamaldehyde, it showed higher antibacterial activity than cinnamon oil against Gram-positive bacteria [72]. Cinnamaldehyde at 150 μg mL^{-1} inhibited the spore production and mycelial growth of *A. niger* [73] and showed antifungal activity and alternariol reduction at 0.200 μL mL^{-1} against *Alt. alternata* [74].

2.3. Glucosinolate Derivatives

Glucosinolates are the main bioactive compounds of *Brassica* plants. The breakdown of glucosinolates releases nitriles, thiocyanates, and isothiocyanates. In particular, isothiocyanates, largely occurring in cruciferous vegetables, are the most reactive compounds endowed with antimicrobial activity. Allyl-, benzyl-, and 4-methylsulfinylbutyl isothiocyanates are the main compounds with antimicrobial activity against bacterial pathogens and fungi [7]. Allyl-isothiocyanate at the concentration of 1 μL L^{-1} reduced of 4 log cfu g^{-1} the *Sal. thyphimurium* load on lettuce [75], whereas at 0.1% v/w inhibited *L. monocytogenes* growth in chickpea puree stored for 10 days at 4 °C [76]. Allyl-isothiocyanate showed antifungal activity against *A. flavus* in maize and *P. verrucosum* in barley, reducing the aflatoxin B1 and ochratoxin A accumulation, respectively [77,78]. Benzyl-isothiocyanate showed MIC values ranging from 60 to 160 μM against enterotoxigenic *E. coli* [79], and 120 μM against *L. monocytogenes* [80]. Benzyl-isothiocyanate at 25 μg mL^{-1} inhibited the growth of *A. carbonarius* and *A. ochraceus*, whereas *A. niger* was more resistant to both allyl- and benzyl-isothiocyanates than other aspergilli [81]. Other bioactive isothiocyanates are sulforaphane (4-methylsulfinylbutyl isothiocyanate) and phenethyl isothiocyanate. Both compounds showed MIC values of 40–88 mg mL^{-1} against *S. aureus* and *E. coli* [82]. However, their use for applications in the food sector is limited compared to allyl- and benzyl-isothiocyanates. Other isothiocyanates demonstrated an interesting antifungal activity. In particular, the volatile compound 2-phenylethyl isothiocyanate showed a MIC value of 1.2 mM against *Alt. alternata*, and reduced the development of the black spot rot on pear [83], whereas 2-(4-methoxyphenyl)ethyl isothiocyanate showed an EC_{50} value of 4.2 μg mL^{-1} against *A. niger*, and inhibited the spore germination by 95% [84].

2.4. Alkaloids and Thiols

Several plant extracts include alkaloids and thiols as antimicrobial compounds. Alkaloids are PSMs classified based on their chemical structure and natural origin. Although more than 18.000 alkaloids are known, mainly represented by plant alkaloids [85], their use in the food sector is limited due to their well-known toxic and neuroactive effects. Recently, berberine, an isoquinoline alkaloid found in roots and stem-bark of *Berberis* plants, was the most studied alkaloid exploited for its antimicrobial activity against food-related microorganisms and was proposed as a food preservative [86–88]. In particular, *Berberis vulgaris* root and leaf extracts, rich in berberine, showed a MIC value of 150 μg mL^{-1} against *E. coli* and *S. aureus*, and 60–100 μg mL^{-1} against different *Aspergillus* species [86]. Berberine at 1.6 mg mL^{-1} inhibited mycelial growth and spore germination of *P. italicum* [88].

As regards thiols, the main antimicrobial compounds are allicin and its derivatives [29]. Allicin is a sulphur compound occurring in garlic, effective against spoilage yeasts, Gram-positive and Gram-negative foodborne pathogens, with MIC values lower than 30 μg mL^{-1} [89]. The main oxidation derivatives of allicin are diallyl disulphide and diallyl trisulfide. Diallyl disulphide showed antibacterial action against *B. cereus* and a MIC value of 120 μg mL^{-1} [90], whereas diallyl trisulfide treatment reduced, by 1.5 log cfu g^{-1}, the *Campylobacter jejuni* count on chicken [91].

2.5. Modes of Action

PSMs described in the previous sections have multiple mechanisms of antimicrobial action (Figure 1). In particular, different cell targets are affected by exposure to polyphenolic substances, essential oil compounds, isothiocyanates, alkaloids, and thiols. As regards polyphenols, the three main mechanisms of action are the modification of the membrane permeability, the intracellular enzyme inactivation, and the modification of fungal morphology. Additional mechanisms of antimicrobial action of polyphenols are the modification of intracellular pH, the interference with the ATP-generating system, and the inhibition of DNA synthesis [1].

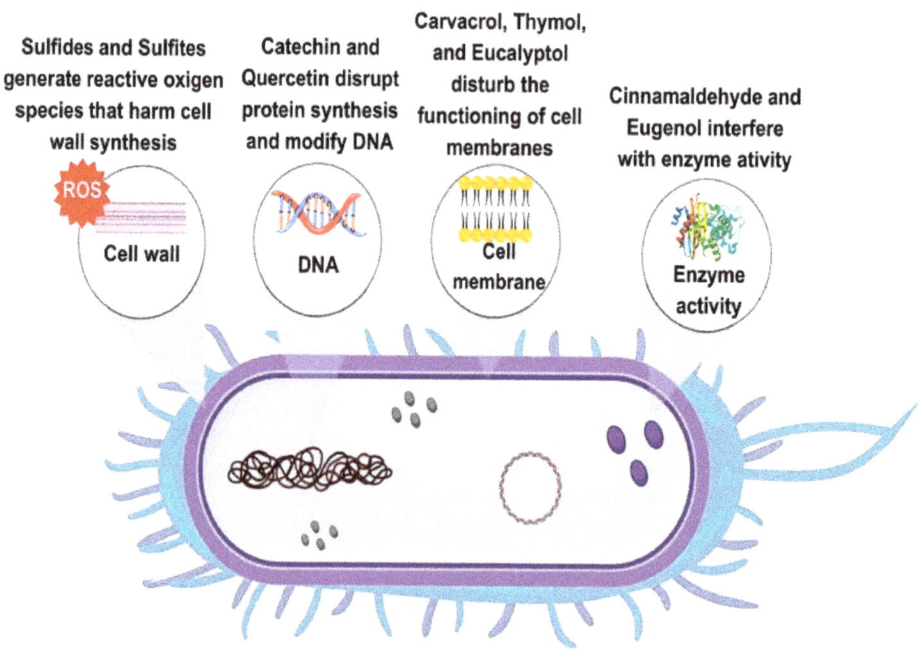

Figure 1. Mechanisms of action of plant antimicrobials against foodborne bacteria.

Different polyphenolic classes have specific mechanisms of action. Phenolic acids mainly interact with the cell membrane intercalating the phospholipid layer, or crossing the membrane, decreasing the intracellular pH, and/or interacting with cellular constituents [92]. The antibacterial action of phenolic acids against *L. monocytogenes* depends on their dissociated/undissociated form. In particular, chlorogenic acid and gallic acid reduced extracellular pH, caffeic acid, p-hydroxybenzoic acid, protocatechuic acid, and vanillic acid were active in their undissociated form, and p-coumaric acid and ferulic acid showed antibacterial action in both dissociated and undissociated form [92]. In *Sal. enteritidis*, chlorogenic acid treatment damaged intracellular and outer membranes and inactivated key enzymes of the tricarboxylic acid cycle (TCA) [93]. Phenolic acid esters showed multiple mechanisms of antibacterial action, such as the damage of bacterial membranes, changes in the conformation of protein membranes, formation of complexes with bacterial DNA, and oxidative damage [35,36]. As regards the antifungal mechanism of phenolic acids, it is well known that these compounds produce oxidative stress and disorganization of the wall or membrane of the hyphae [94], but, as in the case of *B. cinerea*, they can also affect the ATP synthesis and cellular metabolism acting as an uncoupler of oxidative phosphorylation [95]. Resveratrol inhibits ATP synthesis, hydrolysis, and cell division in *E. coli* [40]. Pterostilbene treatment induces incomplete sporangia, membrane rupture, and downregulation of the biosynthetic genes of the OTA production in *A. carbonarius* [41]. The disruption of cell membranes and functions is the primary mode of antibacterial action of tannins. However, the inhibition of microbial enzymes, the deprivation of the nutrients required for the microbial growth, and the inhibition of oxidative phosphorylation were also suggested [42]. In *P. digitatum*, tannins disrupted the cell wall and caused the leakage of intracellular content [96]. The antibacterial modes of action of lignin are the damage of the cell membrane through its phenolic compounds, the decrease in intracellular pH, and the increase in osmotic pressure [97]. Flavonoids have multiple modes of antimicrobial action. Quercetin inhibited DNA gyrase, increased membrane permeability, and prevented ATP synthesis in *E. coli* [29], whereas in *S. aureus*, it inhibited key enzymes necessary for the protein synthesis [31]. Flavonols such as quercetin, rutin, morin, rhamnetin, and flavones such as acacetin and apigenin have membrane-disrupting activity. Conversely, flavanones naringenin and sophoraflavanone G reduce the fluidity in regions of both inner and outer cellular membranes [18]. Catechins such as epigallocatechin gallate, at high concentration, generated reactive oxygen species (ROS), causing membrane damage [18]. As regards the antifungal mechanism of flavonoids, a recent study showed that quercetin downregulated genes involved in the conidial and mycelial development, while reducing the production of aflatoxin probably by lowering levels of ROS [51]. Flavonoids from the medicinal–edible plant *Sedum aizoon* L. damaged the cell membrane and the cell wall, and interfered with the mitochondrial respiratory metabolism, the protein biosynthesis, and the amino acid metabolism in *P. italicum* [98].

As regards the antibacterial mechanism of action of terpenoids and essential oils, these compounds can disrupt cell walls and cytoplasmic membranes, increasing their permeability. Essential oils can also solidify the cytoplasm, damage lipids and proteins in the cell, and inhibit bacterial enzymes [29]. Specifically, terpenoids such as carvacrol disrupted the cell membranes and inhibited the respiratory activity in *L. monocytogenes* [99], while it increased the cell permeability and reduced the ATP levels in *E. coli* [100]. In *Sal. enteritidis*, the antibacterial action of oregano essential oil was mainly attributed to thymol rather than its isomer carvacrol, with changes in the protein regulation and the DNA synthesis [101]. The antibacterial action of terpenes such as limonene was associated with increased cell permeability, inhibition of the ATP synthesis, dysfunction of the respiratory chain complex, and inhibition of the transcription of nucleic acids [102–104]. Phenylpropenes such as eugenol altered the membrane permeability in *E. coli* [105], whereas they increased reactive oxygen species, depolarized the membrane potential, and decreased the ATP content in *Shigella flexneri* [106].

The antifungal mechanisms of action of essential oil compounds such as thymol and carvacrol are related to changes in the morphology of hyphae, the increase in membrane permeability, and the reduction in total lipids and ergosterol content [107,108]. As regards p-cymene, the antibacterial mode of action is related to the expansion of the cytoplasmic membrane and a moderate generation of ROS [63]. Other essential oil compounds, such as citral and geraniol, showed distinctively antifungal mechanisms of action. In particular, citral downregulated the sporulation- and growth-related genes in *A. flavus* and *A. ochraceus*, whereas geraniol determined intracellular ROS accumulation in *A. flavus* and increased cell membrane permeability in *A. ochraceus* [70]. Oxidative stress was partially responsible for the antifungal action of cinnamaldehyde against *A. niger*, causing cell damage and increasing membrane permeability [73]. Citral, limonene, and eugenol damaged the cell membranes and destroyed the yeast proteins in *Zygosaccharomyces rouxii* [109]. In *A. carbonarius*, eugenol determined the leakage of cytoplasmic contents, increased the lipid peroxidation, decreased the ergosterol content, increased the membrane permeability, and induced oxidative stress [110].

Luciano and Holley [111] demonstrated that allyl isothiocyanate inhibited thioredoxin reductase and acetate kinase in *E. coli* O157:H7. The bacteriostatic/fungistatic effects of benzyl isothiocyanate against *E. coli*, *B. subtilis*, *Sal. enterica*, *S. aureus*, *A. niger*, and *P. citrinum* were associated with interferences with the ATP production, enzymes and coenzymes of the energy metabolism [112]. Conversely, in *B. cinerea*, benzyl isothiocyanate disrupted the plasma membrane integrity and induced ROS accumulation in the spores, inhibiting their germination [113]. Other glucosinolate derivatives, such as sulforaphane and phenethyl isothiocyanate, are effective against different pathogenic bacteria by inhibiting the synthesis of nucleic acids or disrupting the membrane integrity depending on bacterial species [83].

The alkaloid berberine binds to the FtsZ protein, causing the inhibition of bacterial cell division [114]. In *Sal. typhimurium*, it reduces the number of type I fimbriae and prevents biofilm formation [115]. In fungi, berberine damages the plasma membrane integrity and reduces the contents of soluble proteins and reducing sugars. In addition, a high H_2O_2 content was found in berberine-treated *P. italicum* mycelia [88]. Thiols such as allicin display antimicrobial action due to the rapid reaction of thiosulfinates with thiol groups of key enzymes [89].

3. Plant Antimicrobials for Food Quality and Safety

In the past, the use of plant material during traditional food processing was defined empirically to improve the sensory characteristics of the food and the food safety and quality levels. It should be considered that several spices, obtained from different plant species, often include antimicrobial molecules and are usually supplemented to foods as flavouring agents. For this reason, the use of plant compounds as food preservatives is close to traditional recipes and, therefore, highly accepted by consumers. This section presents the direct application of plant antimicrobials in different foods, highlighting the antimicrobial action against spoilage and pathogenic microorganisms.

3.1. Plant Antimicrobials as Food Preservatives

Plant antimicrobials were exploited as preservatives in several foods to control the microbial growth of food spoilage microorganisms or foodborne pathogens [116]. This section summarizes recent published results, focusing on the direct application of plant extracts or their bioactive compounds as preservatives in food products (Table 1).

3.1.1. Applications in Plant Foods

This section presents the applications of whole plant extracts or their antimicrobial compounds against spoilage microorganisms of fresh fruit and vegetables, ready-to-eat vegetables, and fruit juices.

With regard to the application of plant antimicrobials on fresh fruits and vegetables, pomegranate (*Punica granatum* L.) peel extract (PPE), rich in polyphenols such as punicalagin and ellagic acid, reduced the growth of post-harvest fungi belonging to the genera *Penicillium*, *Botrytis*, *Monilinia*, and *Colletotrichum* on various fruits including lemon, strawberry, grape, apple, grapefruit, orange, and capsicum. In addition, the PPE ethanolic or aqueous extracts can preserve foods by dipping treatments or using edible coatings [117]. In this context, the use of ethanolic pomegranate peel extract (PPE) was found to significantly

($p \leq 0.05$) reduce the lesion diameter and infection rate in mandarins contaminated with *P. italicum* and *P. digitatum* [118]. In addition, other plant-based extracts were also found to be effective in controlling spoilage microorganisms. A mango kernel extract, rich in mangiferin, chlorogenic acid, and myricetin, inhibited anthracnose development caused by *Colletotrichum brevisporum* on mangoes [119]. A sweet orange (*Citrus sinensis* L.) peel extract, rich in ferulic acid, showed antifungal activity against *M. fructicola* and *Alt. alternata* in a peach-based medium [120].

Other plant antimicrobial extracts with antimicrobial activity against spoilage microorganisms on fresh fruit and vegetables are the essential oils or their main compounds. Mint (*Mentha* × *piperita* L.), basil (*Ocimum basilicum* L.), lavender (*Lavandula angustifolia* Mill.), and thyme EOs in the vapour phase were used for the post-harvest preservation of strawberry, peach, orange, and lemon [121–124]. In particular, as recently reported by Pinto et al. [123], the in-package application of red thyme oil vapours reduced the percentage of infected wounds, the mycelium development, and the production of spores by *Penicillium* strains on oranges during 12 days of cold storage. Dipping in cinnamon essential oil microemulsion at 0.3% v/v eradicated *P. fluorescens* from iceberg lettuce during 28 days of cold storage [124]. As regards other plant antimicrobials, methyl, allyl, and ethyl isothiocyanate (8–12 μL L^{-1}) completely inhibited citrus sour-rot caused by *Geotrichum citriaurantii* [125], whereas berberine at 3 mg mL^{-1} reduced the development of *P. italicum* and natural decay on citrus fruit [88].

Plant antimicrobials were extensively used to control the spoilage microorganisms on ready-to-eat fruits and vegetables [126]. Dipping of fresh-cut pineapple in *Centella asiatica* extract, rich in quercetin and kaempferol, reduced the *A. niger* load during cold storage [127]. In-package application of trans-anethole in ready-to-eat organic lettuce reduced total coliforms during cold storage [128], whereas the addition of β-caryophyllene-rich pepper EOs in salad dressing decreased *P. fluorescens* development and spoilage activity on fresh-cut lettuce [129]. Pomegranate arils coated with savoury essential oil-loaded chitosan showed a reduction in total mesophilic bacteria and total yeasts and moulds of 1 log CFU g^{-1} after 18 days of storage [130]. Peppermint and tea tree (*Melaleuca alternifolia* Cheel) oils controlled the growth of total aerobic bacteria, yeasts, and moulds on fresh-cut green bean pods stored for 9 days at 5 °C [131].

Other applications of plant antimicrobials in plant-based foods concern fruit juices and smoothies [132]. In this context, essential oils and their compounds are the most used antimicrobials. Indeed, *Mentha piperita* L. EO inclusion (7.50 μL mL^{-1}) in cashew and guava juice caused >5 log reductions in counts of the spoilage yeast *Pichia anomala* [133]. Thymol in concentrated apple juice showed higher antimicrobial activity than carvacrol and trans-cinnamaldehyde against *Z. rouxii* [134]. Lee et al. [135] found a synergism between oregano and thyme EOs, at 0.156 μL mL^{-1}, in inhibiting *Leuconostoc citreum* in tomato juice. As regards the applications of plant antimicrobials in smoothies, the addition of beet (*Beta vulgaris* L.) leaf extract (30% w/v) in a vegetable smoothie reduced significantly ($p \leq 0.05$) total mesophilic bacteria, enterobacteria, and total yeasts and moulds throughout 21 days of cold storage [136].

The use of plant antimicrobials in plant-based food products, specifically fruit juices and fresh and ready-to-eat vegetables, effectively reduces spoilage and increases the shelf life of these products. Essential oils and their bioactive compounds, such as *Mentha piperita* L. EO, thymol, carvacrol, trans-cinnamaldehyde, oregano and thyme EOs, demonstrated antimicrobial activity against different spoilage microorganisms. These findings suggest that plant antimicrobials have the potential to play a crucial role in preserving the quality of plant-based foods.

3.1.2. Applications in Animal-Based Foods

This section presents some applications of whole plant extracts or their antimicrobial compounds against spoilage microorganisms contaminating animal-based foods (e.g., meat, seafood, and dairy products). In this context, the addition of 200 mg kg^{-1} of tannic acid or catechin in camel meat decreased total mesophilic and psychrophilic bacterial counts by one order of magnitude after 9 days of refrigeration [137], as well as Nowak et al. [138] demonstrated that sour cherry (*Prunus cerasus* Scop.) leaf extract, rich in coumaric acid, and blackcurrant (*Ribes nigrum* L.) leaf extract, rich in gallic acid and quercetin derivatives, delayed the growth of *Pseudomonas* spp. in pork sausages, but not that of *Brochothrix* spp. and *Enterobacteriaceae*. On the contrary, Casaburi et al. [139] reduced the growth of *Brochotrix* spp. and *Enterobacteriaceae*, but not that of *Pseudomonas* spp., of grounded beef meat during cold storage, adding 5% of a freeze-dried myrtle (*Myrtus communis* L.) extract, rich in phenolic compounds. These results highlight that the effectiveness of phenolics can vary depending on the specific bacterial species, the concentration of phenolics, and other factors such as the food matrix, the presence of other preservatives, and the storage conditions.

The addition of the ethanolic extract of cranberry (*Vaccinium oxycoccos* L.) pomace, characterized by great amounts of anthocyanins, chlorogenic acid, and myricetin and quercetin derivatives, inhibited the growth of *Brochothrix thermospacta* and *P. putida* on pork burgers during the first days of cold storage [140]. As regards the application of essential oils or their compounds on meat products, ethanolic extracts of rosemary and clove (1% v/w) reduced *Pseudomonas* spp. counts in raw chicken meat during cold storage [141]. The use of *Ziziphora clinopodioides* essential oil (0.2% v/w), rich in carvacrol and thymol, reduced the *Enterobacteriaceae* and psychrotrophic bacteria loads of raw beef patties during cold storage by 2–3 log cfu g^{-1} [142]. Thymol or carvacrol at 0.4% w/w in marinated beef significantly reduced the mesophilic total viable count, lactic acid bacteria, *Broch. thermosphacta*, *Pseudomonas* spp., and total coliforms, extending the microbiological shelf life by three days [143]. In this context, only some terpene compounds showed a broad spectrum of activity against various bacterial species, making them effective preservatives for meat products, independently of the source of plant origin, and able to extend the shelf life of some meat products.

Likewise, the reduction in fish spoilage bacteria can be achieved using plant antimicrobials, specifically polyphenolic extracts and essential oils [11]. The use of ethanolic Noni (*Morinda citrifolia* L.) leaf extract, rich in rutin and kaempferol derivatives, was shown to extend the shelf life of striped catfish slices and maintain the acceptable levels of total viable bacteria and psychrophilic bacteria during storage, with loads remaining below 6 log cfu g^{-1} [144]. Similarly, the growth of *Pseudomonas* spp. in Pacific white shrimps was delayed by adding ethanolic guava (*Psidium gujava* L.) leaf extracts, rich in phenolic compounds such as piceatannol 4′-galloylglucoside, epicatechin, epigallocatechin, procyanidin B2, ellagic acid, quercetin 3′-o-glucuronide, and quercetin 3-galactoside [145]. Grape seed extract, containing high levels of phenolic acids, catechins, and proanthocyanidins, decreased the presence of *Aeromonas* spp. in snakehead fillets during cold storage. This reduction limited the release of soluble peptides and biogenic amines and increased the shelf life of snakehead fillets by three days [146]. The application of essential oils, such as cinnamon, oregano, and thyme, as marinades was evaluated in salmon and scampi by Van Haute et al. [147]. The immersion of these products in cinnamon essential oil at 1% w/w inhibited the growth of yeasts and moulds. Similarly, cinnamon essential oil at 0.1% w/v effectively inhibited *Aeromonas* spp. in vacuum-packed carp and extended its shelf life by two days [148]. However, the direct application of essential oils in fish products can cause bitterness, off-flavours, and yellowing of the tissue [11]. The inclusion of essential oils in active packaging or nano-emulsions is recommended to mitigate these effects.

Building on the findings of previous studies on the application of plant antimicrobials in meat and fish products, the use of plant polyphenols, essential oils, and other plant-based compounds in milk and dairy products to control spoilage microorganisms and extend their shelf life is also of interest. For instance, the addition of olive mill wastewater in the governing liquid of "Fior di Latte" cheese (500 μg mL^{-1} of phenols) resulted in a four-day extension of shelf life due to the increase in the lag phase of *P. fluorescens* and *Enterobacteriaceae* [149]. A recent study by Derbassi et al. [150] evaluated the preservative effect of *Arbutus unedo* L. leaf extracts on the microbiological characteristics of quark cheese during storage. They found that incorporating the dry macerated leaf extract into the cheese resulted in higher efficacy against aerobic mesophiles and yeasts than the use of potassium sorbate after 8 days of storage. Milanović et al. [151] investigated the efficacy of seven essential oils against 74 spoilage yeasts. In a yoghurt model, lemongrass and cinnamon EOs demonstrated the highest antifungal activity in vitro. However, it should be noted that cinnamon EO inhibited lactic acid bacteria, while lemongrass EO displayed species-specific antifungal activity. These findings suggest that further research is needed to fully understand the application of plant antimicrobials in the dairy sector to control spoilage microorganisms.

The direct addition of natural plant antimicrobials in animal-based foods, such as meat, fish, and dairy products, shows the potential to control spoilage microorganisms and extend shelf life. Studies demonstrated the effectiveness of compounds such as phenolic acids, catechins, proanthocyanidins, and EOs in inhibiting the growth of spoilage bacteria. However, more research is necessary to fully understand the mode of action of these natural compounds and optimize their application in animal-based foods. Additionally, it is essential to consider the potential drawbacks, such as the development of off-flavours or bitterness, and address them through alternative delivery methods, such as nano-emulsions or active packaging.

Table 1. Applications of plant antimicrobials on plant-based and animal-based foods against spoilage microorganisms.

Food Matrix	Plant Antimicrobial	Concentration/Conditions	Antimicrobial Effect	Data from Ref. *
Mandarins	Pomegranate peel extract	Dipping in 25 g L^{-1} extract for 2 min	Reduction of lesion diameter and infection rate (80–90%) caused by *P. italicum* and *P. digitatum*	[118]
Fresh-cut lettuce	Pepper EO	3–5 μL mL^{-1} addition in salad dressing	Reduction of *P. fluorescens* biomass by 30–40%	[129]
Concentrated apple juice	Thymol, carvacrol	MIC of 0.1–0.16 mM, treatment time 9 days	Reduction of *Z. rouxii* load by 99%	[134]
Pork burgers	Ethanolic extract of cranberry pomace	2% extract-16 days of storage	Bacteriostatic effect on *B. thermospacta* and *P. putida* during cold storage	[140]
Snakehead fillets	Grape seed extract	0.52 mg GAE mL^{-1} for 20 min	Decrease of *Aeromonas* spp. abundance by 37% and reduction of 1 log cfu g^{-1} of total viable counts during cold storage	[146]
Quark cheese	*Arbutus unedo* L. leaf extracts	0.1 g 100 g^{-1} cheese, 8 days of cold storage	Reduction of total aerobic mesophilic bacteria and yeasts by 2–3 log cfu g^{-1}	[150]

* as cited in the text.

3.2. Use of Plant Antimicrobials for Food Safety

The use of natural compounds derived from plants has numerous benefits, including the potential to provide safer, more sustainable and practical solutions for preserving food safety [152]. These plant-derived compounds showed high antimicrobial activity, making them ideal candidates as natural food preservatives. In particular, the correct use of these natural antimicrobials can fight emerging problems such as the spread of multidrug-resistant pathogens, biofilm-producing strains, and microbial toxins through the food chain.

3.2.1. Effect on Viability of Foodborne Pathogens

Foodborne bacteria are a significant public health concern since they can cause gastrointestinal illness, food poisoning, chronic diseases, economic losses, and the spread of antibiotic-resistant bacteria. Multiple foodborne illnesses were caused by various pathogens such as *Sal. enteritidis*, *L. monocytogenes*, *E. coli* toxigenic strains, *Cam. jejuni*, *Cronobacter sakazakii*, and *S. aureus*. Foodborne outbreaks underline the need for more efficient methods to control foodborne pathogens. Symptoms of foodborne illness can range from mild to severe, including nausea, vomiting, diarrhoea, abdominal cramps, and fever [153]. The outbreak of foodborne illnesses can have significant economic consequences, including loss of income for food producers, increased healthcare costs, and decreased consumer confidence in the food industry. Addressing the issue of foodborne bacteria is crucial to ensure the safety and quality of the food supply, protect public health, and minimize the economic impact of foodborne illnesses.

Several phytochemicals showed antibacterial activity against various foodborne pathogens. For example, studies demonstrated that plant compounds such as carvacrol and thymol, found in essential oils extracted from herbs and spices, have high antibacterial activity against *Sal. enteritidis*, *E. coli*, and *L. monocytogenes* [154]. Similarly, compounds such as cinnamaldehyde and eugenol, present in cinnamon EO and clove EO, respectively, inhibited the growth of foodborne pathogens such as *L. monocytogenes* and *S. aureus* [155,156]. These findings provide evidence of the potential of plant-based antimicrobials in controlling foodborne pathogens and improving food safety.

Specific applications of EOs or their compounds were described in plant and animal-based foods to ensure food safety. As regards animal-based foods, cinnamaldehyde inactivated *L. monocytogenes* at 4 °C in ground pork, reducing its viability by 4 log cfu g^{-1} in 5 days [155]. Similarly, thymol reduced, by 3 log cfu g^{-1}, the load of *S. aureus*, *E. coli*, and *C. perfringens* on a sausage product during 4 weeks of storage [157]. In dairy products, myrtle EO (31.25 μL mL^{-1}) reduced, by 1–2 log cfu g^{-1}, the load of *L. monocytogenes* ATCC 679 on sheep cheese during ripening [158], whereas ginger (*Zingiber officinale* R.) and thyme EOs totally inactivated *S. aureus* (6 log cfu g^{-1}) on a fresh soft cheese after two weeks of storage [159]. In plant foods, EOs or their compounds were proposed as sanitizers of fresh-cut vegetables and natural preservatives of fruit juices. Rossi et al. [160] treated fresh-cut lettuce contaminated with a cocktail of *Salmonella* spp. strains, with 5 μL mL^{-1} of cinnamon EO, reducing the attached cells by 0.6–0.8 log cfu cm^{-2}. Cinnamon EO was also successfully used to control *Sal. typhimurium* and *L. monocytogenes* on celery, with a reduction of 2–4 orders of magnitude after 7 days at 4 °C depending on the initial contamination level [161]. As regards the application of EOs in fruit juices, *Litsea cubeba* Pers. EO reduced 3–4 log cfu mL^{-1} of the load of *E. coli* O157:H7 in four vegetable juices after 4 days of storage, and inhibited the respiratory metabolism, the topoisomerase activity, the transcription of virulence genes, and the nucleic acid replication [162]. In watermelon juice, *Melissa officinalis* L. EO reduced the viability of *L. monocytogenes* from 2 to 7 days of storage [163]. In some cases, plant antimicrobials can induce tolerance to environmental stresses in bacteria, and cross-resistance to common antibiotics. The use of *Melissa officinalis* L. EO at subinhibitory levels (0.125 μL mL^{-1}) did not induce high tolerance to stresses (such as high temperature, low pH, osmotic stress, and desiccation) or cross-resistance with antibiotics in *L. monocytogenes* [163].

Plant phenolic compounds are naturally occurring compounds found in plants used as food preservatives due to their high antimicrobial activity against foodborne pathogenic bacteria [92]. Some of the most commonly used plant phenolic compounds in food include quercetin, and derivatives of cinnamic acid and gallic acid. Grape skin pomace extracts from different cultivars, rich in phenolic acids and flavonoids, showed higher antibacterial activity against Gram-positive strains than Gram-negative ones [164]. The addition of cranberry pomace extracts, rich in quinic and chlorogenic acids, procyanidin B3, myricetin and quercetin derivatives, delayed the growth of *L. monocytogenes* in cooked ham during cold storage [144]. *Yersinia enterocolitica* load was reduced by two logarithmic cycles in pork meat containing 5 mg g^{-1} of gallic acid [165]. Phuong et al. [166] evaluated the antibacterial activity of rambutan (*Nephelium lappaceum* L.) peel extracts, rich in geraniin, ellagic acid, rutin, quercetin, and corilagin as main phenolic compounds. The phenolic extract inhibited the growth of *Sal. Enteritidis* in raw chicken and that of *Vibrio parahaemolyticus* in fish during cold storage. The application of polyphenolic extracts or single polyphenols reduced the growth of foodborne pathogens in fresh-cut fruits, as demonstrated by using pomegranate peel extract or ferulic acid against *L. monocytogenes* on fresh-cut pear, apple, and melon [167,168]. The dipping of fresh-cut potatoes and fresh-cut lettuce in *Centella asiatica* L. extract significantly reduced the load of *B. cereus* and *E. coli* O157:H7 [131]. The glabridin, a prenylated isoflavonoid, reduced, by at least 1 log cfu g^{-1}, the load of *L. monocytogenes* on fresh-cut cantaloupe during 4 days of cold storage [169].

As regards the application of the glucosinolate derivatives against food pathogens, the (4-[(4′-O-acetyl-α-L-rhamnosyloxy)benzyl] isothiocyanate) from *Moringa oleifera* seeds reduced the viable load of *Cro. sakazakii* and *B. cereus* in goat milk by three orders of magnitude [170].

In addition to the effect on cell viability, plant antimicrobials improved the thermal sensitivity of foodborne pathogens in the food matrix. In particular, the use of oregano EO in combination with citric acid enhanced the thermal inactivation of *L. monocytogenes* in sous-vide salmon cooked at 60 °C [171], whereas vanillin and emulsified citral improved the heat-sensitization of *E. coli* at 58 °C in a blended carrot-orange juice [172]. However, in certain conditions, plant antimicrobials can induce a viable but not culturable (VBNC) state in foodborne pathogens, as demonstrated for the application of citral and *trans*-cinnamaldehyde in a meat-based broth against *S. aureus* [173].

In contrast with many of the above-reported papers, the methanolic extract of spices mixtures employed to confer typical pungency and a hot taste to 'Nduja, a traditional Calabrian sausage produced with about 20% of different spices, showed a limited inhibitory spectrum against ten common foodborne bacteria. Authors concluded that these spice mixtures, rich in hundreds of potentially antimicrobial compounds, can not exert an antimicrobial effect under normal processing conditions, due to the limited release of the bioactive compounds from the plant tissue [174]. In conclusion, the inclusion of plant antimicrobials in real food model systems can control the growth of foodborne pathogens, representing a valuable option to replace synthetic preservatives, even though their efficacy needs to be carefully evaluated under real production conditions.

3.2.2. Effect on Biofilm-Producing Strains

Bacterial biofilms are communities of microorganisms encased in a self-produced extracellular matrix and attached to a surface [22]. Biofilms are prevalent in many natural and artificial environments, including food processing facilities and equipment. Bacterial biofilms can cause serious problems in the food industry by contaminating food products, leading to foodborne illness and decreasing the product's quality [115]. Biofilms can harbour pathogenic bacteria and provide a protective environment for these microorganisms, making them resistant to cleaning and disinfection procedures. This can result in a persistent contamination and the spread of foodborne illnesses. In addition, biofilm growing on the equipment surfaces can cause clogging, formation of corrosion, and degradation of the equipment surfaces, leading to increased maintenance costs and decreased productivity [175]. Bacterial biofilms are a significant concern in the food industry due to their impact on food safety and quality and the performance and efficiency of food processing equipment. The food industry needs to implement effective strategies to prevent and control the formation of bacterial biofilms to maintain a safe and efficient food processing environment.

The use of plant-derived antimicrobial compounds in food preservation gained attention due to their efficacy against foodborne biofilm-producing strains of bacteria [176]. These compounds act through various mechanisms, such as interference with metabolic processes, oxidative stress, and membrane disruption, and can also exert positive effects in inhibiting the growth and replication of biofilm-producing bacteria. The first anti-biofilm mechanism of action is the inhibition of the bacteria's attachment to the surfaces. Phenolic compounds, such as phenolic acids, catechins, and quercetin, were found to reduce the adhesion of bacteria affecting flagellum, fimbria, and adhesins, delaying the formation of biofilms [12]. Red Globe and Carignan grape stem extracts, rich in caffeic, ferulic and gallic acids, catechin and rutin, inhibited the adhesion of *L. monocytogenes* to stainless steel and polypropylene surfaces by inhibiting motility and reducing the adhesion potential [177], as well as quercetin inhibited the early attachment of *L. monocytogenes* on stainless steel surface by increasing the cell permeability and reducing the superficial cell charge [178,179]. Quercetin also reduced the swimming and swarming motility of *Sal. enterica* at sub-MIC levels [180].

EOs or their compounds inhibit biofilm formation by different mechanisms. Cinnamon EO inhibited the adhesion of *L. monocytogenes* on polystyrene, but its efficacy was low on pre-formed biofilm [22]. Additionally, some terpenes, such as eugenol, carvacrol, and thymol, were demonstrated to suppress the production of exopolysaccharides in *Salmonella* spp., which are key components of bacterial biofilms [175], whereas citral and geraniol decreased the glucan production in *E. coli* O157:H7 [181]. Eugenol showed similar effectiveness against sessile and planktonic cells of *S. aureus*, showing a lower resistance coefficient, the ratio of concentrations required to achieve the same log reductions in both populations ($C_{biofilm}/C_{planktonic}$), as compared to conventional disinfectants [182]. Carvacrol and oregano EO effectively inhibited biofilm formation by *S. aureus* on stainless steel surfaces, but the long-term exposure to a sub-MIC concentration of the oregano EO showed an inductive biofilm formation effect [183]. Another mode of action of plant antimicrobials against foodborne biofilm-producing bacteria is destabilizing the biofilm matrix. Compounds such as sulphides, including allicin and diallyl sulfide, and sulfites were shown to penetrate the biofilm and disrupt its stability, causing the release of bacteria from the biofilm [184], as demonstrated in uropathogenic *E. coli* [185].

On the other hand, plant-derived antimicrobial agents were shown to possess anti-biofilm activity by disrupting the quorum sensing process [186]. Quorum sensing is a communication mechanism that bacteria utilize to coordinate the expression of certain genes, including those involved in biofilm formation. Phytochemicals such as flavonoids (quercetin and kaempferol) and terpenoids (carvacrol and thymol) were demonstrated to interfere with the quorum-sensing by inhibiting the production and the activity of autoinducers (e.g., acyl-homoserine lactone), which play a key role in the quorum-sensing process [187,188]. *Lippia origanoides* K. EO (thymol-carvacrol chemotype) inhibited the expression of the *sdiA*, *luxS*, and *luxR* genes, which were implicated in the quorum-sensing of *Sal. enteritidis*. This effect could be related to the inhibition of the biosynthesis of autoinducers or the interference with the reception of acyl-homoserine lactone [189]. Aqueous pomegranate extract showed anti-quorum sensing activity, reducing the violacein production, the quorum-sensing system's product, in *Chromobacterium violaceum* [186]. *Curcuma longa* L. extract, with curcumin and curcumin derivatives as main compounds, showed anti-quorum sensing activity inhibiting the violacein production in *C. violaceum*, probably disrupting the signal reception or the absorption of the acyl-homoserine lactone. However, this extract showed lower anti-biofilm activity against food pathogens than *Camellia sinensis* L. extract, rich in epigallocatechin and epicatechin [190].

The use of plant antimicrobials as a strategy to control the biofilm formation in foodborne pathogens gained increasing attention in recent years. Phytochemicals such as phenolic acids, tannins, sulphur compounds, and terpenoids (Figure 2) showed anti-biofilm activity by interfering with the quorum-sensing process of bacteria [191]. Despite these promising results, further research is needed to fully understand how these compounds exhibit antibacterial and anti-biofilm activity and to develop effective strategies for controlling biofilm formation during food processing.

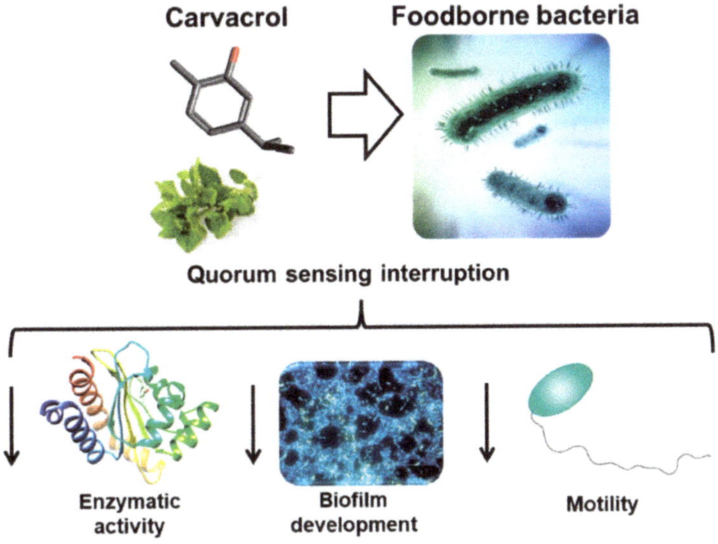

Figure 2. Anti-quorum sensing potential of carvacrol in foodborne bacteria.

3.2.3. Effect on Microbial Toxins

Microbial toxins (e.g., bacterial toxins and mycotoxins) harm human health. Plant antimicrobials were evaluated to reduce toxin production by foodborne bacteria and mycotoxins by filamentous fungi.

Bacterial exotoxins are proteins that damage host cells and are important for the pathogenesis of many bacterial pathogens, such as *Clostridium* spp., *E. coli*, *L. monocytogenes*, and *S. aureus* [192]. The use of plant antimicrobials can attenuate the virulence of these foodborne pathogens. In particular, different flavonoids suppressed the toxin production in different foodborne pathogens. Genistein inhibited the exotoxin produced by *S. aureus*, kaempferol, kaempferol-3-O-rutinoside, quercetin glycoside inhibited the neurotoxin production from *Cl. botulinum*, and green tea catechins inhibited the release of verotoxin from enterohemorrhagic *E. coli* [16]. Recent findings showed that the water-soluble fraction of the *Eucalyptus camaldulensis* Dehnh. leaf extract significantly reduced the listeriolysin O-induced haemolysis in *L. monocytogenes* at sub-inhibitory concentrations [193]. A witch-hazel extract, with hamamelitannin as the main phenolic compound, inhibited the production of the staphylococcal enterotoxin A in *S. aureus* at non-inhibitory concentrations for microbial cells [194]. As regards the EO and their compounds, sub-inhibitory concentrations of tea tree EO downregulated the transcription of genes encoding α-hemolysin, staphylococcal enterotoxin A, and staphylococcal enterotoxin B in *S. aureus*, inhibited their production, and the hemolytic activity [195]. Zhang et al. demonstrated that citronellal significantly reduced the production of enterotoxins in *S. aureus*-contaminated pork meat without reducing the viable cell load [196]. Other EO compounds, such as carvacrol and trans-cinnamaldehyde, reduced the production of TcdA and TcdB toxins produced by *Cl. difficile* in in vitro conditions [197]. Organic sulphur compounds such as the diallyl disulphide, at sub-inhibitory concentrations, reduced the production of the *B. cereus* enterotoxins Nhe and Hbl [91].

Plant antimicrobials also showed the ability to control the mycotoxin production by filamentous fungi. The mechanisms of action are the inhibition of the fungal growth and the induction of xenobiotic detoxification and/or the activation of biotransformation pathways [19]. In the first case, turmeric, rosemary and clove EOs demonstrated great efficacy in controlling the growth of mycotoxigenic *A. flavus* through the inhibition of ergosterol biosynthesis, the disruption of the fungal cell membrane, and the production of reactive oxygen species (ROS). In some cases, essential oils showed anti-aflatoxigenic activity at concentrations inhibiting or completely suppressing fungal growth. In contrast, in other cases, the anti-aflatoxigenic activity was detected at non-inhibiting concentrations. However, in a few cases, plant antimicrobials stimulated the production of secondary metabolites, including mycotoxins, in *Aspergillus* species [19]. Natural flavonoids such as baicalein, flavone, hispidulin, kaempferol, and liquiritigenin reduced the aflatoxin production in maize kernels contaminated with *A. flavus* by 50–67% [198], whereas a ternary mixture of naringin, neohesperidin, and quercetin reduced the aflatoxin accumulation in maize contaminated with *A. parasiticus* by more than 85% [199]. In sausages, the combined application of *Salvia farinacea* Benth. and *Azadirachta indica* A.Juss. extract at 2 mg mL^{-1} suppressed the production of ochratoxin A and aflatoxin B1 produced by *A. ochraceous* and *A. parasiticus*, respectively [200]. The degradation of aflatoxin B1 treated with the leaf extract from rosemary reached 60% after 48 h of incubation. Araçá (*Psidium cattleianum* S.) and oregano extracts produce less degradation than rosemary extract. Substances such as alkaloids and enzymes occurring in the plant extract might be involved in the structural modification of aflatoxin B1 [201]. Although the effect of plant antimicrobials on mycotoxin accumulation in food products is promising, more in-depth information regarding the toxicity of the resulting compounds from the degradation activity is required.

In conclusion, plant antimicrobials can reduce or suppress the production of bacterial and fungal toxins by reducing microbial growth or downregulating toxin gene expression. Further research is necessary to understand the modes of action of different plant extracts and their bioactive compounds on toxin production to exploit their potential to improve food safety under real contamination conditions.

4. Stabilization Techniques

Plant antimicrobials can have limited stability under processing or storage conditions of foods. The efficacy of plant antimicrobials is affected by several factors such as pH, the temperature, and the concentration. Caffeic, chlorogenic, and gallic acids are not stable at high pH values, whereas chlorogenic acid is stable at low pH values and heat [202]. Some phenolic compounds and EOs, and their compounds, are thermolabile. *Achillea* sp., rosemary, sage (*Salvia officinalis* L.), and thyme EOs were more effective at low pH and low temperature against pathogenic bacteria [203], whereas carvacrol and cymene showed higher antibacterial activity in carrot juice at 25 °C than at 4 °C and 15 °C [204]. Several plant antimicrobials show a dose-dependent effect against spoilage and pathogenic microorganisms. The stabilization techniques described in this section can help to protect plant antimicrobials and, in some cases, reduce the concentration necessary to exert their antimicrobial activity. The direct addition of plant extracts or their bioactive compounds in foods is the most common method of food preservation. However, the direct addition of plant extracts is often responsible for changes in sensory properties such as flavour and texture. In addition, the bioavailability of these compounds and their effectiveness in improving food safety can be affected by the interaction with the macronutrients and ingredients. For these reasons, several stabilization techniques were proposed to enhance stability, drive the release of bioactive compounds during storage, and reduce the negative effects of plant extracts on the sensory characteristics of foods.

4.1. Nano-Emulsions

The encapsulation of plant antimicrobials into edible colloidal delivery systems is a promising method to enhance the efficacy of these substances and reduce the negative effects due to the interaction with food ingredients. In particular, encapsulation in small particles increases water dispersibility and resistance to environmental conditions enhancing plant antimicrobials' efficacy [6]. Oil-in-water nano-emulsions containing lipid nanoparticles dispersed in water are currently the most common delivery system for plant antimicrobials. These nano-emulsions can be manufactured from food-grade ingredients, such as plant-based emulsifiers and different stabilizers, using common processing methods, such as mixing (low-energy emulsification), sonication, and homogenization (high-energy emulsification) [6].

Different studies investigated the efficacy of nano-emulsions against foodborne pathogens. The plant antimicrobials most used to prepare nano-emulsions are the EOs and their compounds. Lemongrass, clove, thyme, or palmarosa (*Cymbopogon martini* Will. Watson)-loaded EOs nano-emulsions, prepared after micro fluidization of the primary emulsion, inactivated *E. coli* by 3–4 log cfu mL^{-1}. The use of alginate in the aqueous phase is useful for applying these nano-emulsions in the coating material of fruits and vegetables [205]. Anise (*Pimpinella anisum* L.) oil nano-emulsions showed the same MIC (1% v/v) of the bulk EO and coarse emulsion against *L. monocytogenes* and *E. coli* O157:H7. However, the anise oil nano-emulsion displayed the highest physical stability and antibacterial efficacy [206]. More recently, other plant antimicrobials were used to prepare nano-emulsions to control the growth of bacterial pathogens. Anise seed extract, with anethole, naringenin, and taxifolin as main compounds, was used to develop an antibacterial nano-emulsion using the ultrasound emulsification method. The nano-emulsion was active against *E. coli* and *Sal. thyphimurium*, whose growth was not affected by the bulk extract [207]. Ghazy et al. [208] evaluated the antimicrobial action of henna (*Lawsonia inermis* L.) extract as a nano-emulsion against seven pathogenic bacteria. The nano-emulsion, rich in catechin, methyl gallate, ellagic acid, and coumaric acid, displayed higher antimicrobial activity against *E. coli*, and *B. cereus*, than the course emulsion. Regarding the application of nano-emulsions, including plant essential oils, to control pathogens in plant foods, oregano oil nano-emulsion at 0.1% reduced the load of *L. monocytogenes*, *Sal. typhimurium*, and *E. coli* O157:H7 on lettuce by 3 log cfu g^{-1} [209]. Cinnamon oil nano-emulsion at 0.5% determined more than five log reductions in *L. monocytogenes* and *Salmonella* spp. on melon [210]. Lemongrass and mandarin (*Citrus reticulata* Blanco) EO nano-emulsions inactivated *E. coli* in apple juice, but when the nano-emulsions were prepared directly in the apple medium as a continuous phase, the antibacterial efficacy was reduced in comparison to the use of water [211]. Citral nano-emulsions at 0.15 μL mL^{-1} inactivated *L. monocytogenes* (5 log cfu g^{-1} reduction) on fresh-cut melon and papaya during cold storage [212].

Regarding the efficacy of nano-emulsions including plant antimicrobials against spoilage microorganisms, thyme EO nano-emulsion showed lower efficacy than bulk EO against fish spoilage bacteria, except for *Serratia liquefascens* [213]. For this spoilage bacteria, laurel (*Laurus nobilis* L.) and grapefruit (*Citrus paradisi* Macfad.) EO nano-emulsions showed lower MIC values than the corresponding EOs [214,215]. Ginger EO nano-emulsion, prepared with zein and sodium caseinate as co-emulsifiers, showed higher bactericidal activity against total viable counts of chicken breasts than the bulk EO, extending the shelf life of the product by 6 days [216]. As regards the antifungal activity of plant antimicrobial nano-emulsions, cinnamaldehyde, eugenol, and carvacrol nano-emulsion showed a dose-dependent effect against the spore germination and mycelial growth of *P. digitatum*, with a MIC value of 0.125 mg mL^{-1} [217]. Gundewadi et al. [218] found that basil EO nano-emulsion displayed lower lethal concentration values (LC_{50}) than course emulsion against *P. chrysogenum* and *A. flavus* during 8 days of incubation. Oregano and clove EOs nano-emulsions, at 1.95 mg g^{-1}, showed fungicidal activity against *Z. bailii* in a salad dressing after 4 days of storage [219]. EOs nano-emulsions also showed anti-mycotoxigenic activity. Indeed, lemongrass EO nano-emulsion reduced by 99.5% the deoxynivalenol content in rice contaminated with *F. graminearum*. The lemongrass EO nano-emulsion showed better anti-mycotoxigenic activity than the bulk EO, but the efficacy was strain-specific [220]. Oregano

EO encapsulated into chitosan nano-emulsion suppressed the production of aflatoxin B1 by *A. flavus* in maize [221].

Given these results, the antimicrobial action of plant antimicrobial nano-emulsions depends on the chemical composition of the plant extract, the emulsion droplet size, and the target microbial species. In addition, many studies demonstrated higher efficacy of nano-emulsion than course emulsion and bulk plant extract.

4.2. Spray-Drying and Encapsulation

Spray-drying and encapsulation are techniques commonly used to improve the stability and functionality of plant antimicrobials in food products. Spray-drying is a process in which a liquid solution or suspension is atomized into a hot air stream, causing the rapid evaporation of droplets, resulting in a dry powder. This process can produce dry powders of plant antimicrobials that are more stable and easier to handle than the liquid form. Spray-drying can also encapsulate the plant antimicrobials in a protective matrix, improving their stability and functionality. Encapsulation is a process in which a natural antimicrobial is surrounded by a protective matrix, such as a polymer or lipid, to improve its stability and functionality. Encapsulation can enhance the natural antimicrobials' shelf life and protect them from degradation induced by light, heat, or moisture. Additionally, encapsulation can improve the solubility and dispersibility of natural antimicrobials, making them easier to incorporate into foods. These techniques can help to preserve the antimicrobial activity of the natural antimicrobials and improve their effectiveness in controlling the growth of spoilage and pathogenic microorganisms.

4.2.1. Spray-Drying Process

Spray-drying and encapsulation techniques provides numerous benefits in handling, storage, and transportation of plant antimicrobials. Powdered antimicrobials are more suitable for various applications within the food industry [222]. Powdered antimicrobials minimize the risk of spillage and waste during handling and processing, as they can be easily measured and transferred without causing mess or loss of material. This ensures a more efficient use of resources and reduced operational costs. In some cases, the spray-drying process conditions can have pros and cons related to the stability and functionality of plant extracts, as briefly pointed out in Table 2.

Table 2. Advantages and disadvantages of the spray-drying process for producing stable and functional plant antimicrobial powders.

Plant Antimicrobial	Spray Drying Inlet Temperature (°C)	Protective Matrix	Microbial Targets	Advantages	Disadvantages	Data from Ref.
Eugenol and thymol	105	Zein/casein	*E. coli* O157:H7, *L. monocytogenes* Scott A	Good dispersion in water and good stability during storage	not reported	[223]
Eugenol	180	Whey protein/ maltodextrin/ chitosan	*E. coli*, *L. innocua*	High encapsulation efficiency and thermal stability	Chitosan inclusion negatively affects thermal stability, releasing and antimicrobial properties of the powder	[224]
Carvacrol	100–190	Pectin/sodium alginate	*E. coli* K12	Better thermal stability	High inlet temperature affects dissolution time and hygroscopicity	[225]

Table 2. *Cont.*

Plant Antimicrobial	Spray Drying Inlet Temperature (°C)	Protective Matrix	Microbial Targets	Advantages	Disadvantages	Data from Ref.
Oregano EO	100	Whey protein/ maltodextrin	*E. coli, S. aureus*	Low residence time, high yield, low inlet temperature	Low throughput, extended processing hours, high production cost	[226]
Green tea extract	150	Maltodextrin	-	High thermal stability and reduced weight loss	not reported	[227]

In the study of Chen et al. [223], eugenol and thymol were co-encapsulated into zein-casein nano-capsules through spray-drying. The resulting powders showed good water hydration, stability during storage, controlled release during 24 h, and bactericidal and bacteriostatic effects against *E. coli* O157:H7 and *L. monocytogenes* in milk whey, respectively. Thyme EO encapsulated by spray-drying, with casein and maltodextrin as wall materials, showed antibacterial action against thermotolerant coliforms and *E. coli* in meat burgers [228]. The wall material employed to protect plant antimicrobials can affect their antibacterial action. Indeed, the inclusion of chitosan in a whey protein/maltodextrin blend reduced the antibacterial action of eugenol against *E. coli* and *L. innocua*. A low inlet temperature used in the spray-drying of pectin/sodium alginate capsules including carvacrol, increased the antibacterial activity against *E. coli* K12 (Table 2). The use of nano spray-drying, a novel process to produce plant antimicrobial powders, was evaluated to obtain whey protein/maltodextrin capsules, including oregano EO. The capsules showed enhanced antibacterial action against *E. coli* and *S. aureus* compared to pure EO. However, this process has drawbacks such as high production costs, high processing time, and reduced spraying effectiveness of viscous solutions [226]. These examples illustrate the potential benefits and drawbacks of the spray-drying technique to produce plant antimicrobial powders. The specific advantages and disadvantages observed depend on the antimicrobial compound, the spray-drying conditions, and the choice of the protective matrix.

The increased stability of powdered antimicrobials extends their shelf life, as demonstrated for the encapsulated peanut (*Arachis hypogaea* L.) skin extracts [229]. It maintains the efficacy of plant antimicrobials throughout storage, reducing the need for frequent replacements and ensuring consistent antimicrobial action.

Powdered antimicrobials generally have lower storage requirements than their liquid counterparts, as they do not require refrigeration or specific storage conditions to maintain their stability. This reduces energy consumption and storage costs for food manufacturers. Additionally, the nature, weight, and form of powdered antimicrobials facilitates more efficient transportation and shipping, as they occupy less space and require less protective packaging than liquid antimicrobials [222]. Powdered antimicrobials can be more easily integrated into various food matrices, as their fine and uniform particles allow a more homogeneous distribution throughout the product. This ensures consistent antimicrobial protection across the whole food matrix, enhancing food safety and quality.

4.2.2. Other Encapsulation Techniques of Plant Antimicrobials

Encapsulation of plant antimicrobials, a method of entrapment of a core material within another solid or liquid immiscible substance, allows the production of capsules or spheres in micrometre to millimetre in size [230]. Encapsulation can involve various types of protective matrices that impact the stability and functionality of the antimicrobial agents. Some common matrices for encapsulation include polysaccharides, lipids, and proteins. Polysaccharides such as alginate, chitosan, and maltodextrin are widely used as encapsulating agents due to their biocompatibility, non-toxicity, and excellent film-forming properties. A study by de Araújo et al. [231] demonstrated that using maltodextrin/gelatine mixtures as a protective matrix for encapsulating the sweet orange EO positively affected the thermo-oxidative stability of bioactive compounds and maintained its antibacterial properties. The encapsulation of plant antimicrobials can be obtained through the formation of inclusion complexes using the β-cyclodextrins, cyclic oligosaccharides with amphipathic properties. These complexes can stabilize the guest molecule against the degradation, mask off-flavours, and control the release of the encapsulated compounds [232]. Thyme EO microcapsules exerted a bacteriostatic effect over *Enterobacteriaceae*, mesophilic bacteria, and psychrotrophic bacteria on lettuce [232], whereas inclusion complexes with rosemary EO showed better antimicrobial activity against *Saccharomyces pastorianus* than free EO in pasteurised tomato juice [233]. Coriander (*Coriandrum sativum* L.) EO encapsulated in β-cyclodextrin nano-sponge showed bactericidal activity against *L. monocytogenes*, *Y. enterocolitica*, and *Cam. jejuni* in aqueous media [234]. Black pepper (*Piper nigrum* L.) oleoresin was stabilized in β-cyclodextrins using the kneading method, a method in which the β-cyclodextrins and the guest compound are mixed with small amounts of ethanol or water using a kneader for a specific time, showing antimicrobial activity against *L. monocytogenes* and improved thermal stability [235].

Lipid-based encapsulation systems, such as solid lipid nanoparticles, the use of nanostructured lipid carriers, and liposomes, are also employed for encapsulating plant-derived antimicrobials. These systems can improve the stability of the encapsulated compounds, their bioavailability, and ensure a controlled release. The study by Lin et al. [236] reported that encapsulating chrysanthemum (*Chrysanthemum flosculosum* L.) EO in triple-layer liposomes led to long-term antimicrobial activity against *Cam. jejuni* in chicken.

Protein-based matrices, such as gelatine, soy protein, and whey protein, can also encapsulate plant-derived antimicrobials. These matrices offer advantages in biodegradability, biocompatibility, and the ability to form stable complexes with antimicrobial agents. Recently, the microencapsulation of cinnamon EO using chitosan and whey protein isolate showed enhanced thermal stability and long-term antimicrobial effect against *S. aureus*, *E. coli*, *P. fragi*, and *Shewanella putrefaciens* [237]. The selection of the most suitable matrix depends on the specific antimicrobial compound, the target application, and the desired release characteristics.

4.2.3. Challenges Associated with Spray-Drying and Encapsulation Techniques

Although spray-drying and encapsulation techniques offer various advantages for the stabilization and incorporation of plant-derived antimicrobials into food systems, there are challenges associated with these processes that require further research. One of the issues associated with spray-drying is the potential degradation or loss of activity of heat-sensitive plant compounds during the drying process, as high temperatures are often involved [222]. This can lead to reduced antimicrobial efficacy or the modification of sensory properties. Additional and specific research studies need to be carried out to explore alternative drying techniques, such as freeze-drying or nano spray-drying, that might better preserve heat-sensitive compounds. Another challenge is the selection of the most appropriate encapsulation matrix to ensure optimal protection, release, and stability of the encapsulated antimicrobial compound. The choice of the encapsulation material can greatly affect the effectiveness and shelf life of the antimicrobial agent for food applications [238]. Further research is needed to understand the interactions between various wall materials and plant-derived antimicrobials, and to optimize the encapsulation processes for specific food systems. Additionally, scaling up from lab-scale to industrial-scale production of encapsulated plant-derived antimicrobials poses challenges related to the encapsulation efficiency, the product stability, and the process economics [239]. More research is also required to develop cost-effective and efficient methods for large-scale production, maintaining the quality and functionality of the encapsulated antimicrobials.

4.3. Active Packaging

Active packaging involves the deliberate inclusion of subsidiary constituents in or on either the packaging material or the package headspace to enhance the performance of the package system. Active packaging can preserve the food quality and can extend the product's shelf life through the direct interaction between the food and bioactive substances intentionally incorporated into the package [240]. Antimicrobial packaging is one type of active packaging, in which the antimicrobial activity strongly depends on the migration rate of the biologically active molecule incorporated into the polymer matrix [240]. Recently, the market was oriented to replace packaging produced with fossil fuels with more sustainable materials such as biopolymers. In this context, the use of biopolymers, including plant antimicrobials, is very attractive to develop active films or coatings, to overcome the thermal/oxidative instability of these compounds during the manufacturing of the polymers or storage of the final product, and to mask undesirable sensorial aspects of some plant extracts. In addition, biopolymers have a lower environmental impact compared to standard food packaging polymers such as polyethylene or polypropylene. Plant antimicrobials can be added directly into the biopolymer or loaded into clays or nanocarriers, as demonstrated for EO compounds [241,242] and polyphenols [243,244]. This strategy ensures, in most cases, their controlled release and, in some cases, an improvement in the mechanical and physical properties of the film.

The most used biopolymers for including plant antimicrobials are chitosan, starch, carrageenan, cellulose, and alginate. However, other polymers used for this purpose are polyvinyl alcohol (PVA), poly lactic acid (PLA), poly butylene-succinate-co-adipate (PBSA), poly butylene-adipate-co-terephthalate (PBAT), poly(hydroxybutyrate)s (PHBs), and poly (ε-caprolactone) (PCL).

Regarding active chitosan films, the inclusion of apple peel polyphenols (1%) into chitosan film enhanced the antibacterial activity against *B. cereus*, *E. coli*, *Sal. typhimurium*, and *S. aureus* [245]. A composite film based on grapefruit seed extract-loaded poly(ε-caprolactone)/chitosan reduced the *E. coli* population on salmon by more than 2 log cfu g^{-1} after 6 days at 4 °C compared to the packaging into polyethylene or poly(ε-caprolactone)/chitosan films, and suppressed mould development on bread stored for 7 days at 24 °C [246]. Surendhiran et al. [247] developed active nanofibers based on chitosan/Poly (ethylene oxide) loaded with pomegranate peel extract. The nanofibers reduced by 3 log cfu g^{-1} the *E. coli* O157:H7 population in raw beef stored at 4 °C for 10 days. The coating of fresh cucumber with chitosan loaded with oregano EO reduced the viability of total mesophilic bacteria and total yeasts and moulds during storage at 10 °C for 15 days [248]. Starch films were also enriched with plant antimicrobials as demonstrated by Saberi et al. [249] that developed pea starch-guar gum films including epigallocatechin-3-gallate and blueberry ash fruit (*Elaeocarpus reticulatus* Sm.) and macadamia (*Macadamia tetraphylla*) skin extracts. Active films showed antimicrobial activity against spoilage bacteria and fungi, and pathogenic bacteria, with a reduction in microbial load in the range of 40–80% for the films loaded with epigallocatechin-3-gallate and blueberry ash fruit skin extracts at the MIC level (ranging from 93 to 1500 µg mL^{-1}). A bio-composite film made with cassava starch and whey protein loaded with rambutan peel extract and clove oil slightly inhibited *B. cereus*, *E. coli*, and *S. aureus* in in vitro conditions, and reduced total viable count of salami stored for 10 days [250]. A sweet potato starch-based film activated with montmorillonite nano-clay and thyme EO reduced the load of *E. coli* and *S. thypimurium* on fresh spinach leaves during 8 days of cold storage [251]. Other biopolymers used to manufacture active films/coatings with plant antimicrobials are alginate and carrageenan. A sodium alginate film loaded with the gallnut extract (*Quercus infectoria* Oliv.), rich in gallotannins, ellagic acid, and gallic acid, showed antibacterial activity against *S. aureus* and *E. coli* [252], whereas the coating of apples and pears with alginate loaded with cinnamon EO at 0.9% v/v inhibited the *A. carbonarius* growth and the OTA production [253]. Compared with the control film, a carrageenan film containing 3% rosemary extract displayed >99% inhibition against *B. cereus*, *E. coli*, *P. aeruginosa*, and *S. aureus*, reducing by 2–4 orders of magnitude the microbial load [254]. He and Wang [255] recently demonstrated that a K-carrageenan coating enriched with cinnamon EO delayed the growth of total viable count, lactic acid bacteria, and H_2S-producing bacteria in pork meat. Active films incorporating plant antimicrobials can be also produced using proteins. Indeed, zein nanofibers loaded with 1,8-cineol rich extracts reduced the load of *L. monocytogenes* and *S. aureus* on cheese slices by two orders of magnitude during 28 days of cold storage [256].

Other active films, including plant antimicrobials, can be manufactured using PLA and PHBs. A PLA/PBAT composite film including 7% *w*/*w* of grapefruit seed extract showed bactericidal activity against *L. monocytogenes* and a bacteriostatic effect against *E. coli* [257]. Additionally, a PLA/PBSA blend including 6% *w*/*w* thymol delayed the mould development on bread compared to polypropylene or neat PLA [258]. PHBV films loaded with eugenol and carvacrol showed antibacterial action against *E. coli* in cheese and pumpkin but not in melon, where the highest release of the active compounds from the films was observed [259]. This result highlights that the interaction of antimicrobial compounds with the food components and their diffusion within the food matrix play an important role in the antimicrobial activity of active films including plant antimicrobials.

Plant antimicrobials can be incorporated into active packaging using encapsulated extracts or nano-emulsions. Pabast et al. [260] developed a chitosan film in which the EO extracted from *Satureja khuzestanica* J. was encapsulated into nanoliposomes. The film delayed the growth of total mesophilic bacteria, *Pseudomonas* spp., and lactic acid bacteria of lamb meat stored for 20 days at 4 °C. Interestingly, the antimicrobial effect was higher than that displayed by the film including free *Satureja khuzestanica* J. EO [260]. Chitosan film loaded with microcapsules, including basil EO, slightly reduces total mesophilic bacteria, enterobacteria, and lactic acid bacteria on cooked ham during storage [261]. A PVA loaded with cinnamon EO encapsulated in β-cyclodextrin showed a bacteriostatic effect against *S. aureus* and *E. coli* [262]. In addition, a PVA/starch film including β-cyclodextrin inclusion complex embedding lemongrass EO showed antibacterial action against *She. putrefaciens* [263]. Nano-emulsions of essential oils were used for the inclusion in active packaging. Lee et al. [264] developed hydroxypropyl methylcellulose-based films incorporating oregano EO nano-emulsions. The active film showed inhibition zones against *Sal. thyphimurium*, *E. coli*, *L. monocytogenes*, *B. cereus*, and *S. aureus*. Chitosan-*Ferulago angulata* essential oil nano-emulsion showed lower MIC and MBC values against the fish-spoilage bacteria *P. fluorescens* and *She. putrefaciens* than the corresponding coating emulsions. Moreover, the coatings, including the nano-emulsion, reduced the total viable and psychrotrophic counts of rainbow trout fillets by 3 log cfu g^{-1} after 16 days of storage at 4 °C [265]. The incorporation of *Zataria multiflora* Boiss. EO and cinnamaldehyde in the form of nano-emulsions into starch coatings reduced the growth of *L. monocytogenes*, psychrotrophic bacteria, and *Enterobacteriaceae* in chicken during cold storage [266].

5. Combining Effects and Hurdle Technologies

In this section, the more recent studies demonstrating the possible improvement in food safety and shelf life by combination of bioactive compounds from different plant-based extracts or by their combination with non-thermal or mild food technologies are briefly reported.

5.1. Additive or Synergistic Effects

The use of combinations of antimicrobial plant extracts and their compounds showed additive or synergistic effects against spoilage and pathogenic microorganisms. This approach is cost-efficient for the food industry and adheres to the hurdle technology in inhibiting the proliferation of undesirable microorganisms, improving the preservative effects of plant antimicrobials and reducing the negative sensory effects of single plant extracts [2].

Additive or synergistic effects are found in combinations of plant extracts and their compounds, reducing the MIC of the plant antimicrobials. As regards the additive effects, cinnamon EO with clove EO showed an additive effect against *L. monocytogenes* [67], whereas the combination of cinnamaldehyde with 2-hydroxycinnamic acid showed additive effects against *L. monocytogenes* and *Sal. enteritidis* under in vitro conditions, but it was not effective in contaminated cooked ham [267]. Regarding the synergism among plant antimicrobials, thyme EO with cinnamon EO (0.312 g L^{-1}), cinnamon EO (0.156 g L^{-1}) with rosemary EO (0.625 g L^{-1}) and thyme EO (0.078 g L^{-1}) showed synergistic effects in inhibiting *Alt. alternata* and *P. expansum* on jujube fruit [268]. Cinnamon EO with clove EO showed synergistic antibacterial activity against *S. aureus*, *L. monocytogenes*, and *Sal. typhimurium* [269], and against *L. monocytogenes* when vanillin was combined with both EOs [67]. Synergistic effects were found using combinations of EOs compounds or isothiocyanates with phenolic acids against bacterial pathogens. The combination of thymol with gallic acid determined a synergistic effect at sub-inhibitory concentrations against *E. coli* O157:H7 and *S. aureus* on fresh-cut tomatoes [270]. Allyl isothiocyanate with o-coumaric acid showed synergism, obtaining 2 log reduction in *E. coli* O157:H7 in a dry-fermented sausage when the antimicrobial compounds were added at the concentration of 6.25 μL and 750 mg per 100 g fresh weight, respectively [271].

As regards the antifungal interactive effect of plant antimicrobials, a triple combination of thyme EO, cinnamon EO, and rosemary EO showed a synergistic antifungal effect against *B. cinerea* and *P. expansum*, reducing their development on pear [272]. Pinto et al. [64] demonstrated a synergistic effect in the vapour phase between thymol and γ-terpinene in binary combinations and between p-cymene, γ-terpinene, and thymol in ternary combinations against the strain *P. digitatum* ITEM 9569, which is resistant to single thyme EO exposure. The use of combinations of plant antimicrobials to control the development of spoilage and pathogenic microorganisms in foods is a research area showing rapid development. The use of combinations of plant antimicrobials can reduce the concentration of plant antimicrobials added in foods, minimizing the negative impact of these compounds on the sensory properties of foods as previously demonstrated [269,270]. Studies related to this topic are expected to increase in the future, paying attention to the effect of the combination of plant antimicrobial compounds in real food matrices, and the elucidation of the modes of action.

5.2. Hurdle Technologies

The "hurdle approach" in the food sector refers to the successive or simultaneous application of two or more food preservation techniques for enhancing food safety and quality using lower individual treatment intensities and for achieving multi-target, mild, and reliable preservation effects [273]. This approach was followed to reduce the dose of chemical preservatives used to control the development of spoilage microorganisms in foods [274]. Recently, several studies investigated the application of mild or non-thermal technologies in combination with the use of plant antimicrobials to control the growth of spoilage and pathogenic bacteria in food and to extend the shelf life.

The combined application of high-pressure homogenization (HPH) and nano-emulsions of hexanal and trans-2-hexanal inactivated *S. cerevisiae* in apple juice up to 22 days of storage, with better performance as compared to individual treatments [275]. Citral (1% *w*/*w*) combined with high-pressure processing reduced the viability of a cocktail of *E. coli* STEC in ground beef by 4–7 log cfu g^{-1}, depending on the pressure level applied [276]. The treatment of ground chicken meat with 320 MPa for 23 min at 4 °C with allyl isothiocyanate and acetic acid at ca. 0.2% *w*/*w* achieved a 5-log reduction in *E. coli* O157:H7, with a better inactivation compared to single treatments [277]. As regards the application of cold plasma technology with plant antimicrobials, González-González et al. [278] found that the combined application of cold plasma and linalool nano-emulsion reduced by 3 log cfu g^{-1} the load of *E. coli* O157:H7 and *Sal. enterica* in chicken meat, while individual treatments showed limited efficacy. The effect of plant antimicrobials combined with food technologies against *E. coli* O157:H7 on chicken meat is depicted in Figure 3.

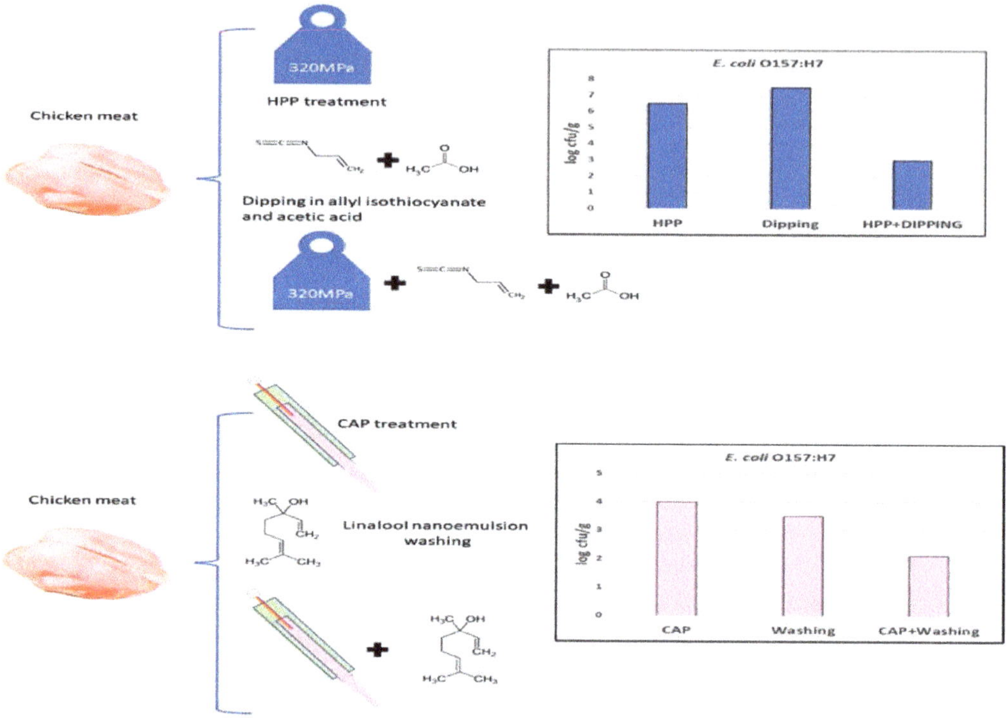

Figure 3. Hurdle effect of plant bioactive compounds against *E. coli* O157:H7 contaminating chicken meat when combined with high pressure or cold plasma treatments. The picture was made by drawing the main results from references [277,278].

Sea bass slices packed under modified atmosphere packaging and pre-treated with cold plasma and a liposomal ethanolic coconut husk extract showed the lowest increase in *Pseudomonas* spp. and *Enterobacteriaceae* during 18 days at 4 °C in comparison to cold plasma treatment alone or the application of the liposomal ethanolic coconut husk extract [279]. On the contrary, the combined use of cinnamon EO and modified atmosphere packaging showed limited efficacy against spoilage bacteria of lean pork meat or salmon during cold storage [280]. The pulsed electric field pre-treatment of Pacific white shrimp, followed by the soaking in 1% of Chamuang (*Garcinia cowa* Roxb.) leaf extract, showed a lower increase in mesophilic, psychrophilic, *Pseudomonas* spp., *Enterobacteriaceae*, and H_2S producing bacterial counts in comparison to individual treatments and the application of sodium metabisulfite during cold storage [281]. A synergistic effect in reducing the microbial load of *E. coli* O157:H7 was found between thyme EO nano-emulsion treatment and the sonoporation induced by ultrasounds [282]. Microwave heating at 915 MHz with carvacrol showed a synergistic effect against *E. coli* O157:H7, *Sal. typhimurium* and *L. monocytogenes* in buffered peptone water but not in hot chilli sauce [283].

Light technologies were also combined with the use of plant antimicrobials to control the contamination by foodborne pathogens. In reconstituted powdered infant formula, the load of *Cro. sakazakii* was reduced by 6.5 log cfu mL^{-1} following the combined 405 nm light-emitting diode and 9 μL mL^{-1} citral treatment for 90 min compared with untreated samples [284]. Silva-Espinosa et al. [285] found that the combined application of UV-C light and clove EO on stainless steel surface achieved a complete bacterial reduction (6.8 log cm^{-2}) on biofilms of *Sal. thyphimurium*.

The use of chemical compounds was associated with plant extracts and their compounds. The nano-emulsion of thyme EO enhanced the antimicrobial effect of slightly acidic electrolyzed water against foodborne pathogens, suggesting the formation of complexes probably through hydrophobic interactions [286]. Based on total viable counts, a shelf-life extension of 7 days was observed in fresh fish fillets treated with gaseous ozone and coated with alginate, including different EOs and citrus extract, as compared to single treatments [287].

The combination of mild or non-thermal technologies with plant antimicrobials is a recent research trend. Many of these studies showed synergistic effects in reducing the microbial load of spoilage and pathogenic microorganisms on foods. These effects can be explained by the exposure of target microorganisms to multiple hurdles and stresses. Even though more studies are necessary to understand the modes of action of these combined approaches, the enhanced antimicrobial activity of plant extracts and their compounds, when coupled with innovative technologies, allows to reduce their concentration in different foods, mitigating some of their drawbacks such as the modification of sensory characteristics.

6. Regulation and Safety Issues of Plant Extracts

Plant extracts can be contaminated by different dangerous compounds, such as heavy metals, crop-protection residues, and mycotoxins. The concentration of these compounds depends on the cultivation practices employed, the geographical location of the cultivation site, the application of crop protection products, and the extraction method. In this section, data related to the contamination levels of plant extracts employed for the production of plant antimicrobials for food purposes are reported.

6.1. Heavy Metals and Crop-Protection Residues

Heavy metals can contaminate plant extracts. In plants that produce EOs, the uptake of metals is associated with soil contamination, and their transfer to EOs depends on the extraction technology. Moreover, the storage of EOs in metallic containers can promote the transfer of metals into the oil. As, Cd, Pb, and Hg cause toxic effects at relatively low levels. For Cd, Hg, and Pb, recommendations for safety limits in medicinal plants are imposed by European Pharmacopoeia, FAO, and the World Health Organization (WHO). Iordache et al. [25] evaluated the heavy metal content of EOs from different sites. High levels of Hg, Cr, Pb, Cu contamination were found in *Mentha* × *pipperita* L. EO. However, the authors concluded that the analyzed EOs could be safely consumed in the doses recommended by the manufacturers, and the content of heavy metals does not pose a significant risk to the consumer's health. High levels of Cr, Cd, and Pb in thyme and oregano plant samples were found by Reinholds et al. [288].

Regarding the contamination levels of crop-protection residues in EOs, they depended on the agricultural practices. Conventional orange EOs contained 17 pesticides and a total concentration of 5.1 mg L^{-1}, whereas organic orange EOs contained only 4 pesticides and a total concentration of 0.087 mg L^{-1} [27]. Organophosphorus and organochlorine pesticide residues were found in citrus EOs, especially those produced by cold-pressing and conventional agriculture practice, albeit with a concentration lower than 1 mg L^{-1}. Tebuconazole and propiconazole co-distillated in peppermint EO with a different degree depending on the vapour pressure [289]. Cymoxanyl, dimethoate, and tebuconazole residues exceeded the maximum residue level set by the European Union in thyme samples from Poland [288]. In conventional grape skin extracts, Boscalid, Fludioxinil, Mycobutanil, and Pyraclostrobin levels exceeded the maximum residue level. However, the concentration of these crop-protection products was lower than the detection limit or the maximum residue level in organic grape skin extracts [290].

6.2. Mycotoxins

Mycotoxins can contaminate plant extracts. Different studies evaluated the mycotoxin contamination of plants such as *Mentha* sp. and *Zingiber officinale* R., which are generally used to produce EOs. Although the plant material showed levels of mycotoxins within the EU regulation limits, in some cases, the aflatoxin concentration exceeded acceptable standards [26]. In addition, the fungal contamination and mycotoxin accumulation in *Z. officinale* R. decreased the bioactive compounds of ginger [291]. Dried thyme herbs from Lebanon showed 75% of samples exceeding the limit of aflatoxin B1 for spices according to the European regulation. Similarly, the OTA level exceeded the maximum limits for Lebanese thyme and thyme mixes in 13% of the samples [292]. Zearalenone and deoxynivalenol contamination (range 10–209 $\mu g\ kg^{-1}$) was detected in thyme samples from Poland [288]. Additional data are required to monitor the mycotoxin contamination of plant extracts (e.g., essential oils, phenolic extracts) and to establish regulatory limits.

6.3. Regulation

Comprehensive toxicological studies are necessary to obtain the approval of plant antimicrobials as food preservatives by the European Food Safety Authority (EFSA), the Food and Drug Administration (FDA), and the China Food Additives Association (CFAA). Many plant antimicrobials have the GRAS status for specific food applications, but their use in other food applications is not expressly approved. Indeed, plant phenolics are actually absent in the positive list of food preservatives. Regulatory authorities' approval of plant extracts as food additives is essential to ensure consumer safety and confidence. Such authorization must be based on comprehensive safety assessments, including toxicological studies, exposure assessments, and evaluations of factors such as purity, stability, and potential allergenicity. Regulatory approval helps to guarantee that these plant extracts are safe for human consumption and meet specific quality standards. However, there can be challenges and barriers in the regulatory approval process. The complexity of plant antimicrobial mixtures and the need for extensive safety data may lead to time-consuming and costly approval processes. Global policy inequities significantly impact the approval of plant extracts as food additives. Addressing these inequities requires international cooperation, knowledge sharing, and capacity building to develop robust regulatory frameworks, protect and promote traditional knowledge, enhance research capacity in developing countries, and ensure fair and equitable access to the global market for plant-derived products. In order to improve and expedite the approval of plant extracts as food additives by authorities, several actions could be taken: encourage international collaboration and communication to harmonize regulations, establishing consistent guidelines and standards, increase investment in research and development to generate robust scientific data on plant extracts' safety, efficacy, and potential applications, strengthen developing countries' research and regulatory capacities through training, resources, and technical assistance, facilitate knowledge sharing and collaboration among various stakeholders, including researchers, industry professionals, and regulatory authorities, and employ advanced technologies such as data science for data-driven decision-making in the approval process. In turn, this will ensure consumer safety, promote the sustainable use of plant resources, and contribute to the growth of the global market for plant-derived products.

7. Conclusions and Future Directions

Plant antimicrobials gained considerable attention as promising alternatives to synthetic preservatives in the food industry, offering numerous benefits such as enhanced safety, extended shelf life, and increased consumer acceptance. The observed additive or synergistic effects between plant extracts, essential oils, and their compounds, and the successful integration of hurdle technologies, contributed to the scouting of novel and mild food preservation methods. Despite the significant progress made in the field, several research areas warrant further investigation. A deeper understanding of modes of action of plant antimicrobials and their combinations, including their effects at molecular and cellular levels on target microorganisms, is crucial for optimizing their application and enhancing their efficacy. The development of optimized formulations and delivery systems, such as nano-emulsions, encapsulation, or edible coatings, will improve stability, bioavailability, and targeted delivery of plant antimicrobials and, as a consequence, will increase their use in various food systems. A thorough examination of the impact of plant antimicrobials on the sensory properties of food is important to ensure consumer acceptance, focusing on minimizing adverse effects on taste, aroma, and texture while maintaining antimicrobial effectiveness.

Conducting comprehensive toxicological studies on plant antimicrobials, their derivatives, and combinations is essential to establish safe consumption levels and guarantee consumer safety. Additionally, addressing the challenges and barriers to regulatory approval for plant antimicrobials as food additives will accelerate their broader application in the food industry. The development of sustainable and environmentally friendly methods for extracting and producing plant antimicrobials aligns with the overall sustainability goals of the food industry. Given the knowledge on the chemical composition of plant extracts and the antimicrobial activity of plant compounds, plant antimicrobials could be recovered from agri-food by-products or from food waste, creating new value chains. Moreover, since non-edible wild plant species are often a good source of plant antimicrobial compounds, these species could be cultivated in marginal areas, promoting new cultivation practices in depressed rural territories, and also increasing the sustainable production of plant additives for the food industry. Increasing consumer awareness of plant antimicrobials' benefits and their role in food safety and preservation will enhance market acceptance and drive demand for such products.

The description of the results allows us to decontextualize results reported in single studies for a wider comprehension. The analysis of studies suggests that the application of plant antimicrobials, also when combined with other technologies, still needs to overcome some critical aspects. As with all reviews, this study reported a selection of case studies among many others and some of the drawbacks highlighted herein are actually under examination.

In conclusion, we believe that this review provided useful updated information promoting a scientific evidence-based approach for researchers, aimed to understand if the use of plant antimicrobials can be scaled up from laboratory trials towards industrial applications. Further research and development efforts in these areas will help to overcome the current challenges and pave the way towards a widespread adoption of plant antimicrobials in the food industry, contributing to safer, more sustainable, and consumer-friendly foods.

Author Contributions: Conceptualization, L.P. and J.F.A.-Z.; writing—original draft preparation, L.P., M.R.T.-R. and J.F.A.-Z.; writing—review and editing, L.P., F.B. and J.F.A.-Z.; visualization, L.P. and M.R.T.-R.; supervision, J.F.A.-Z. All authors have read and agreed to the published version of the manuscript.

Funding: The authors greatly acknowledge the Project "Eco-sustainable flexible plastic films with high breathability endowed with antibacterial activity for vegetable foods—ECOATTIVE (Ministerial decree n. 0002441/28-07-2022, CUP: B29J21032720008—COR: 9200716)" and Carmela Magno (Vibac S.p.a, Ticineto (AL), Italy) for the stimulating discussion and useful suggestions.

Data Availability Statement: Not applicable.

Conflicts of Interest: The authors declare no conflict of interest.

References

1. Bouarab Chibane, L.; Degraeve, P.; Ferhout, H.; Bouajila, J.; Oulahal, N. Plant antimicrobial polyphenols as potential natural food preservatives. *J. Sci. Food Agric.* **2019**, *99*, 1457–1474. [CrossRef] [PubMed]
2. Olszewska, M.A.; Gędas, A.; Simões, M. Antimicrobial polyphenol-rich extracts: Applications and limitations in the food industry. *Food Res. Int.* **2020**, *134*, 109214. [CrossRef]
3. Pisoschi, A.M.; Pop, A.; Georgescu, C.; Turcuş, V.; Olah, N.K.; Mathe, E. An overview of natural antimicrobials role in food. *Eur. J. Med. Chem.* **2018**, *143*, 922–935. [CrossRef] [PubMed]

4. Petruzzi, L.; Corbo, M.R.; Sinigaglia, M.; Bevilacqua, A. Microbial spoilage of foods: Fundamentals. In *The Microbiological Quality of Food*; Bevilacqua, A., Corbo, M.R., Sinigaglia, M., Eds.; Woodhead Publishing: Sawston, UK, 2017; pp. 1–21. [CrossRef]
5. Joardder, M.U.; Masud, M.H. *Food Preservation in Developing Countries: Challenges and Solutions*; Springer: Berlin, Germany, 2019; pp. 1–239.
6. McClements, D.J.; Das, A.K.; Dhar, P.; Nanda, P.K.; Chatterjee, N. Nanoemulsion-based technologies for delivering natural plant-based antimicrobials in foods. *Front. Sustain. Food Syst.* **2021**, *5*, 643208. [CrossRef]
7. Favela-González, K.M.; Hernández-Almanza, A.Y.; De la Fuente-Salcido, N.M. The value of bioactive compounds of cruciferous vegetables (*Brassica*) as antimicrobials and antioxidants: A review. *J. Food Biochem.* **2020**, *44*, e13414. [CrossRef] [PubMed]
8. Saeed, F.; Afzaal, M.; Tufail, T.; Ahmad, A. Use of natural antimicrobial agents: A safe preservation approach. In *Active Antimicrobial Food Packaging*; Var, I., Uzunlu, S., Eds.; IntechOpen: London, UK, 2019; Volume 18, pp. 7–24. [CrossRef]
9. EFSA Panel on Food Additives and Nutrient Sources Added to Food (EFSA ANS Panel); Younes, M.; Aggett, P.; Aguilar, F.; Crebelli, R.; Dusemund, B.; Filipič, M.; Frutos, M.J.; Galtier, P.; Gott, D.; et al. Refined Exposure Assessment of Extracts of Rosemary (E 392) from Its Use as Food Additive. *EFSA J.* **2018**, *16*, e05373.
10. Christopoulou, S.D.; Androutsopoulou, C.; Hahalis, P.; Kotsalou, C.; Vantarakis, A.; Lamari, F.N. Rosemary extract and essential oil as drink ingredients: An evaluation of their chemical composition, genotoxicity, antimicrobial, antiviral, and antioxidant properties. *Foods* **2021**, *10*, 3143. [CrossRef]
11. Rathod, N.B.; Ranveer, R.C.; Benjakul, S.; Kim, S.K.; Pagarkar, A.U.; Patange, S.; Ozogul, F. Recent developments of natural antimicrobials and antioxidants on fish and fishery food products. *Compr. Rev. Food Sci. Food Saf.* **2021**, *20*, 4182–4210. [CrossRef]
12. Papuc, C.; Goran, G.V.; Predescu, C.N.; Nicorescu, V.; Stefan, G. Plant polyphenols as antioxidant and antibacterial agents for shelf-life extension of meat and meat products: Classification, structures, sources, and action mechanisms. *Compr. Rev. Food Sci. Food Saf.* **2017**, *16*, 1243–1268. [CrossRef]
13. Pateiro, M.; Munekata, P.E.; Sant'Ana, A.S.; Domínguez, R.; Rodríguez-Lázaro, D.; Lorenzo, J.M. Application of essential oils as antimicrobial agents against spoilage and pathogenic microorganisms in meat products. *Int. J. Food Microbiol.* **2021**, *337*, 108966. [CrossRef]
14. Patrignani, F.; Siroli, L.; Serrazanetti, D.I.; Gardini, F.; Lanciotti, R. Innovative strategies based on the use of essential oils and their components to improve safety, shelf-life and quality of minimally processed fruits and vegetables. *Trends Food Sci. Technol.* **2015**, *46*, 311–319. [CrossRef]
15. Ceruso, M.; Clement, J.A.; Todd, M.J.; Zhang, F.; Huang, Z.; Anastasio, A.; Pepe, T.; Liu, Y. The inhibitory effect of plant extracts on growth of the foodborne pathogen, *Listeria monocytogenes*. *Antibiotics* **2020**, *9*, 319. [CrossRef] [PubMed]
16. Kalogianni, A.I.; Lazou, T.; Bossis, I.; Gelasakis, A.I. Natural phenolic compounds for the control of oxidation, bacterial spoilage, and foodborne pathogens in meat. *Foods* **2020**, *9*, 794. [CrossRef] [PubMed]
17. Ju, J.; Xie, Y.; Guo, Y.; Cheng, Y.; Qian, H.; Yao, W. The inhibitory effect of plant essential oils on foodborne pathogenic bacteria in food. *Crit. Rev. Food Sci. Nutr.* **2019**, *59*, 3281–3292. [CrossRef] [PubMed]
18. Górniak, I.; Bartoszewski, R.; Króliczewski, J. Comprehensive review of antimicrobial activities of plant flavonoids. *Phytochem. Rev.* **2019**, *18*, 241–272. [CrossRef]
19. Makhuvele, R.; Naidu, K.; Gbashi, S.; Thipe, V.C.; Adebo, O.A.; Njobeh, P.B. The use of plant extracts and their phytochemicals for control of toxigenic fungi and mycotoxins. *Heliyon* **2020**, *6*, e05291. [CrossRef] [PubMed]
20. Sakarikou, C.; Kostoglou, D.; Simões, M.; Giaouris, E. Exploitation of plant extracts and phytochemicals against resistant *Salmonella* spp. in biofilms. *Food Res. Int.* **2020**, *128*, 108806. [CrossRef]
21. Qian, W.; Liu, M.; Fu, Y.; Zhang, J.; Liu, W.; Li, J.; Li, X.; Li, Y.; Wang, T. Antimicrobial mechanism of luteolin against *Staphylococcus aureus* and *Listeria monocytogenes* and its antibiofilm properties. *Microb. Pathogen.* **2020**, *142*, 104056. [CrossRef]
22. Somrani, M.; Inglés, M.-C.; Debbabi, H.; Abidi, F.; Palop, A. Garlic, onion, and cinnamon essential oil anti-biofilms' effect against *Listeria monocytogenes*. *Foods* **2020**, *9*, 567. [CrossRef]
23. Castro-Rosas, J.; Ferreira-Grosso, C.R.; Gómez-Aldapa, C.A.; Rangel-Vargas, E.; Rodríguez-Marín, M.L.; Guzmán-Ortiz, F.A.; Falfan-Cortes, R.N. Recent advances in microencapsulation of natural sources of antimicrobial compounds used in food—A review. *Food Res. Int.* **2017**, *102*, 575–587. [CrossRef]
24. Pinto, L.; Bonifacio, M.A.; De Giglio, E.; Santovito, E.; Cometa, S.; Bevilacqua, A.; Baruzzi, F. Biopolymer hybrid materials: Development, characterization, and food packaging applications. *Food Packag. Shelf Life* **2021**, *28*, 100676. [CrossRef]
25. Iordache, A.M.; Nechita, C.; Voica, C.; Roba, C.; Botoran, O.R.; Ionete, R.E. Assessing the health risk and the metal content of thirty-four plant essential oils using the ICP-MS technique. *Nutrients* **2022**, *14*, 2363. [CrossRef] [PubMed]
26. Ałtyn, I.; Twarużek, M. Mycotoxin contamination concerns of herbs and medicinal plants. *Toxins* **2020**, *12*, 182. [CrossRef] [PubMed]
27. Fillâtre, Y.; Gray, F.X.; Roy, C. Pesticides in essential oils: Occurrence and concentration in organic and conventional orange essential oils from eleven geographical origins. *Anal. Chim. Acta* **2017**, *992*, 55–66. [CrossRef] [PubMed]
28. Wink, M. Current understanding of modes of action of multicomponent bioactive phytochemicals: Potential for nutraceuticals and antimicrobials. *Annu. Rev. Food Sci. Technol.* **2022**, *13*, 337–359. [CrossRef]
29. Gutiérrez-del-Río, I.; Fernández, J.; Lombó, F. Plant nutraceuticals as antimicrobial agents in food preservation: Terpenoids, polyphenols and thiols. *Int. J. Antimicrob. Agents* **2018**, *52*, 309–315. [CrossRef]

30. Villalobos-Delgado, L.H.; Nevárez-Moorillon, G.V.; Caro, I.; Quinto, E.J.; Mateo, J. Natural antimicrobial agents to improve foods shelf-life. In *Food Quality and Shelf-Life*; Galanakis, C.M., Ed.; Academic Press: Cambridge, MA, USA, 2019; pp. 125–157. [CrossRef]
31. Bae, J.Y.; Seo, Y.H.; Oh, S.W. Antibacterial activities of polyphenols against foodborne pathogens and their application as antibacterial agents. *Food Sci. Biotechnol.* **2022**, *31*, 985–997. [CrossRef]
32. Borges, A.; Ferreira, C.; Saavedra, M.J.; Simões, M. Antibacterial activity and mode of action of ferulic and gallic acids against pathogenic bacteria. *Microb. Drug Resist.* **2013**, *19*, 256–265. [CrossRef]
33. Xu, J.G.; Hu, H.X.; Chen, J.Y.; Xue, Y.S.; Kodirkhonov, B.; Han, B.Z. Comparative study on inhibitory effects of ferulic acid and p-coumaric acid on *Salmonella Enteritidis* biofilm formation. *World J. Microbiol. Biotechnol.* **2022**, *38*, 136. [CrossRef]
34. Li, J.; Zhao, N.; Xu, R.; Li, G.; Dong, H.; Wang, B.; Li, Z.; Fan, M.; Wei, X. Deciphering the antibacterial activity and mechanism of p-coumaric acid against *Alicyclobacillus acidoterrestris* and its application in apple juice. *Int. J. Food Microbiol.* **2022**, *378*, 109822. [CrossRef]
35. Shi, Y.G.; Zhu, Y.J.; Shao, S.Y.; Zhang, R.R.; Wu, Y.; Zhu, C.M.; Liang, X.R.; Cai, W.Q. Alkyl ferulate esters as multifunctional food additives: Antibacterial activity and mode of action against *Escherichia coli* in vitro. *J. Agric. Food Chem.* **2018**, *66*, 12088–12101. [CrossRef] [PubMed]
36. Shi, Y.G.; Zhang, R.R.; Zhu, C.M.; Liang, X.R.; Ettelaie, R.; Jiang, L.; Lin, S. On the mechanism behind enhanced antibacterial activity of alkyl gallate esters against foodborne pathogens and its application in Chinese icefish preservation. *Food Microbiol.* **2021**, *99*, 103817. [CrossRef] [PubMed]
37. Patzke, H.; Schieber, A. Growth-inhibitory activity of phenolic compounds applied in an emulsifiable concentrate-ferulic acid as a natural pesticide against *Botrytis cinerea*. *Food Res. Int.* **2018**, *113*, 18–23. [CrossRef] [PubMed]
38. Yuan, S.; Li, W.; Li, Q.; Wang, L.; Cao, J.; Jiang, W. Defense responses, induced by p-coumaric acid and methyl p-coumarate, of jujube (*Ziziphus jujuba* Mill.) fruit against black spot rot caused by *Alternaria alternata*. *J. Agric. Food Chem.* **2019**, *67*, 2801–2810. [CrossRef]
39. Li, W.; Yuan, S.; Sun, J.; Li, Q.; Jiang, W.; Cao, J. Ethyl p-coumarate exerts antifungal activity in vitro and in vivo against fruit *Alternaria alternata* via membrane-targeted mechanism. *Int. J. Food Microbiol.* **2018**, *278*, 26–35. [CrossRef]
40. Vestergaard, M.; Ingmer, H. Antibacterial and antifungal properties of resveratrol. *Int. J. Antimicrob. Agents* **2019**, *53*, 716–723. [CrossRef]
41. Cai, X.; Qi, J.; Xu, Z.; Huang, L.; Li, Y.; Ren, X.; Kong, Q. Three stilbenes make difference to the antifungal effects on ochratoxin A and its precursor production of *Aspergillus carbonarius*. *Food Microbiol.* **2022**, *103*, 103967. [CrossRef]
42. Puljula, E.; Walton, G.; Woodward, M.J.; Karonen, M. Antimicrobial activities of ellagitannins against *Clostridiales perfringens*, *Escherichia coli*, *Lactobacillus plantarum* and *Staphylococcus aureus*. *Molecules* **2020**, *25*, 3714. [CrossRef]
43. Wang, R.; Zhang, Y.; Jia, Y.; Zhang, M.; Huang, Y.; Li, C.; Li, K. Persimmon oligomeric proanthocyanidins exert antibacterial activity through damaging the cell membrane and disrupting the energy metabolism of *Staphylococcus aureus*. *ACS Food Sci. Technol.* **2020**, *1*, 35–44. [CrossRef]
44. Peng, M.; Jiang, C.; Jing, H.; Du, X.; Fan, X.; Zhang, Y.; Wang, H. Comparison of different extraction methods on yield, purity, antioxidant, and antibacterial activities of proanthocyanidins from chokeberry (*Aronia melanocarpa*). *J. Food Meas. Charact.* **2022**, *16*, 2049–2059. [CrossRef]
45. Wang, G.; Pang, T.; Xia, Y.; Liu, X.; Li, S.; Parvez, A.M.; Kong, F.; Si, C. Subdivision of bamboo kraft lignin by one-step ethanol fractionation to enhance its water-solubility and antibacterial performance. *Int. J. Biol. Macromol.* **2019**, *133*, 156–164. [CrossRef] [PubMed]
46. Du, W.; Zhou, M.; Liu, Z.; Chen, Y.; Li, R. Inhibition effects of low concentrations of epigallocatechin gallate on the biofilm formation and hemolytic activity of *Listeria monocytogenes*. *Food Control* **2018**, *85*, 119–126. [CrossRef]
47. Xiong, L.G.; Chen, Y.J.; Tong, J.W.; Huang, J.A.; Li, J.; Gong, Y.S.; Liu, Z.H. Tea polyphenol epigallocatechin gallate inhibits *Escherichia coli* by increasing endogenous oxidative stress. *Food Chem.* **2017**, *217*, 196–204. [CrossRef] [PubMed]
48. Cetin-Karaca, H.; Newman, M.C. Antimicrobial efficacy of plant phenolic compounds against *Salmonella* and *Escherichia coli*. *Food Biosci.* **2015**, *11*, 8–16. [CrossRef]
49. Amin, M.U.; Khurram, M.; Khattak, B.; Khan, J. Antibiotic additive and synergistic action of rutin, morin and quercetin against methicillin resistant *Staphylococcus aureus*. *BMC Complement. Altern. Med.* **2015**, *15*, 59. [CrossRef] [PubMed]
50. Zhang, M.; Xu, L.; Zhang, L.; Guo, Y.; Qi, X.; He, L. Effects of quercetin on postharvest blue mold control in kiwifruit. *Sci. Hortic.* **2018**, *228*, 18–25. [CrossRef]
51. Li, X.M.; Li, Z.Y.; Wang, Y.D.; Wang, J.Q.; Yang, P.L. Quercetin inhibits the proliferation and aflatoxins biosynthesis of *Aspergillus flavus*. *Toxins* **2019**, *11*, 154. [CrossRef]
52. Cha, J.D.; Moon, S.E.; Kim, J.Y.; Jung, E.K.; Lee, Y.S. Antibacterial activity of sophoraflavanone G isolated from the roots of *Sophora flavescens* against methicillin-resistant *Staphylococcus aureus*. *Phytother. Res.* **2009**, *23*, 1326–1331. [CrossRef]
53. Kalli, S.; Araya-Cloutier, C.; Chapman, J.; Sanders, J.W.; Vincken, J.P. Prenylated (iso) flavonoids as antifungal agents against the food spoiler *Zygosaccharomyces parabailii*. *Food Control* **2022**, *132*, 108434. [CrossRef]
54. Batiha, G.E.S.; Hussein, D.E.; Algammal, A.M.; George, T.T.; Jeandet, P.; Al-Snafi, A.E.; Tiwari, A.; Pagnossa, G.P.; Lima, C.M.; Thorat, N.D.; et al. Application of natural antimicrobials in food preservation: Recent views. *Food Control* **2021**, *126*, 108066. [CrossRef]

55. Bajalan, I.; Rouzbahani, R.; Pirbalouti, A.G.; Maggi, F. Antioxidant and antibacterial activities of the essential oils obtained from seven Iranian populations of *Rosmarinus officinalis*. *Ind. Crops Prod.* **2017**, *107*, 305–311. [CrossRef]
56. Rathore, S.; Mukhia, S.; Kapoor, S.; Bhatt, V.; Kumar, R.; Kumar, R. Seasonal variability in essential oil composition and biological activity of *Rosmarinus officinalis* L. accessions in the western Himalaya. *Sci. Rep.* **2022**, *12*, 3305. [CrossRef] [PubMed]
57. Micić, D.; Đurović, S.; Riabov, P.; Tomić, A.; Šovljanski, O.; Filip, S.; Tosti, T.; Dojčinović, B.; Božović, R.; Jovanović, D.; et al. Rosemary essential oils as a promising source of bioactive compounds: Chemical composition, thermal properties, biological activity, and gastronomical perspectives. *Foods* **2021**, *10*, 2734. [CrossRef] [PubMed]
58. Dammak, I.; Hamdi, Z.; El Euch, S.K.; Zemni, H.; Mliki, A.; Hassouna, M.; Lasram, S. Evaluation of antifungal and anti-ochratoxigenic activities of *Salvia officinalis*, *Lavandula dentata* and *Laurus nobilis* essential oils and a major monoterpene constituent 1, 8-cineole against *Aspergillus carbonarius*. *Ind. Crops Prod.* **2019**, *128*, 85–93. [CrossRef]
59. Sakkas, H.; Economou, V.; Gousia, P.; Bozidis, P.; Sakkas, V.A.; Petsios, S.; Mpekoulis, G.; Ilia, A.; Papadopoulou, C. Antibacterial efficacy of commercially available essential oils tested against drug-resistant Gram-positive pathogens. *Appl. Sci.* **2018**, *8*, 2201. [CrossRef]
60. Valdivieso-Ugarte, M.; Plaza-Diaz, J.; Gomez-Llorente, C.; Gómez, E.L.; Sabés-Alsina, M.; Gil, Á. In vitro examination of antibacterial and immunomodulatory activities of cinnamon, white thyme, and clove essential oils. *J. Funct. Foods* **2021**, *81*, 104436. [CrossRef]
61. Reyes-Jurado, F.; Cervantes-Rincón, T.; Bach, H.; López-Malo, A.; Palou, E. Antimicrobial activity of Mexican oregano (*Lippia berlandieri*), thyme (*Thymus vulgaris*), and mustard (*Brassica nigra*) essential oils in gaseous phase. *Ind. Crops Prod.* **2019**, *131*, 90–95. [CrossRef]
62. Císarová, M.; Hleba, L.; Medo, J.; Tančinová, D.; Mašková, Z.; Čuboň, J.; Kováčik, A.; Foltinová, D.; Božike, M.; Klouček, P. The in vitro and in situ effect of selected essential oils in vapour phase against bread spoilage toxicogenic aspergilli. *Food Control* **2020**, *110*, 107007. [CrossRef]
63. Marchese, A.; Arciola, C.R.; Barbieri, R.; Silva, A.S.; Nabavi, S.F.; Tsetegho Sokeng, A.J.; Izadi, M.; Jafari, N.J.; Suntar, I.; Daglia, M.; et al. Update on monoterpenes as antimicrobial agents: A particular focus on p-cymene. *Materials* **2017**, *10*, 947. [CrossRef]
64. Pinto, L.; Bonifacio, M.A.; De Giglio, E.; Cometa, S.; Logrieco, A.F.; Baruzzi, F. Unravelling the antifungal effect of red thyme oil (*Thymus vulgaris* L.) compounds in vapor phase. *Molecules* **2020**, *25*, 4761. [CrossRef]
65. Zhang, L.L.; Zhang, L.F.; Xu, J.G.; Hu, Q.P. Comparison study on antioxidant, DNA damage protective and antibacterial activities of eugenol and isoeugenol against several foodborne pathogens. *Food Nutr. Res.* **2017**, *61*, 1353356. [CrossRef] [PubMed]
66. Chen, C.; Cai, N.; Chen, J.; Wan, C. Clove essential oil as an alternative approach to control postharvest blue mold caused by *Penicillium italicum* in citrus fruit. *Biomolecules* **2019**, *9*, 197. [CrossRef] [PubMed]
67. Cava-Roda, R.; Taboada-Rodríguez, A.; López-Gómez, A.; Martínez-Hernández, G.B.; Marín-Iniesta, F. Synergistic antimicrobial activities of combinations of vanillin and essential oils of cinnamon bark, cinnamon leaves, and cloves. *Foods* **2021**, *10*, 1406. [CrossRef] [PubMed]
68. Yoplac, I.; Vargas, L.; Robert, P.; Hidalgo, A. Characterization and antimicrobial activity of microencapsulated citral with dextrin by spray drying. *Heliyon* **2021**, *7*, e06737. [CrossRef]
69. Zheng, S.; Jing, G.; Wang, X.; Ouyang, Q.; Jia, L.; Tao, N. Citral exerts its antifungal activity against *Penicillium digitatum* by affecting the mitochondrial morphology and function. *Food Chem.* **2015**, *178*, 76–81. [CrossRef]
70. Tang, X.; Shao, Y.L.; Tang, Y.J.; Zhou, W.W. Antifungal activity of essential oil compounds (geraniol and citral) and inhibitory mechanisms on grain pathogens (*Aspergillus flavus* and *Aspergillus ochraceus*). *Molecules* **2018**, *23*, 2108. [CrossRef]
71. Quintieri, L.; Fancello, F.; Caputo, L.; Sorrentino, A.; Zara, S.; Lippolis, V.; Cervellieri, S.; Fanelli, F.; Corvino, A.; Pace, B.; et al. Effect of gaseous citral on table grapes contaminated by *Rhizopus oryzae* ITEM 18876. *Foods* **2022**, *11*, 2478. [CrossRef]
72. Siddiqua, S.; Anusha, B.A.; Ashwini, L.S.; Negi, P.S. Antibacterial activity of cinnamaldehyde and clove oil: Effect on selected foodborne pathogens in model food systems and watermelon juice. *J. Food Sci. Technol.* **2015**, *52*, 5834–5841. [CrossRef]
73. Sun, Q.; Li, J.; Sun, Y.; Chen, Q.; Zhang, L.; Le, T. The antifungal effects of cinnamaldehyde against *Aspergillus niger* and its application in bread preservation. *Food Chem.* **2020**, *317*, 126405. [CrossRef]
74. Xu, L.; Tao, N.; Yang, W.; Jing, G. Cinnamaldehyde damaged the cell membrane of *Alternaria alternata* and induced the degradation of mycotoxins in vivo. *Ind. Crops Prod.* **2018**, *112*, 427–433. [CrossRef]
75. Song, H.J.; Ku, K.M. Optimization of allyl isothiocyanate sanitizing concentration for inactivation of *Salmonella typhimurium* on lettuce based on its phenotypic and metabolome changes. *Food Chem.* **2021**, *364*, 130438. [CrossRef] [PubMed]
76. Olaimat, A.N.; Al-Holy, M.A.; Ghoush, M.A.; Al-Nabulsi, A.A.; Holley, R.A. Control of *Salmonella enterica* and *Listeria monocytogenes* in hummus using allyl isothiocyanate. *Int. J. Food Microbiol.* **2018**, *278*, 73–80. [CrossRef]
77. de Melo Nazareth, T.; Alonso-Garrido, M.; Stanciu, O.; Mañes, J.; Manyes, L.; Meca, G. Effect of allyl isothiocyanate on transcriptional profile, aflatoxin synthesis, and *Aspergillus flavus* growth. *Food Res. Int.* **2020**, *128*, 108786. [CrossRef] [PubMed]
78. de Melo Nazareth, T.; Quiles, J.M.; Torrijos, R.; Luciano, F.B.; Manes, J.; Meca, G. Antifungal and antimycotoxigenic activity of allyl isothiocyanate on barley under different storage conditions. *LWT* **2019**, *112*, 108237. [CrossRef]
79. Yang, C.X.; Wu, H.T.; Li, X.X.; Wu, H.Y.; Niu, T.X.; Wang, X.N.; Lian, R.; Zhang, G.L.; Hou, H.M. Comparison of the inhibitory potential of benzyl isothiocyanate and phenethyl isothiocyanate on Shiga toxin-producing and enterotoxigenic *Escherichia coli*. *LWT* **2020**, *118*, 108806. [CrossRef]

80. Wu, H.Y.; Xu, Y.H.; Wei, L.N.; Bi, J.R.; Hou, H.M.; Hao, H.S.; Zhang, G.L. Inhibitory effects of 3-(methylthio) propyl isothiocyanate in comparison with benzyl isothiocyanate on *Listeria monocytogenes*. *J. Food Meas. Charact.* **2022**, *16*, 1768–1775. [CrossRef]
81. Yang, B.; Li, L.; Geng, H.; Zhang, C.; Wang, G.; Yang, S.; Gao, S.; Zhao, Y.; Xing, F. Inhibitory effect of allyl and benzyl isothiocyanates on ochratoxin a producing fungi in grape and maize. *Food Microbiol.* **2021**, *100*, 103865. [CrossRef]
82. Nowicki, D.; Maciąg-Dorszyńska, M.; Bogucka, K.; Szalewska-Pałasz, A.; Herman-Antosiewicz, A. Various modes of action of dietary phytochemicals, sulforaphane and phenethyl isothiocyanate, on pathogenic bacteria. *Sci. Rep.* **2019**, *9*, 1–12. [CrossRef]
83. Zhang, M.; Li, Y.; Bi, Y.; Wang, T.; Dong, Y.; Yang, Q.; Zhang, T. 2-Phenylethyl isothiocyanate exerts antifungal activity against *Alternaria alternata* by affecting membrane integrity and mycotoxin production. *Toxins* **2020**, *12*, 124. [CrossRef]
84. Wu, T.L.; Hu, Y.M.; Sun, Y.; Zhang, Z.J.; Wu, Z.R.; Zhao, W.B.; Tang, C.; Du, S.S.; He, Y.H.; Ma, Y.; et al. Insights into the mode of action of 2-(4-methoxyphenyl) ethyl isothiocyanate on *Aspergillus niger*. *Food Control* **2022**, *136*, 108871. [CrossRef]
85. Othman, L.; Sleiman, A.; Abdel-Massih, R.M. Antimicrobial activity of polyphenols and alkaloids in middle eastern plants. *Front. Microbiol.* **2019**, *10*, 911. [CrossRef]
86. El-Zahar, K.M.; Al-Jamaan, M.E.; Al-Mutairi, F.R.; Al-Hudiab, A.M.; Al-Einzi, M.S.; Mohamed, A.A.Z. Antioxidant, antibacterial, and antifungal activities of the ethanolic extract obtained from *Berberis vulgaris* roots and leaves. *Molecules* **2022**, *27*, 6114. [CrossRef] [PubMed]
87. Ma, K.; Zhe, T.; Li, F.; Zhang, Y.; Yu, M.; Li, R.; Wang, L. Sustainable films containing AIE-active berberine-based nanoparticles: A promising antibacterial food packaging. *Food Hydrocoll.* **2022**, *123*, 107147. [CrossRef]
88. He, M.; Wang, Y.; Hong, M.; Li, T. Berberine as a promising natural compound to control *Penicillium italicum* causing blue mold of citrus fruit. *Sci. Hortic.* **2022**, *305*, 111370. [CrossRef]
89. Marchese, A.; Barbieri, R.; Sanches-Silva, A.; Daglia, M.; Nabavi, S.F.; Jafari, N.J.; Izadi, M.; Ajami, M.; Nabavi, S.M. Antifungal and antibacterial activities of allicin: A review. *Trends Food Sci. Technol.* **2016**, *52*, 49–56. [CrossRef]
90. Jin, Z.; Li, L.; Zheng, Y.; An, P. Diallyl disulfide, the antibacterial component of garlic essential oil, inhibits the toxicity of *Bacillus cereus* ATCC 14579 at sub-inhibitory concentrations. *Food Control* **2021**, *126*, 108090. [CrossRef]
91. Tang, Y.; Li, F.; Gu, D.; Wang, W.; Huang, J.; Jiao, X. Antimicrobial effect and the mechanism of diallyl trisulfide against *Campylobacter jejuni*. *Antibiotics* **2021**, *10*, 246. [CrossRef]
92. Pernin, A.; Guillier, L.; Dubois-Brissonnet, F. Inhibitory activity of phenolic acids against *Listeria monocytogenes*: Deciphering the mechanisms of action using three different models. *Food Microbiol.* **2019**, *80*, 18–24. [CrossRef]
93. Sun, Z.; Zhang, X.; Wu, H.; Wang, H.; Bian, H.; Zhu, Y.; Xu, W.; Liu, F.; Wang, D.; Fu, L. Antibacterial activity and action mode of chlorogenic acid against *Salmonella enteritidis*, a foodborne pathogen in chilled fresh chicken. *World J. Microbiol. Biotechnol.* **2020**, *36*, 24. [CrossRef]
94. Ansari, M.A.; Anurag, A.; Fatima, Z.; Hameed, S. Natural phenolic compounds: A potential antifungal agent. In *Microbial Pathogens and Strategies for Combating Them: Science, Technology and Education*, 1st ed.; Méndez-Vilas, A., Ed.; Formatex Research Center: Badajoz, Spain, 2013; pp. 1189–1195.
95. Morales, J.; Mendoza, L.; Cotoras, M. Alteration of oxidative phosphorylation as a possible mechanism of the antifungal action of p-coumaric acid against *Botrytis cinerea*. *J. Appl. Microbiol.* **2017**, *123*, 969–976. [CrossRef]
96. Zhu, C.; Lei, M.; Andargie, M.; Zeng, J.; Li, J. Antifungal activity and mechanism of action of tannic acid against *Penicillium digitatum*. *Physiol. Mol. Plant Pathol.* **2019**, *107*, 46–50. [CrossRef]
97. You, S.; Xie, Y.; Zhuang, X.; Chen, H.; Qin, Y.; Cao, J.; Lan, T. Effect of high antioxidant activity on bacteriostasis of lignin from sugarcane bagasse. *Biochem. Eng. J.* **2022**, *180*, 108335. [CrossRef]
98. Luo, J.; Xu, F.; Zhang, X.; Shao, X.; Wei, Y.; Wang, H. Transcriptome analysis of *Penicillium italicum* in response to the flavonoids from *Sedum aizoon* L. *World J. Microbiol. Biotechnol.* **2020**, *36*, 62. [CrossRef] [PubMed]
99. Churklam, W.; Chaturongakul, S.; Ngamwongsatit, B.; Aunpad, R. The mechanisms of action of carvacrol and its synergism with nisin against *Listeria monocytogenes* on sliced bologna sausage. *Food Control* **2020**, *108*, 106864. [CrossRef]
100. Stratakos, A.C.; Sima, F.; Ward, P.; Linton, M.; Kelly, C.; Pinkerton, L.; Stef, L.; Pet, I.; Corcionivoschi, N. The in vitro effect of carvacrol, a food additive, on the pathogenicity of O157 and non-O157 Shiga-toxin producing *Escherichia coli*. *Food Control* **2018**, *84*, 290–296. [CrossRef]
101. Barbosa, L.N.; Alves, F.C.B.; Andrade, B.F.M.T.; Albano, M.; Rall, V.L.M.; Fernandes, A.A.H.; Buzalaf, M.A.R.; de Lima Leite, A.; de Pontes, L.G.; dos Santos, L.D.; et al. Proteomic analysis and antibacterial resistance mechanisms of *Salmonella enteritidis* submitted to the inhibitory effect of *Origanum vulgare* essential oil, thymol and carvacrol. *J. Proteomics* **2020**, *214*, 103625. [CrossRef]
102. Han, Y.; Sun, Z.; Chen, W. Antimicrobial susceptibility and antibacterial mechanism of limonene against *Listeria monocytogenes*. *Molecules* **2019**, *25*, 33. [CrossRef]
103. Gupta, A.; Jeyakumar, E.; Lawrence, R. Strategic approach of multifaceted antibacterial mechanism of limonene traced in *Escherichia coli*. *Sci. Rep.* **2021**, *11*, 13816. [CrossRef]
104. Han, Y.; Chen, W.; Sun, Z. Antimicrobial activity and mechanism of limonene against *Staphylococcus aureus*. *J. Food Saf.* **2021**, *41*, e12918. [CrossRef]
105. Jeyakumar, G.E.; Lawrence, R. Mechanisms of bactericidal action of eugenol against *Escherichia coli*. *J. Herb. Med.* **2021**, *26*, 100406. [CrossRef]
106. Bai, X.; Li, X.; Liu, X.; Xing, Z.; Su, R.; Wang, Y.; Xia, X.; Shi, C. Antibacterial effect of eugenol on *Shigella flexneri* and its mechanism. *Foods* **2022**, *11*, 2565. [CrossRef]

107. Zhang, J.; Ma, S.; Du, S.; Chen, S.; Sun, H. Antifungal activity of thymol and carvacrol against postharvest pathogens *Botrytis cinerea*. *J. Food Sci. Technol.* **2019**, *56*, 2611–2620. [CrossRef] [PubMed]
108. Qu, C.; Li, Z.; Wang, X. UHPLC-HRMS-based untargeted lipidomics reveal mechanism of antifungal activity of carvacrol against *Aspergillus flavus*. *Foods* **2022**, *11*, 93. [CrossRef] [PubMed]
109. Cai, R.; Hu, M.; Zhang, Y.; Niu, C.; Yue, T.; Yuan, Y.; Wang, Z. Antifungal activity and mechanism of citral, limonene and eugenol against *Zygosaccharomyces rouxii*. *LWT* **2019**, *106*, 50–56. [CrossRef]
110. Jiang, N.; Wang, L.; Jiang, D.; Wang, M.; Liu, H.; Yu, H.; Yao, W. Transcriptomic analysis of inhibition by eugenol of ochratoxin A biosynthesis and growth of *Aspergillus carbonarius*. *Food Control* **2022**, *135*, 108788. [CrossRef]
111. Luciano, F.B.; Holley, R.A. Enzymatic inhibition by allyl isothiocyanate and factors affecting its antimicrobial action against *Escherichia coli* O157: H7. *Int. J. Food Microbiol.* **2009**, *131*, 240–245. [CrossRef]
112. Li, P.; Zhao, Y.M.; Wang, C.; Zhu, H.P. Antibacterial activity and main action pathway of benzyl isothiocyanate extracted from papaya seeds. *J. Food Sci.* **2021**, *86*, 169–176. [CrossRef]
113. Sun, Y.; Wang, Y.; Xu, Y.; Chen, T.; Li, B.; Zhang, Z.; Tian, S. Application and mechanism of benzyl-isothiocyanate, a natural antimicrobial agent from cruciferous vegetables, in controlling postharvest decay of strawberry. *Postharvest Biol. Technol.* **2021**, *180*, 111604. [CrossRef]
114. Boberek, J.M.; Stach, J.; Good, L. Genetic evidence for inhibition of bacterial division protein FtsZ by berberine. *PLoS ONE* **2010**, *5*, e13745. [CrossRef]
115. Xu, C.; Wang, F.; Huang, F.; Yang, M.; He, D.; Deng, L. Targeting effect of berberine on type I fimbriae of *Salmonella typhimurium* and its effective inhibition of biofilm. *Appl. Microbiol. Biotechnol.* **2021**, *105*, 1563–1573. [CrossRef]
116. Arshad, M.S.; Batool, S.A. Natural antimicrobials, their sources and food safety. In *Food Additives*; Karunaratne, D.N., Pamunuwa, G., Eds.; IntechOpen: London, UK, 2017; Volume 87, pp. 87–104. [CrossRef]
117. Belgacem, I.; Li Destri Nicosia, M.G.; Pangallo, S.; Abdelfattah, A.; Benuzzi, M.; Agosteo, G.E.; Schena, L. Pomegranate peel extracts as safe natural treatments to control plant diseases and increase the shelf-life and safety of fresh fruits and vegetables. *Plants* **2021**, *10*, 453. [CrossRef] [PubMed]
118. Givi, F.; Gholami, M.; Massah, A. Application of pomegranate peel extract and essential oil as a safe botanical preservative for the control of postharvest decay caused by *Penicillium italicum* and *Penicillium digitatum* on "Satsuma" mandarin. *J. Food Saf.* **2019**, *39*, e12639. [CrossRef]
119. Gómez-Maldonado, D.; Lobato-Calleros, C.; Aguirre-Mandujano, E.; Leyva-Mir, S.G.; Robles-Yerena, L.; Vernon-Carter, E.J. Antifungal activity of mango kernel polyphenols on mango fruit infected by anthracnose. *LWT* **2020**, *126*, 109337. [CrossRef]
120. Hernández, A.; Ruiz-Moyano, S.; Galván, A.I.; Merchán, A.V.; Nevado, F.P.; Aranda, E.; Serradilla, M.J.; de Guía Córdoba, M.; Martín, A. Anti-fungal activity of phenolic sweet orange peel extract for controlling fungi responsible for post-harvest fruit decay. *Fungal Biol.* **2021**, *125*, 143–152. [CrossRef] [PubMed]
121. Yan, J.; Wu, H.; Shi, F.; Wang, H.; Chen, K.; Feng, J.; Jia, W. Antifungal activity screening for mint and thyme essential oils against *Rhizopus stolonifer* and their application in postharvest preservation of strawberry and peach fruits. *J. Appl. Microbiol.* **2021**, *130*, 1993–2007. [CrossRef] [PubMed]
122. Sumalan, R.M.; Kuganov, R.; Obistioiu, D.; Popescu, I.; Radulov, I.; Alexa, E.; Negrea, M.; Salimzoda, A.F.; Sumalan, R.L.; Cocan, I. Assessment of mint, basil, and lavender essential oil vapor-phase in antifungal protection and lemon fruit quality. *Molecules* **2020**, *25*, 1831. [CrossRef]
123. Pinto, L.; Cefola, M.; Bonifacio, M.A.; Cometa, S.; Bocchino, C.; Pace, B.; De Giglio, E.; Palumbo, M.; Sada, A.; Logrieco, A.F.; et al. Effect of red thyme oil (*Thymus vulgaris* L.) vapours on fungal decay, quality parameters and shelf-life of oranges during cold storage. *Food Chem.* **2021**, *336*, 127590. [CrossRef]
124. Arellano, S.; Zhu, L.; Dev Kumar, G.; Law, B.; Friedman, M.; Ravishankar, S. Essential oil microemulsions inactivate antibiotic-resistant bacteria on iceberg lettuce during 28-day storage at 4 °C. *Molecules* **2022**, *27*, 6699. [CrossRef]
125. Kara, M.; Soylu, E.M. Assessment of glucosinolate-derived isothiocyanates as potential natural antifungal compounds against citrus sour rot disease agent *Geotrichum citri-aurantii*. *J. Phytopathol.* **2020**, *168*, 279–289. [CrossRef]
126. Santos, M.I.S.; Marques, C.; Mota, J.; Pedroso, L.; Lima, A. Applications of essential oils as antibacterial agents in minimally processed fruits and vegetables—A review. *Microorganisms* **2022**, *10*, 760. [CrossRef]
127. Wong, J.X.; Ramli, S.; Desa, S.; Chen, S.N. Use of *Centella asiatica* extract in reducing microbial contamination and browning effect in fresh cut fruits and vegetables during storage: A potential alternative of synthetic preservatives. *LWT* **2021**, *151*, 112229. [CrossRef]
128. Wieczyńska, J.; Cavoski, I. Antimicrobial, antioxidant and sensory features of eugenol, carvacrol and trans-anethole in active packaging for organic ready-to-eat iceberg lettuce. *Food Chem.* **2018**, *259*, 251–260. [CrossRef] [PubMed]
129. Myszka, K.; Schmidt, M.T.; Majcher, M.; Juzwa, W.; Czaczyk, K. β-Caryophyllene-rich pepper essential oils suppress spoilage activity of *Pseudomonas fluorescens* KM06 in fresh-cut lettuce. *LWT Food Sci. Technol.* **2017**, *83*, 118–126. [CrossRef]
130. Amiri, A.; Ramezanian, A.; Mortazavi, S.M.H.; Hosseini, S.M.H.; Yahia, E. Shelf-life extension of pomegranate arils using chitosan nanoparticles loaded with *Satureja hortensis* essential oil. *J. Sci. Food Agric.* **2021**, *101*, 3778–3786. [CrossRef]
131. Awad, A.H.R.; Parmar, A.; Ali, M.R.; El-Mogy, M.M.; Abdelgawad, K.F. Extending the shelf-life of fresh-cut green bean pods by ethanol, ascorbic Acid, and essential oils. *Foods* **2021**, *10*, 1103. [CrossRef] [PubMed]

132. Maleš, I.; Pedisić, S.; Zorić, Z.; Elez-Garofulić, I.; Repajić, M.; You, L.; Vladimir-Knežević, S.; Butorac, D.; Dragović-Uzelac, V. The medicinal and aromatic plants as ingredients in functional beverage production. *J. Funct. Foods* **2022**, *96*, 105210. [CrossRef]
133. da Cruz Almeida, E.T.; de Souza, G.T.; de Sousa Guedes, J.P.; Barbosa, I.M.; de Sousa, C.P.; Castellano, L.R.C.; Magnani, M.; de Souza, E.L. *Mentha piperita* L. essential oil inactivates spoilage yeasts in fruit juices through the perturbation of different physiological functions in yeast cells. *Food Microbiol.* **2019**, *82*, 20–29. [CrossRef] [PubMed]
134. Wang, H.; Sun, H. Assessment of different antimicrobials to inhibit the growth of *Zygosaccharomyces rouxii* cocktail in concentrated apple juice. *Food Microbiol.* **2020**, *91*, 103549. [CrossRef]
135. Lee, S.; Kim, H.; Beuchat, L.R.; Kim, Y.; Ryu, J.H. Synergistic antimicrobial activity of oregano and thyme thymol essential oils against *Leuconostoc citreum* in a laboratory medium and tomato juice. *Food Microbiol.* **2020**, *90*, 103489. [CrossRef]
136. Fernandez, M.V.; Bengardino, M.; Jagus, R.J.; Agüero, M.V. Enrichment and preservation of a vegetable smoothie with an antioxidant and antimicrobial extract obtained from beet by-products. *LWT* **2020**, *117*, 108622. [CrossRef]
137. Maqsood, S.; Abushelaibi, A.; Manheem, K.; Al Rashedi, A.; Kadim, I.T. Lipid oxidation, protein degradation, microbial and sensorial quality of camel meat as influenced by phenolic compounds. *LWT Food Sci. Technol.* **2015**, *63*, 953–959. [CrossRef]
138. Nowak, A.; Czyzowska, A.; Efenberger, M.; Krala, L. Polyphenolic extracts of cherry (*Prunus cerasus* L.) and blackcurrant (*Ribes nigrum* L.) leaves as natural preservatives in meat products. *Food Microbiol.* **2016**, *59*, 142–149. [CrossRef] [PubMed]
139. Casaburi, A.; Di Martino, V.; Ercolini, D.; Parente, E.; Villani, F. Antimicrobial activity of *Myrtus communis* L. water-ethanol extract against meat spoilage strains of *Brochothrix thermosphacta* and *Pseudomonas fragi* in vitro and in meat. *Ann. Microbiol.* **2015**, *65*, 841–850. [CrossRef]
140. Tamkutė, L.; Gil, B.M.; Carballido, J.R.; Pukalskienė, M.; Venskutonis, P.R. Effect of cranberry pomace extracts isolated by pressurized ethanol and water on the inhibition of food pathogenic/spoilage bacteria and the quality of pork products. *Food Res. Int.* **2019**, *120*, 38–51. [CrossRef]
141. Zhang, H.; Wu, J.; Guo, X. Effects of antimicrobial and antioxidant activities of spice extracts on raw chicken meat quality. *Food Sci. Hum. Wellness* **2016**, *5*, 39–48. [CrossRef]
142. Shahbazi, Y.; Shavisi, N.; Mohebi, E. Effects of *Ziziphora clinopodioides* essential oil and nisin, both separately and in combination, to extend shelf-life and control *Escherichia coli* O157:H7 and *Staphylococcus aureus* in raw beef patty during refrigerated storage. *J. Food Saf.* **2016**, *36*, 227–236. [CrossRef]
143. Karam, L.; Chehab, R.; Osaili, T.M.; Savvaidis, I.N. Antimicrobial effect of thymol and carvacrol added to a vinegar-based marinade for controlling spoilage of marinated beef (*Shawarma*) stored in air or vacuum packaging. *Int. J. Food Microbiol.* **2020**, *332*, 108769. [CrossRef] [PubMed]
144. Olatunde, O.O.; Benjakul, S.; Huda, N.; Zhang, B.; Deng, S. Ethanolic Noni (*Morinda citrifolia* L.) leaf extract dechlorophyllised using sedimentation process: Antioxidant, antibacterial properties and efficacy in extending the shelf-life of striped catfish slices. *Int. J. Food Sci. Technol.* **2021**, *56*, 2804–2819. [CrossRef]
145. Olatunde, O.O.; Della Tan, S.L.; Shiekh, K.A.; Benjakul, S.; Nirmal, N.P. Ethanolic guava leaf extracts with different chlorophyll removal processes: Anti-melanosis, antibacterial properties and the impact on qualities of Pacific white shrimp during refrigerated storage. *Food Chem.* **2021**, *341*, 128251. [CrossRef]
146. Li, Y.; Zhuang, S.; Liu, Y.; Zhang, L.; Liu, X.; Cheng, H.; Liu, J.; Shu, R.; Luo, Y. Effect of grape seed extract on quality and microbiota community of container-cultured snakehead (*Channa argus*) fillets during chilled storage. *Food Microbiol.* **2020**, *91*, 103492. [CrossRef]
147. Van Haute, S.; Raes, K.; Van Der Meeren, P.; Sampers, I. The effect of cinnamon, oregano and thyme essential oils in marinade on the microbial shelf-life of fish and meat products. *Food Control* **2016**, *68*, 30–39. [CrossRef]
148. Zhang, Y.; Li, D.; Lv, J.; Li, Q.; Kong, C.; Luo, Y. Effect of cinnamon essential oil on bacterial diversity and shelf-life in vacuum-packaged common carp (*Cyprinus carpio*) during refrigerated storage. *Int. J. Food Microbiol.* **2017**, *249*, 1–8. [CrossRef] [PubMed]
149. Roila, R.; Valiani, A.; Ranucci, D.; Ortenzi, R.; Servili, M.; Veneziani, G.; Branciari, R. Antimicrobial efficacy of a polyphenolic extract from olive oil by-product against "Fior di latte" cheese spoilage bacteria. *Int. J. Food Microbiol.* **2019**, *295*, 49–53. [CrossRef]
150. Derbassi, N.; Pedrosa, M.C.; Heleno, S.; Fernandes, F.; Dias, M.I.; Calhelha, R.C.; Rodrigues, P.; Carocho, M.; Ferreira, I.C.F.R.; Barros, L. *Arbutus unedo* leaf extracts as potential dairy preservatives: Case study on quark cheese. *Food Funct.* **2022**, *13*, 5442–5454. [CrossRef] [PubMed]
151. Milanović, V.; Sabbatini, R.; Garofalo, C.; Cardinali, F.; Pasquini, M.; Aquilanti, L.; Osimani, A. Evaluation of the inhibitory activity of essential oils against spoilage yeasts and their potential application in yogurt. *Int. J. Food Microbiol.* **2021**, *341*, 109048. [CrossRef] [PubMed]
152. Novais, C.; Molina, A.K.; Abreu, R.M.; Santo-Buelga, C.; Ferreira, I.C.; Pereira, C.; Barros, L. Natural food colorants and preservatives: A review, a demand, and a challenge. *J. Agric. Food Chem.* **2022**, *70*, 2789–2805. [CrossRef]
153. Schirone, M.; Visciano, P.; Tofalo, R.; Suzzi, G. Foodborne pathogens: Hygiene and safety. *Front. Microbiol.* **2019**, *10*, 1974. [CrossRef]
154. Kachur, K.; Suntres, Z. The antibacterial properties of phenolic isomers, carvacrol and thymol. *Crit. Rev. Food Sci. Nutr.* **2020**, *60*, 3042–3053. [CrossRef]
155. Guan, P.; Wang, X.; Dong, Z.; Song, M.; Zhu, H.; Suo, B. Cinnamaldehyde inactivates Listeria monocytogenes at a low temperature in ground pork by disturbing the expression of stress regulatory genes. *Food Biosci.* **2023**, *51*, 102277. [CrossRef]

156. Bai, J.; Li, J.; Chen, Z.; Bai, X.; Yang, Z.; Wang, Z.; Yang, Y. Antibacterial activity and mechanism of clove essential oil against foodborne pathogens. *LWT* **2023**, *173*, 114249. [CrossRef]
157. Sepahvand, S.; Amiri, S.; Radi, M.; Akhavan, H.R. Antimicrobial activity of thymol and thymol-nanoemulsion against three food-borne pathogens inoculated in a sausage model. *Food Bioproc. Tech.* **2021**, *14*, 1936–1945. [CrossRef]
158. Saraiva, C.; Silva, A.C.; García-Díez, J.; Cenci-Goga, B.; Grispoldi, L.; Silva, A.F.; Almeida, J.M. Antimicrobial activity of *Myrtus communis* L. and *Rosmarinus officinalis* L. essential oils against *Listeria monocytogenes* in cheese. *Foods* **2021**, *10*, 1106. [CrossRef]
159. Ahmed, L.I.; Ibrahim, N.; Abdel-Salam, A.B.; Fahim, K.M. Potential application of ginger, clove and thyme essential oils to improve soft cheese microbial safety and sensory characteristics. *Food Biosci.* **2021**, *42*, 101177. [CrossRef]
160. Rossi, C.; Chaves-López, C.; Možina, S.S.; Di Mattia, C.; Scuota, S.; Luzzi, I.; Jenič, T.; Paparella, A.; Serio, A. *Salmonella enterica* adhesion: Effect of *Cinnamomum zeylanicum* essential oil on lettuce. *LWT* **2019**, *111*, 16–22. [CrossRef]
161. Brnawi, W.I.; Hettiarachchy, N.S.; Horax, R.; Kumar-Phillips, G.; Ricke, S. Antimicrobial activity of leaf and bark cinnamon essential oils against *Listeria monocytogenes* and *Salmonella typhimurium* in broth system and on celery. *J. Food Process. Preserv.* **2019**, *43*, e13888. [CrossRef]
162. Dai, J.; Li, C.; Cui, H.; Lin, L. Unraveling the anti-bacterial mechanism of *Litsea cubeba* essential oil against *E. coli* O157: H7 and its application in vegetable juices. *Int. J. Food Microbiol.* **2021**, *338*, 108989. [CrossRef] [PubMed]
163. Carvalho, F.; Coimbra, A.T.; Silva, L.; Duarte, A.P.; Ferreira, S. *Melissa officinalis* essential oil as an antimicrobial agent against *Listeria monocytogenes* in watermelon juice. *Food Microbiol.* **2023**, *109*, 104105. [CrossRef]
164. Gerardi, C.; Pinto, L.; Baruzzi, F.; Giovinazzo, G. Comparison of antibacterial and antioxidant properties of red (cv. Negramaro) and white (cv. Fiano) skin pomace extracts. *Molecules* **2021**, *26*, 5918. [CrossRef]
165. Tian, L.; Fu, J.; Wu, M.; Liao, S.; Jia, X.; Wang, J.; Yang, S.; Liu, Z.; Liu, Z.; Xue, Z.; et al. Evaluation of gallic acid on membrane damage of *Yersinia enterocolitica* and its application as a food preservative in pork. *Int. J. Food Microbiol.* **2022**, *374*, 109720. [CrossRef]
166. Phuong, N.N.M.; Le, T.T.; Van Camp, J.; Raes, K. Evaluation of antimicrobial activity of rambutan (*Nephelium lappaceum* L.) peel extracts. *Int. J. Food Microbiol.* **2020**, *321*, 108539. [CrossRef]
167. Belgacem, I.; Schena, L.; Teixidó, N.; Romeo, F.V.; Ballistreri, G.; Abadias, M. Effectiveness of a pomegranate peel extract (PGE) in reducing *Listeria monocytogenes* in vitro and on fresh-cut pear, apple and melon. *Eur. Food Res. Technol.* **2020**, *246*, 1765–1772. [CrossRef]
168. Nicolau-Lapeña, I.; Aguiló-Aguayo, I.; Bobo, G.; Viñas, I.; Anguera, M.; Abadias, M. Ferulic acid application to control growth *Listeria monocytogenes* and *Salmonella enterica* on fresh-cut apples and melon, and its effect in quality parameters. *Postharvest Biol. Technol.* **2022**, *186*, 111831. [CrossRef]
169. Bombelli, A.; Araya-Cloutier, C.; Vincken, J.P.; Abee, T.; den Besten, H.M. Impact of food-relevant conditions and food matrix on the efficacy of prenylated isoflavonoids glabridin and 6, 8-diprenylgenistein as potential natural preservatives against *Listeria monocytogenes*. *Int. J. Food Microbiol.* **2023**, *390*, 110109. [CrossRef] [PubMed]
170. Wang, S.; Liu, S.; Hao, G.; Zhao, L.; Lü, X.; Wang, H.; Wang, L.; Zhang, J.; Ge, W. Antimicrobial activity and mechanism of isothiocyanate from *Moringa oleifera* seeds against *Bacillus cereus* and *Cronobacter sakazakii* and its application in goat milk. *Food Control* **2022**, *139*, 109067. [CrossRef]
171. Dogruyol, H.; Mol, S.; Cosansu, S. Increased thermal sensitivity of *Listeria monocytogenes* in sous-vide salmon by oregano essential oil and citric acid. *Food Microbiol.* **2020**, *90*, 103496. [CrossRef]
172. Orizano-Ponce, E.; Char, C.; Sepúlveda, F.; Ortiz-Viedma, J. Heat sensitization of *Escherichia coli* by the natural antimicrobials vanillin and emulsified citral in blended carrot-orange juice. *Food Microbiol.* **2022**, *107*, 104058. [CrossRef]
173. Cheng, S.; Su, R.; Song, L.; Bai, X.; Yang, H.; Li, Z.; Li, Z.; Zhan, X.; Xia, X.; Lü, X.; et al. Citral and trans-cinnamaldehyde, two plant-derived antimicrobial agents can induce *Staphylococcus aureus* into VBNC state with different characteristics. *Food Microbiol.* **2023**, *112*, 104241. [CrossRef]
174. de Candia, S.; Quintieri, L.; Caputo, L.; Baruzzi, F. Antimicrobial activity of processed spices used in traditional Southern Italian sausage processing. *J. Food Process. Preserv.* **2017**, *41*, e13022. [CrossRef]
175. Palomares-Navarro, J.J.; Bernal-Mercado, A.T.; González-Aguilar, G.A.; Ortega-Ramirez, L.A.; Martínez-Téllez, M.A.; Ayala-Zavala, J.F. Antibiofilm action of plant terpenes in *Salmonella* strains: Potential inhibitors of the synthesis of extracellular polymeric substances. *Pathogens* **2023**, *12*, 35. [CrossRef]
176. Silva, L.N.; Zimmer, K.R.; Macedo, A.J.; Trentin, D.S. Plant natural products targeting bacterial virulence factors. *Chem. Rev.* **2016**, *116*, 9162–9236. [CrossRef]
177. Vazquez-Armenta, F.J.; Bernal-Mercado, A.T.; Lizardi-Mendoza, J.; Silva-Espinoza, B.A.; Cruz-Valenzuela, M.R.; Gonzalez-Aguilar, G.A.; Nazzaro, F.; Fratianni, F.; Ayala-Zavala, J.F. Phenolic extracts from grape stems inhibit *Listeria monocytogenes* motility and adhesion to food contact surfaces. *J. Adhes. Sci. Technol.* **2018**, *32*, 889–907. [CrossRef]
178. Vazquez-Armenta, F.J.; Bernal-Mercado, A.T.; Tapia-Rodriguez, M.R.; Gonzalez-Aguilar, G.A.; Lopez-Zavala, A.A.; Martinez-Tellez, M.A.; Hernandez-Oñate, M.A.; Ayala-Zavala, J.F. Quercetin reduces adhesion and inhibits biofilm development by *Listeria monocytogenes* by reducing the amount of extracellular proteins. *Food Control* **2018**, *90*, 266–273. [CrossRef]
179. Vazquez-Armenta, F.J.; Hernandez-Oñate, M.A.; Martinez-Tellez, M.A.; Lopez-Zavala, A.A.; Gonzalez-Aguilar, G.A.; Gutierrez-Pacheco, M.M.; Ayala-Zavala, J.F. Quercetin repressed the stress response factor (*sigB*) and virulence genes (*prfA*, *actA*, *inlA*, and *inlC*), lower the adhesion, and biofilm development of *L. monocytogenes*. *Food Microbiol.* **2020**, *87*, 103377. [CrossRef]

180. Kim, Y.K.; Roy, P.K.; Ashrafudoulla, M.; Nahar, S.; Toushik, S.H.; Hossain, M.I.; Ha, S.D. Antibiofilm effects of quercetin against *Salmonella enterica* biofilm formation and virulence, stress response, and quorum-sensing gene expression. *Food Control* **2022**, *137*, 108964. [CrossRef]
181. Ortega-Ramirez, L.A.; Gutiérrez-Pacheco, M.M.; Vargas-Arispuro, I.; González-Aguilar, G.A.; Martínez-Téllez, M.A.; Ayala-Zavala, J.F. Inhibition of glucosyltransferase activity and glucan production as an antibiofilm mechanism of lemongrass essential oil against *Escherichia coli* O157: H7. *Antibiotics* **2020**, *9*, 102. [CrossRef] [PubMed]
182. Kostoglou, D.; Protopappas, I.; Giaouris, E. Common plant-derived terpenoids present increased anti-biofilm potential against *Staphylococcus* bacteria compared to a quaternary ammonium biocide. *Foods* **2020**, *9*, 697. [CrossRef]
183. dos Santos Rodrigues, J.B.; de Carvalho, R.J.; de Souza, N.T.; de Sousa Oliveira, K.; Franco, O.L.; Schaffner, D.; de Souza, E.L.; Magnani, M. Effects of oregano essential oil and carvacrol on biofilms of *Staphylococcus aureus* from food-contact surfaces. *Food Control* **2017**, *73*, 1237–1246. [CrossRef]
184. Ozma, M.A.; Abbasi, A.; Ahangarzadeh Rezaee, M.; Hosseini, H.; Hosseinzadeh, N.; Sabahi, S.; Kafil, H.S. A critical review on the nutritional and medicinal profiles of garlic's (*Allium sativum* L.) bioactive compounds. *Food Rev. Int.* **2022**, 1–38. [CrossRef]
185. Yang, X.; Sha, K.; Xu, G.; Tian, H.; Wang, X.; Chen, S.; Wang, Y.; Li, J.; Chen, J.; Huang, N. Subinhibitory concentrations of allicin decrease uropathogenic *Escherichia coli* (UPEC) biofilm formation, adhesion ability, and swimming motility. *Int. J. Mol. Sci.* **2016**, *17*, 979. [CrossRef]
186. Cruz-Valenzuela, M.R.; Ayala-Soto, R.E.; Ayala-Zavala, J.F.; Espinoza-Silva, B.A.; González-Aguilar, G.A.; Martín-Belloso, O.; Soliva-Fortuny, R.; Nazzaro, F.; Fratianni, F.; Tapia-Rodriguez, M.T.; et al. Pomegranate (*Punica granatum* L.) peel extracts as antimicrobial and antioxidant additives used in alfalfa sprouts. *Foods* **2022**, *11*, 2588. [CrossRef]
187. Bouyahya, A.; Chamkhi, I.; Balahbib, A.; Rebezov, M.; Shariati, M.A.; Wilairatana, P.; El Omari, N. Mechanisms, anti-quorum-sensing actions, and clinical trials of medicinal plant bioactive compounds against bacteria: A comprehensive review. *Molecules* **2022**, *27*, 1484. [CrossRef] [PubMed]
188. Fimbres-García, J.O.; Flores-Sauceda, M.; Othon-Díaz, E.D.; García-Galaz, A.; Tapia-Rodríguez, M.R.; Silva-Espinoza, B.A.; Ayala-Zavala, J.F. Facing resistant bacteria with plant essential oils: Reviewing the oregano case. *Antibiotics* **2022**, *11*, 1777. [CrossRef] [PubMed]
189. Guillín, Y.; Cáceres, M.; Torres, R.; Stashenko, E.; Ortiz, C. Effect of essential oils on the inhibition of biofilm and quorum sensing in *Salmonella enteritidis* 13076 and *Salmonella typhimurium* 14028. *Antibiotics* **2021**, *10*, 1191. [CrossRef] [PubMed]
190. Tamfu, A.N.; Ceylan, O.; Kucukaydin, S.; Duru, M.E. HPLC-DAD phenolic profiles, antibiofilm, anti-quorum sensing and enzyme inhibitory potentials of *Camellia sinensis* (L.) O. Kuntze and *Curcuma longa* L. *LWT* **2020**, *133*, 110150. [CrossRef]
191. Tapia-Rodriguez, M.R.; Bernal-Mercado, A.T.; Gutierrez-Pacheco, M.M.; Vazquez-Armenta, F.J.; Hernandez-Mendoza, A.; Gonzalez-Aguilar, G.A.; Martinez-Tellez, M.A.; Nazzaro, F.; Ayala-Zavala, J.F. Virulence of *Pseudomonas aeruginosa* exposed to carvacrol: Alterations of the Quorum sensing at enzymatic and gene levels. *J. Cell Commun. Signal.* **2019**, *13*, 531–537. [CrossRef] [PubMed]
192. Forbes, J.D. Clinically important toxins in bacterial infection: Utility of laboratory detection. *Clin. Microbiol. Newsl.* **2020**, *42*, 163–170. [CrossRef] [PubMed]
193. Nwabor, O.F.; Singh, S.; Syukri, D.M.; Voravuthikunchai, S.P. Bioactive fractions of *Eucalyptus camaldulensis* inhibit important foodborne pathogens, reduce listeriolysin O-induced haemolysis, and ameliorate hydrogen peroxide-induced oxidative stress on human embryonic colon cells. *Food Chem.* **2021**, *344*, 128571. [CrossRef]
194. Rasooly, R.; Molnar, A.; Choi, H.-Y.; Do, P.; Racicot, K.; Apostolidis, E. In-vitro inhibition of staphylococcal pathogenesis by witch-hazel and green tea extracts. *Antibiotics* **2019**, *8*, 244. [CrossRef] [PubMed]
195. Shi, C.; Zhao, X.; Yan, H.; Meng, R.; Zhang, Y.; Li, W.; Liu, Z.; Guo, N. Effect of tea tree oil on *Staphylococcus aureus* growth and enterotoxin production. *Food Control* **2016**, *62*, 257–263. [CrossRef]
196. Zhang, M.; Li, H.; Agyekumwaa, A.K.; Yu, Y.; Xiao, X. Effects of citronellal on growth and enterotoxins production in *Staphylococcus aureus* ATCC 29213. *Toxicon* **2022**, *213*, 92–98. [CrossRef]
197. Mooyottu, S.; Kollanoor-Johny, A.; Flock, G.; Bouillaut, L.; Upadhyay, A.; Sonenshein, A.L.; Venkitanarayanan, K. Carvacrol and trans-cinnamaldehyde reduce *Clostridium difficile* toxin production and cytotoxicity in vitro. *Int. J. Mol. Sci.* **2014**, *15*, 4415–4430. [CrossRef]
198. Tian, F.; Woo, S.Y.; Lee, S.Y.; Park, S.B.; Im, J.H.; Chun, H.S. Plant-based natural flavonoids show strong inhibition of aflatoxin production and related gene expressions correlated with chemical structure. *Food Microbiol.* **2023**, *109*, 104141. [CrossRef]
199. Pok, P.S.; Londoño, V.A.G.; Vicente, S.; Romero, S.M.; Pacín, A.; Tolaba, M.; Alzamora, S.M.; Resnik, S.L. Evaluation of citrus flavonoids against *Aspergillus parasiticus* in maize: Aflatoxins reduction and ultrastructure alterations. *Food Chem.* **2020**, *318*, 126414. [CrossRef] [PubMed]
200. Hamad, G.M.; Mohdaly, A.A.A.; El-Nogoumy, B.A.; Ramadan, M.F.; Hassan, S.A.; Zeitoun, A.M. Detoxification of aflatoxin B1 and ochratoxin A using *Salvia farinacea* and *Azadirachta indica* water extract and application in meat products. *Appl. Biochem. Biotechnol.* **2021**, *193*, 3098–3120. [CrossRef] [PubMed]
201. Ponzilacqua, B.; Rottinghaus, G.E.; Landers, B.R.; Oliveira, C.A.F.D. Effects of medicinal herb and Brazilian traditional plant extracts on in vitro mycotoxin decontamination. *Food Control* **2019**, *100*, 24–27. [CrossRef]
202. Friedman, M.; Jürgens, H.S. Effect of pH on the stability of plant phenolic compounds. *J. Agric. Food Chem.* **2000**, *48*, 2101–2110. [CrossRef]

203. Hatab, S.; Athanasio, R.; Holley, R.; Rodas-Gonzalez, A.; Narvaez-Bravo, C. Survival and reduction of shiga toxin-producing *Escherichia coli* in a fresh cold-pressed juice treated with antimicrobial plant extracts. *J. Food Sci.* **2016**, *81*, M1987–M1995. [CrossRef] [PubMed]
204. Rattanachaikunsopon, P.; Phumkhachorn, P. Assessment of factors influencing antimicrobial activity of carvacrol and cymene against *Vibrio cholerae* in food. *J. Biosci. Bioeng.* **2010**, *110*, 614–619. [CrossRef]
205. Salvia-Trujillo, L.; Rojas-Graü, A.; Soliva-Fortuny, R.; Martín-Belloso, O. Physicochemical characterization and antimicrobial activity of food-grade emulsions and nanoemulsions incorporating essential oils. *Food Hydrocoll.* **2015**, *43*, 547–556. [CrossRef]
206. Topuz, O.K.; Özvural, E.B.; Zhao, Q.; Huang, Q.; Chikindas, M.; Gölükçü, M. Physical and antimicrobial properties of anise oil loaded nanoemulsions on the survival of foodborne pathogens. *Food Chem.* **2016**, *203*, 117–123. [CrossRef]
207. Ghazy, O.A.; Fouad, M.T.; Saleh, H.H.; Kholif, A.E.; Morsy, T.A. Ultrasound-assisted preparation of anise extract nanoemulsion and its bioactivity against different pathogenic bacteria. *Food Chem.* **2021**, *341*, 128259. [CrossRef] [PubMed]
208. Ghazy, O.A.; Fouad, M.T.; Morsy, T.A.; Kholif, A.E. Nanoemulsion formulation of *Lawsonia inermis* extract and its potential antimicrobial and preservative efficacy against foodborne pathogens. *Food Control* **2023**, *145*, 109458. [CrossRef]
209. Bhargava, K.; Conti, D.S.; da Rocha, S.R.; Zhang, Y. Application of an oregano oil nanoemulsion to the control of foodborne bacteria on fresh lettuce. *Food Microbiol.* **2015**, *47*, 69–73. [CrossRef] [PubMed]
210. Paudel, S.K.; Bhargava, K.; Kotturi, H. Antimicrobial activity of cinnamon oil nanoemulsion against *Listeria monocytogenes* and *Salmonella* spp. on melons. *LWT* **2019**, *111*, 682–687. [CrossRef]
211. Molet-Rodríguez, A.; Turmo-Ibarz, A.; Salvia-Trujillo, L.; Martín-Belloso, O. Incorporation of antimicrobial nanoemulsions into complex foods: A case study in an apple juice-based beverage. *LWT* **2021**, *141*, 110926. [CrossRef]
212. Luciano, W.A.; Pimentel, T.C.; Bezerril, F.F.; Barão, C.E.; Marcolino, V.A.; Carvalho, R.D.S.F.; dos Santos Lima, M.; Martín-Belloso, O.; Magnani, M. Effect of citral nanoemulsion on the inactivation of *Listeria monocytogenes* and sensory properties of fresh-cut melon and papaya during storage. *Int. J. Food Microbiol.* **2023**, *384*, 109959. [CrossRef]
213. Ozogul, Y.; Boğa, E.K.; Akyol, I.; Durmus, M.; Ucar, Y.; Regenstein, J.M.; Köşker, A.R. Antimicrobial activity of thyme essential oil nanoemulsions on spoilage bacteria of fish and food-borne pathogens. *Food Biosci.* **2020**, *36*, 100635. [CrossRef]
214. Özogul, Y.; El Abed, N.; Özogul, F. Antimicrobial effect of laurel essential oil nanoemulsion on food-borne pathogens and fish spoilage bacteria. *Food Chem.* **2022**, *368*, 130831. [CrossRef]
215. Özogul, Y.; Özogul, F.; Kulawik, P. The antimicrobial effect of grapefruit peel essential oil and its nanoemulsion on fish spoilage bacteria and food-borne pathogens. *LWT* **2021**, *136*, 110362. [CrossRef]
216. Tang, M.; Liu, F.; Wang, Q.; Wang, D.; Wang, D.; Zhu, Y.; Sun, Z.; Xu, W. Physicochemical characteristics of ginger essential oil nanoemulsion encapsulated by zein/NaCas and antimicrobial control on chilled chicken. *Food Chem.* **2022**, *374*, 131624. [CrossRef]
217. Yang, R.; Miao, J.; Shen, Y.; Cai, N.; Wan, C.; Zou, L.; Chen, C.; Chen, J. Antifungal effect of cinnamaldehyde, eugenol and carvacrol nanoemulsion against *Penicillium digitatum* and application in postharvest preservation of citrus fruit. *LWT* **2021**, *141*, 110924. [CrossRef]
218. Gundewadi, G.; Sarkar, D.J.; Rudra, S.G.; Singh, D. Preparation of basil oil nanoemulsion using *Sapindus mukorossi* pericarp extract: Physico-chemical properties and antifungal activity against food spoilage pathogens. *Ind. Crops Prod.* **2018**, *125*, 95–104. [CrossRef]
219. Ribes, S.; Fuentes, A.; Barat, J.M. Effect of oregano (*Origanum vulgare* L. ssp. *hirtum*) and clove (*Eugenia* spp.) nanoemulsions on *Zygosaccharomyces bailii* survival in salad dressings. *Food Chem.* **2019**, *295*, 630–636. [CrossRef] [PubMed]
220. Wan, J.; Zhong, S.; Schwarz, P.; Chen, B.; Rao, J. Physical properties, antifungal and mycotoxin inhibitory activities of five essential oil nanoemulsions: Impact of oil compositions and processing parameters. *Food Chem.* **2019**, *291*, 199–206. [CrossRef] [PubMed]
221. Chaudhari, A.K.; Singh, V.K.; Das, S.; Prasad, J.; Dwivedy, A.K.; Dubey, N.K. Improvement of in vitro and in situ antifungal, AFB1 inhibitory and antioxidant activity of *Origanum majorana* L. essential oil through nanoemulsion and recommending as novel food preservative. *Food Chem. Toxicol.* **2020**, *143*, 111536. [CrossRef]
222. Gharsallaoui, A.; Roudaut, G.; Chambin, O.; Voilley, A.; Saurel, R. Applications of spray-drying in microencapsulation of food ingredients: An overview. *Food Res. Int.* **2007**, *40*, 1107–1121. [CrossRef]
223. Chen, H.; Zhang, Y.; Zhong, Q. Physical and antimicrobial properties of spray-dried zein–casein nanocapsules with co-encapsulated eugenol and thymol. *J. Food Eng.* **2015**, *144*, 93–102. [CrossRef]
224. Talón, E.; Lampi, A.M.; Vargas, M.; Chiralt, A.; Jouppila, K.; González-Martínez, C. Encapsulation of eugenol by spray-drying using whey protein isolate or lecithin: Release kinetics, antioxidant and antimicrobial properties. *Food Chem.* **2019**, *295*, 588–598. [CrossRef]
225. Sun, X.; Cameron, R.G.; Bai, J. Effect of spray-drying temperature on physicochemical, antioxidant and antimicrobial properties of pectin/sodium alginate microencapsulated carvacrol. *Food Hydrocoll.* **2020**, *100*, 105420. [CrossRef]
226. Plati, F.; Papi, R.; Paraskevopoulou, A. Characterization of oregano essential oil (*Origanum vulgare* L. subsp. *hirtum*) particles produced by the novel nano spray drying technique. *Foods* **2021**, *10*, 2923. [CrossRef]
227. Cruz-Molina, A.V.D.L.; Ayala Zavala, J.F.; Bernal Mercado, A.T.; Cruz Valenzuela, M.R.; González-Aguilar, G.A.; Lizardi-Mendoza, J.; Brown-Bojorquez, F.; Silva-Espinoza, B.A. Maltodextrin encapsulation improves thermal and pH stability of green tea extract catechins. *J. Food Process. Preserv.* **2021**, *45*, e15729. [CrossRef]

228. Radünz, M.; dos Santos Hackbart, H.C.; Camargo, T.M.; Nunes, C.F.P.; de Barros, F.A.P.; Dal Magro, J.; Filho, P.J.S.; Gandra, E.A.; Radünz, A.L.; da Rosa Zavareze, E. Antimicrobial potential of spray drying encapsulated thyme (*Thymus vulgaris*) essential oil on the conservation of hamburger-like meat products. *Int. J. Food Microbiol.* **2020**, *330*, 108696. [CrossRef] [PubMed]
229. do Valle Calomeni, A.; de Souza, V.B.; Tulini, F.L.; Thomazini, M.; Ostroschi, L.C.; de Alencar, S.M.; Massarioli, A.P.; de Carvalho Balieiro, J.C.; de Carvalho, R.A.; Favaro-Trindade, C.S. Characterization of antioxidant and antimicrobial properties of spray-dried extracts from peanut skins. *Food Bioprod. Process.* **2017**, *105*, 215–223. [CrossRef]
230. Vinceković, M.; Viskić, M.; Jurić, S.; Giacometti, J.; Kovačević, D.B.; Putnik, P.; Donsì, F.; Barba, F.J.; Jambrak, A.R. Innovative technologies for encapsulation of Mediterranean plants extracts. *Trends Food Sci. Technol.* **2017**, *69*, 1–12. [CrossRef]
231. de Araújo, J.S.F.; de Souza, E.L.; Oliveira, J.R.; Gomes, A.C.A.; Kotzebue, L.R.V.; da Silva Agostini, D.L.; de Oliveira, D.L.V.; Mazzetto, S.M.; da Silva, A.L.; Cavalcanti, M.T. Microencapsulation of sweet orange essential oil (*Citrus aurantium* var. *dulcis*) by liophylization using maltodextrin and maltodextrin/gelatin mixtures: Preparation, characterization, antimicrobial and antioxidant activities. *Int. J. Biol. Macromol.* **2020**, *143*, 991–999. [CrossRef]
232. Viacava, G.E.; Ayala-Zavala, J.F.; González-Aguilar, G.A.; Ansorena, M.R. Effect of free and microencapsulated thyme essential oil on quality attributes of minimally processed lettuce. *Postharvest Biol. Technol.* **2018**, *145*, 125–133. [CrossRef]
233. Garcia-Sotelo, D.; Silva-Espinoza, B.; Perez-Tello, M.; Olivas, I.; Alvarez-Parrilla, E.; González-Aguilar, G.A.; Ayala-Zavala, J.F. Antimicrobial activity and thermal stability of rosemary essential oil: β−cyclodextrin capsules applied in tomato juice. *LWT* **2019**, *111*, 837–845. [CrossRef]
234. Silva, F.; Caldera, F.; Trotta, F.; Nerin, C.; Domingues, F.C. Encapsulation of coriander essential oil in cyclodextrin nanosponges: A new strategy to promote its use in controlled-release active packaging. *Innov. Food Sci. Emerg. Technol.* **2019**, *56*, 102177. [CrossRef]
235. Ozdemir, N.; Pola, C.C.; Teixeira, B.N.; Hill, L.E.; Bayrak, A.; Gomes, C.L. Preparation of black pepper oleoresin inclusion complexes based on beta-cyclodextrin for antioxidant and antimicrobial delivery applications using kneading and freeze drying methods: A comparative study. *LWT* **2018**, *91*, 439–445. [CrossRef]
236. Lin, L.; Gu, Y.; Sun, Y.; Cui, H. Characterization of chrysanthemum essential oil triple-layer liposomes and its application against *Campylobacter jejuni* on chicken. *LWT* **2019**, *107*, 16–24. [CrossRef]
237. Yang, K.; Liu, A.; Hu, A.; Li, J.; Zen, Z.; Liu, Y.; Tang, S.; Li, C. Preparation and characterization of cinnamon essential oil nanocapsules and comparison of volatile components and antibacterial ability of cinnamon essential oil before and after encapsulation. *Food Control* **2021**, *123*, 107783. [CrossRef]
238. Augustin, M.A.; Sanguansri, P. Challenges and solutions to incorporation of nutraceuticals in foods. *Annu. Rev. Food Sci. Technol.* **2015**, *3*, 61–83. [CrossRef] [PubMed]
239. Fang, Z.; Bhandari, B. Encapsulation of polyphenols—A review. *Trends Food Sci. Technol.* **2010**, *21*, 510–523. [CrossRef]
240. Arruda, T.R.; Bernardes, P.C.; e Moraes, A.R.F.; Soares, N.D.F.F. Natural bioactives in perspective: The future of active packaging based on essential oils and plant extracts themselves and those complexed by cyclodextrins. *Food Res. Int.* **2022**, *156*, 111160. [CrossRef]
241. Cui, R.; Zhu, B.; Yan, J.; Qin, Y.; Yuan, M.; Cheng, G.; Yuan, M. Development of a sodium alginate-based active package with controlled release of cinnamaldehyde loaded on halloysite nanotubes. *Foods* **2021**, *10*, 1150. [CrossRef]
242. Cometa, S.; Bonifacio, M.A.; Bellissimo, A.; Pinto, L.; Petrella, A.; De Vietro, N.; Iannaccone, G.; Baruzzi, F.; De Giglio, E. A green approach to develop zeolite-thymol antimicrobial composites: Analytical characterization and antimicrobial activity evaluation. *Heliyon* **2022**, *8*, e09551. [CrossRef]
243. Wang, Y.; Yi, S.; Lu, R.; Sameen, D.E.; Ahmed, S.; Dai, J.; Qin, W.; Li, S.; Liu, Y. Preparation, characterization, and 3D printing verification of chitosan/halloysite nanotubes/tea polyphenol nanocomposite films. *Int. J. Biol. Macromol.* **2021**, *166*, 32–44. [CrossRef]
244. Jiang, Y.; Yin, H.; Zhou, X.; Wang, D.; Zhong, Y.; Xia, Q.; Deng, Y.; Zhao, Y. Antimicrobial, antioxidant and physical properties of chitosan film containing *Akebia trifoliata* (Thunb.) Koidz. peel extract/montmorillonite and its application. *Food Chem.* **2021**, *361*, 130111. [CrossRef]
245. Riaz, A.; Lei, S.; Akhtar, H.M.S.; Wan, P.; Chen, D.; Jabbar, S.; Abid, M.; Hashim, M.M.; Zeng, X. Preparation and characterization of chitosan-based antimicrobial active food packaging film incorporated with apple peel polyphenols. *Int. J. Biol. Macromol.* **2018**, *114*, 547–555. [CrossRef]
246. Wang, K.; Lim, P.N.; Tong, S.Y.; San Thian, E. Development of grapefruit seed extract-loaded poly (ε-caprolactone)/chitosan films for antimicrobial food packaging. *Food Packag. Shelf Life* **2019**, *22*, 100396. [CrossRef]
247. Surendhiran, D.; Li, C.; Cui, H.; Lin, L. Fabrication of high stability active nanofibers encapsulated with pomegranate peel extract using chitosan/PEO for meat preservation. *Food Packag. Shelf Life* **2020**, *23*, 100439. [CrossRef]
248. Gutiérrez-Pacheco, M.M.; Ortega-Ramírez, L.A.; Silva-Espinoza, B.A.; Cruz-Valenzuela, M.R.; González-Aguilar, G.A.; Lizardi-Mendoza, J.; Miranda, R.; Ayala-Zavala, J.F. Individual and combined coatings of chitosan and carnauba wax with oregano essential oil to avoid water loss and microbial decay of fresh cucumber. *Coatings* **2020**, *10*, 614. [CrossRef]
249. Saberi, B.; Chockchaisawasdee, S.; Golding, J.B.; Scarlett, C.J.; Stathopoulos, C.E. Characterization of pea starch-guar gum biocomposite edible films enriched by natural antimicrobial agents for active food packaging. *Food Bioprod. Process.* **2017**, *105*, 51–63. [CrossRef]

250. Chollakup, R.; Pongburoos, S.; Boonsong, W.; Khanoonkon, N.; Kongsin, K.; Sothornvit, R.; Sukyai, P.; Sukatta, U.; Harnkarnsujarit, N. Antioxidant and antibacterial activities of cassava starch and whey protein blend films containing rambutan peel extract and cinnamon oil for active packaging. *LWT* **2020**, *130*, 109573. [CrossRef]
251. Issa, A.; Ibrahim, S.A.; Tahergorabi, R. Impact of sweet potato starch-based nanocomposite films activated with thyme essential oil on the shelf-life of baby spinach leaves. *Foods* **2017**, *6*, 43. [CrossRef]
252. Aloui, H.; Deshmukh, A.R.; Khomlaem, C.; Kim, B.S. Novel composite films based on sodium alginate and gallnut extract with enhanced antioxidant, antimicrobial, barrier and mechanical properties. *Food Hydrocoll.* **2021**, *113*, 106508. [CrossRef]
253. Kapetanakou, A.E.; Nestora, S.; Evageliou, V.; Skandamis, P.N. Sodium alginate–cinnamon essential oil coated apples and pears: Variability of *Aspergillus carbonarius* growth and ochratoxin A production. *Food Res. Int.* **2019**, *119*, 876–885. [CrossRef]
254. Nouri, A.; Yaraki, M.T.; Ghorbanpour, M.; Wang, S. Biodegradable κ-carrageenan/nanoclay nanocomposite films containing *Rosmarinus officinalis* L. extract for improved strength and antibacterial performance. *Int. J. Biol. Macromol.* **2018**, *115*, 227–235. [CrossRef]
255. He, S.; Wang, Y. Antimicrobial and antioxidant effects of kappa-carrageenan coatings enriched with cinnamon essential oil in pork meat. *Foods* **2022**, *11*, 2885. [CrossRef]
256. Göksen, G.; Fabra, M.J.; Ekiz, H.I.; López-Rubio, A. Phytochemical-loaded electrospun nanofibers as novel active edible films: Characterization and antibacterial efficiency in cheese slices. *Food Control* **2020**, *112*, 107133. [CrossRef]
257. Shankar, S.; Rhim, J.W. Preparation of antibacterial poly (lactide)/poly (butylene adipate-co-terephthalate) composite films incorporated with grapefruit seed extract. *Int. J. Biol. Macromol.* **2018**, *120*, 846–852. [CrossRef] [PubMed]
258. Suwanamornlert, P.; Kerddonfag, N.; Sane, A.; Chinsirikul, W.; Zhou, W.; Chonhenchob, V. Poly (lactic acid)/poly (butylene-succinate-co-adipate) (PLA/PBSA) blend films containing thymol as alternative to synthetic preservatives for active packaging of bread. *Food Packag. Shelf Life* **2020**, *25*, 100515. [CrossRef]
259. Requena, R.; Vargas, M.; Chiralt, A. Eugenol and carvacrol migration from PHBV films and antibacterial action in different food matrices. *Food Chem.* **2019**, *277*, 38–45. [CrossRef] [PubMed]
260. Pabast, M.; Shariatifar, N.; Beikzadeh, S.; Jahed, G. Effects of chitosan coatings incorporating with free or nano-encapsulated *Satureja* plant essential oil on quality characteristics of lamb meat. *Food Control* **2018**, *91*, 185–192. [CrossRef]
261. Amor, G.; Sabbah, M.; Caputo, L.; Idbella, M.; De Feo, V.; Porta, R.; Fechtali, T.; Mauriello, G. Basil essential oil: Composition, antimicrobial properties, and microencapsulation to produce active chitosan films for food packaging. *Foods* **2021**, *10*, 121. [CrossRef]
262. Pan, J.; Ai, F.; Shao, P.; Chen, H.; Gao, H. Development of polyvinyl alcohol/β-cyclodextrin antimicrobial nanofibers for fresh mushroom packaging. *Food Chem.* **2019**, *300*, 125249. [CrossRef]
263. Chen, Z.; Zong, L.; Chen, C.; Xie, J. Development and characterization of PVA-Starch active films incorporated with β-cyclodextrin inclusion complex embedding lemongrass (*Cymbopogon citratus*) oil. *Food Packag. Shelf Life* **2020**, *26*, 100565. [CrossRef]
264. Lee, J.Y.; Garcia, C.V.; Shin, G.H.; Kim, J.T. Antibacterial and antioxidant properties of hydroxypropyl methylcellulose-based active composite films incorporating oregano essential oil nanoemulsions. *LWT* **2019**, *106*, 164–171. [CrossRef]
265. Shokri, S.; Parastouei, K.; Taghdir, M.; Abbaszadeh, S. Application an edible active coating based on chitosan-*Ferulago angulata* essential oil nanoemulsion to shelf life extension of Rainbow trout fillets stored at 4 °C. *Int. J. Biol. Macromol.* **2020**, *153*, 846–854. [CrossRef]
266. Abbasi, Z.; Aminzare, M.; Hassanzad Azar, H.; Rostamizadeh, K. Effect of corn starch coating incorporated with nanoemulsion of *Zataria multiflora* essential oil fortified with cinnamaldehyde on microbial quality of fresh chicken meat and fate of inoculated *Listeria monocytogenes*. *J. Food Sci. Technol.* **2021**, *58*, 2677–2687. [CrossRef]
267. Corrêa, J.A.F.; dos Santos, J.V.G.; Evangelista, A.G.; Pinto, A.C.S.M.; de Macedo, R.E.F.; Luciano, F.B. Combined application of phenolic acids and essential oil components against *Salmonella Enteritidis* and *Listeria monocytogenes* in vitro and in ready-to-eat cooked ham. *LWT* **2021**, *149*, 111881. [CrossRef]
268. Nikkhah, M.; Hashemi, M. Boosting antifungal effect of essential oils using combination approach as an efficient strategy to control postharvest spoilage and preserving the jujube fruit quality. *Postharvest Biol. Technol.* **2020**, *164*, 111159. [CrossRef]
269. Purkait, S.; Bhattacharya, A.; Bag, A.; Chattopadhyay, R.R. Synergistic antibacterial, antifungal and antioxidant efficacy of cinnamon and clove essential oils in combination. *Arch. Microbiol.* **2020**, *202*, 1439–1448. [CrossRef] [PubMed]
270. Zhang, X.; Zhou, D.; Cao, Y.; Zhang, Y.; Xiao, X.; Liu, F.; Yu, Y. Synergistic inactivation of *Escherichia coli* O157: H7 and *Staphylococcus aureus* by gallic acid and thymol and its potential application on fresh-cut tomatoes. *Food Microbiol.* **2022**, *102*, 103925. [CrossRef]
271. Meira, N.V.; Holley, R.A.; Bordin, K.; de Macedo, R.E.; Luciano, F.B. Combination of essential oil compounds and phenolic acids against *Escherichia coli* O157: H7 in vitro and in dry-fermented sausage production. *Int. J. Food Microbiol.* **2017**, *260*, 59–64. [CrossRef]
272. Nikkhah, M.; Hashemi, M.; Najafi, M.B.H.; Farhoosh, R. Synergistic effects of some essential oils against fungal spoilage on pear fruit. *Int. J. Food Microbiol.* **2017**, *257*, 285–294. [CrossRef]
273. Aaliya, B.; Sunooj, K.V.; Navaf, M.; Akhila, P.P.; Sudheesh, C.; Mir, S.A.; Sabu, S.; Sasidharan, A.; Theingi Hlaing, M.; George, J. Recent trends in bacterial decontamination of food products by hurdle technology: A synergistic approach using thermal and non-thermal processing techniques. *Food Res. Int.* **2021**, *147*, 110514. [CrossRef]

274. Pinto, L.; Baruzzi, F.; Cocolin, L.; Malfeito-Ferreira, M. Emerging technologies to control *Brettanomyces* spp. in wine: Recent advances and future trends. *Trends Food Sci. Technol.* **2020**, *99*, 88–100. [CrossRef]
275. Patrignani, F.; Siroli, L.; Braschi, G.; Lanciotti, R. Combined use of natural antimicrobial based nanoemulsions and ultra high pressure homogenization to increase safety and shelf-life of apple juice. *Food Control* **2020**, *111*, 107051. [CrossRef]
276. Chien, S.Y.; Sheen, S.; Sommers, C.; Sheen, L.Y. Combination effect of high-pressure processing and essential oil (*Melissa officinalis* extracts) or their constituents for the inactivation of *Escherichia coli* in ground beef. *Food Bioproc. Tech.* **2019**, *12*, 359–370. [CrossRef]
277. Sheen, S.; Huang, C.Y.; Chuang, S. Synergistic effect of high hydrostatic pressure, allyl isothiocyanate, and acetic acid on the inactivation and survival of pathogenic *Escherichia coli* in ground chicken. *J. Food Sci.* **2022**, *87*, 5042–5053. [CrossRef] [PubMed]
278. González-González, C.R.; Labo-Popoola, O.; Delgado-Pando, G.; Theodoridou, K.; Doran, O.; Stratakos, A.C. The effect of cold atmospheric plasma and linalool nanoemulsions against *Escherichia coli* O157: H7 and *Salmonella* on ready-to-eat chicken meat. *LWT* **2021**, *149*, 111898. [CrossRef]
279. Olatunde, O.O.; Benjakul, S.; Vongkamjan, K. Cold plasma combined with liposomal ethanolic coconut husk extract: A potential hurdle technology for shelf-life extension of Asian sea bass slices packaged under modified atmosphere. *Innov. Food Sci. Emerg. Technol.* **2020**, *65*, 102448. [CrossRef]
280. Van Haute, S.; Raes, K.; Devlieghere, F.; Sampers, I. Combined use of cinnamon essential oil and MAP/vacuum packaging to increase the microbial and sensorial shelf life of lean pork and salmon. *Food Packag. Shelf Life* **2017**, *12*, 51–58. [CrossRef]
281. Shiekh, K.A.; Benjakul, S. Melanosis and quality changes during refrigerated storage of Pacific white shrimp treated with Chamuang (*Garcinia cowa* Roxb.) leaf extract with the aid of pulsed electric field. *Food Chem.* **2020**, *309*, 125516. [CrossRef]
282. Guo, M.; Zhang, L.; He, Q.; Arabi, S.A.; Zhao, H.; Chen, W.; Ye, X.; Liu, D. Synergistic antibacterial effects of ultrasound and thyme essential oils nanoemulsion against *Escherichia coli* O157: H7. *Ultrason. Sonochem.* **2020**, *66*, 104988. [CrossRef] [PubMed]
283. Kim, W.; Khang, D. Synergistic effects of 915 MHz microwave heating and essential oils on inactivation of foodborne pathogen in hot-chili sauce. *Int. J. Food Microbiol.* **2023**, *398*, 110210. [CrossRef]
284. Yang, Y.; Ma, S.; Guo, K.; Guo, D.; Li, J.; Wang, M.; Wang, Y.; Zhang, C.; Xia, X.; Shi, C. Efficacy of 405-nm LED illumination and citral used alone and in combination for the inactivation of *Cronobacter sakazakii* in reconstituted powdered infant formula. *Food Res. Int.* **2022**, *154*, 111027. [CrossRef]
285. Silva-Espinoza, B.A.; Palomares-Navarro, J.J.; Tapia-Rodriguez, M.R.; Cruz-Valenzuela, M.R.; González-Aguilar, G.A.; Silva-Campa, E.; Pedroza-Montero, M.; Almeida-Lopes, M.; Miranda, R.; Ayala-Zavala, J.F. Combination of ultraviolet light-C and clove essential oil to inactivate *Salmonella typhimurium* biofilms on stainless steel. *J. Food Saf.* **2020**, *40*, e12788. [CrossRef]
286. Chelliah, R.; Jo, K.H.; Yan, P.; Chen, X.; Jo, H.Y.; Madar, I.H.; Sultan, G.; Oh, D.H. Unravelling the sanitization potential of slightly acidic electrolyzed water combined *Thymus vulgaris* based nanoemulsion against foodborne pathogens and its safety assessment. *Food Control* **2023**, *146*, 109527. [CrossRef]
287. Shankar, S.; Danneels, F.; Lacroix, M. Coating with alginate containing a mixture of essential oils and citrus extract in combination with ozonation or gamma irradiation increased the shelf life of *Merluccius* sp. fillets. *Food Packag. Shelf Life* **2019**, *22*, 100434. [CrossRef]
288. Reinholds, I.; Pugajeva, I.; Bavrins, K.; Kuckovska, G.; Bartkevics, V. Mycotoxins, pesticides and toxic metals in commercial spices and herbs. *Food Addit. Contam. Part B* **2017**, *10*, 5–14. [CrossRef] [PubMed]
289. Nikolic, N.; Höferl, M.; Buchbauer, G. Pesticides in essential oils and selected fragrance extracts. Some examples. A review. *Flavour Frag. J.* **2018**, *33*, 373–384. [CrossRef]
290. Corrales, M.; Fernandez, A.; Pinto, M.G.V.; Butz, P.; Franz, C.M.; Schuele, E.; Tauscher, B. Characterization of phenolic content, in vitro biological activity, and pesticide loads of extracts from white grape skins from organic and conventional cultivars. *Food Chem. Toxicol.* **2010**, *48*, 3471–3476. [CrossRef] [PubMed]
291. Yang, Z.; Wang, H.; Ying, G.; Yang, M.; Nian, Y.; Liu, J.; Kong, W. Relationship of mycotoxins accumulation and bioactive components variation in ginger after fungal inoculation. *Front. Pharmacol.* **2017**, *8*, 331. [CrossRef]
292. Hassan, H.F.; Koaik, L.; Khoury, A.E.; Atoui, A.; El Obeid, T.; Karam, L. Dietary exposure and risk assessment of mycotoxins in thyme and thyme-based products marketed in Lebanon. *Toxins* **2022**, *14*, 331. [CrossRef]

foods

MDPI

Article

Fermented Grapevine Leaves: Potential Preserving Agent in Yogurt

Lais Freitas [1,2,3], Miguel Sousa-Dias [2,3], Vanessa B. Paula [1,2,3], Luís G. Dias [2,3] and Leticia M. Estevinho [2,3,*]

1 Doctoral School, University of León (ULE), Campus de Vegazana, 24007 León, Spain; laisfreitas@ipb.pt (L.F.); vanessapaula@ipb.pt (V.B.P.)
2 Centro de Investigação de Montanha (CIMO), Instituto Politécnico de Bragança, Campus de Santa Apolónia, 5300-252 Bragança, Portugal; miglsdias@gmail.com (M.S.-D.); ldias@ipb.pt (L.G.D.)
3 Laboratório para a Sustentabilidade e Tecnologia em Regiões de Montanha, Instituto Politécnico de Bragança, Campus de Santa Apolónia, 5300-253 Bragança, Portugal
* Correspondence: leticia@ipb.pt

Abstract: In this study, we monitored the fermentative process of *Vitis vinifera* L. leaves (grapevine), spontaneously or promoted by *Saccharomyces cerevisiae*, in both solid and liquid media. We also aimed to evaluate the effect on the bioactivity and shelf life of yogurt incorporating fermented and non-fermented grapevine leaves compared to yogurt produced with the preservative potassium sorbate. The results revealed that fermented grapevine leaf extracts increased their bioactive compounds and antioxidant activity, particularly in fermentations in a solid medium. In yogurt samples with incorporation extract from solid spontaneous fermentation and extract from solid yeast fermentation, even in small quantities, they exhibited higher levels of total phenols (1.94 and 2.16 mg GAE/g of yogurt, respectively) and antioxidant activity (5.30 and 5.77 mg TroloxE/g of yogurt; and 1.33 and 1.34 mg Fe(II)E/g of yogurt, respectively) compared to control yogurt (1.44 mg GAE/g of yogurt, 4.00 mg TroloxE/g of yogurt, and 1.01 mg Fe(II)E/g of yogurt). Additionally, yogurts supplemented with fermented grapevine leaves demonstrated the potential to inhibit microbial growth without impairing the multiplication of lactic acid bacteria.

Keywords: *Vitis vinifera*; fermentation; yogurt; preserving; *Saccharomyces cerevisiae*

1. Introduction

Yogurt is a fermented milk product that has become popular over the years for its nutritional profile, stability, natural characteristics and status as a healthy food. Yogurt, as well as other fermented dairy products, is obtained through fermentation by lactic acid bacteria, which are mainly responsible for the product's characteristics, such as its taste, aroma, and texture. This technology consists of inoculating pasteurized milk with lactic acid bacteria, *Streptococcus thermophilus* and *Lactobacillus bulgaricus*. Fermentation begins with the production of lactic acid by *S. thermophilus* and then, *L. bulgaricus* is used to provide the characteristic flavor of yogurt. Studies show that the use of these bacteria has health benefits due to the fact that they stimulate the immune system and regulate the gastrointestinal system [1,2].

During the fermentation process, due to the enzymatic action of lactic acid bacteria, biochemical reactions take place that are responsible for converting organic substances into smaller compounds [3].

Yogurt is a product in which functional additives or those with greater nutritional value can be incorporated to enrich the food and, consequently, have beneficial effects on consumer health. In the case of extending the shelf life of food, additives are used that are able to preserve the food by inhibiting the multiplication of microflora, thus preventing it from deteriorating [4,5].

Sorbic acid is an additive widely used as a preservative in the food industry, as it is mainly capable of inhibiting the growth of molds and yeasts [6]. However, the use of these additives should be moderated as they can pose a risk to consumer health [4]. The maximum limit of potassium sorbate allowed by Regulation (EU) n° 1129/2011 on food additives in yogurt is 1000 mg/kg [7].

In recent years, the demand for healthier foods with fewer chemical additives and preservatives and with functional properties capable of promoting health benefits has increased considerably, as is the case with probiotic foods and natural additives [3,8]. In this context, a recently studied alternative is the use of vine leaves, which are an abundant source of vitamins and phenolic compounds [8].

V. vinifera L., native to Western Asia and Europe, produces fruit that is widely used for both culinary and beverage purposes [9]. Beyond the primary yield of grapes, *V. vinifera* L. offers an array of ancillary products and by-products, such as grape leaves and pomace [10]. Grape leaves, in particular, are rich in phenolic compounds that are known for their powerful antioxidant and anti-inflammatory properties. These bioactive constituents, including polyphenols, resveratrol, quercetin, anthocyanins, flavanols, vitamins, and minerals, offer significant health benefits and can significantly enhance the functional properties of foods [11]. The biological properties of grapevine leaves have been widely recognized, mainly due to their rich phenolic compounds, including phenolic acids, flavonols, tannins, and anthocyanins. These compounds, predominantly found in the leaves, provide liver protection and exhibit a range of beneficial effects, such as their antibacterial, antifungal, antiviral, anti-inflammatory, and antioxidant activities. The use of grape leaves, a by-product of viticulture, promotes sustainability and adds value to agricultural waste, supporting the principles of a circular economy. The extraction of bioactive compounds from *Vitis* leaves represents a promising solution for the management of ecotoxic waste. It also facilitates the development of innovative and valuable natural products with multiple applications in different industries by exploiting their bioactive properties [9]. A major advantage of yogurt is that it can be used as a natural additive to food products. Yogurt, a widely consumed and globally accepted dairy product, offers an ideal medium for introducing new functional ingredients without changing consumers' eating habits. Its matrix is particularly suitable for incorporating bioactive compounds, ensuring controlled release, and improving the bioavailability of these phenolic compounds [12].

Fermentation by yeasts is a processing technique widely used in the food industry as it is responsible for transforming starch and sugars into alcohol and CO_2, resulting in products such as beer, wine, and bread. This process has the advantage of low energy costs and allows enzymes such as lipase, amylase, glucoamylase, and protease to synthesize bioactive compounds present in the substrate [13].

Yeasts are also used for their functional capacity as stabilizers and thickeners. The most commercially used genera are *Saccharomyces* and *Candida*. Each strain requires specific medium, temperature, and aeration conditions for growth [14,15]. Fermentation can cause biological changes in the substrates, producing biologically active secondary metabolites and increasing their antioxidant activity. In previous studies, researchers have observed that fermentation in solid media can increase the antioxidant content and nutritional profile of cereals and legumes [16,17].

The study carried out by Dhull et al. [16] indicated that fermentation conducted by *Aspergillus awamori* in a solid state is an efficient, economical, and reliable method, as it promotes an increase in total phenolic content and condensed tannin content. García and Bianchi [18] indicated that solid-medium fermentation with the fungus *Penicillium purpurogenum* was able to increase the content of phenolic compounds in extracts obtained from coffee husks. According to Kosar et al. [19], the content of total flavonoids, such as quercetin, can be increased by fermenting the leaves.

In view of the above, no studies have been found in the literature regarding the introduction of fermented vine leaf extract for enrichment or as a natural additive in foods.

This work focused on two main objectives. First, it focused on evaluating the ability of vine leaf extracts fermented in solid or liquid media to fully or partially replace potassium sorbate, the conventional preservative of yogurt. Secondly, we investigated whether yogurt enriched with fermented leaf extracts could provide additional health benefits due to their high levels of bioactive compounds. Given the high concentration of bio-active compounds in leaves, it was also investigated whether incorporating leaf extracts into yogurt could enhance its health benefits.

2. Materials and Methods

2.1. Sampling

The leaves of *V. vinifera* L. were harvested at the Agrarian School (ESA) of the Polytechnic Institute of Bragança, Portugal, in June 2020. The leaves were harvested during the vegetative cycle of the vine, when the plant has little water availability. Under these conditions, the plant becomes water stressed and concentrates on the production of high-quality and concentrated by-products, such as the 3-O-glycosides of campherol and quercetin [9].

Undamaged leaves were selected, weighed, and dried in a forced-air oven (VWR® VENTI-Line®, Carnaxide, Portugal; ±0.1 °C) at 40 °C for approximately 72 h to a constant weight. The dried leaves were ground in an IKA TUBE-MILL (Deutschland, Germany) mill until a powder was obtained (grain size 250–800 mesh).

2.2. Reagents

The *Saccharomyces cerevisiae* culture belonged to the microorganism collection of the Escola Superior de Agrária de Bragança, Portugal (ESA).

Yeast extract was obtained from HiMedia Laboratories (Modautal, Germany), bacteriological peptone and glucose were obtained from OXOID LTD (Basingstoke, Hampshire, UK), and glucose (dextrose) monohydrate was obtained from CeaMed, Lda CEIM (Funchal, Portugal). Rose Bengal CAF agar culture medium was supplied by Liofilchem (Roseto degli Abruzzi, Italy), and Man, Rogosa and Sharpe (MRS) agar culture medium was supplied by HiMedia (Mumbai, India). Modified iron sulfite agar culture medium was supplied by HiMedia Laboratories (Modautal, Germany). The SimPlate for Total Plate Count (TPC) kit was provided by Biocontrol® (Bellevue, WA, USA). Baird-Parker medium (BP) was provided by HiMedia Laboratories (Modautal, Germany). Nutrients were provided by Enovit® of the AEB Group (Viseu, Portugal).

Ethanol, 99.8% pure, was obtained from Carlo Erba Reagents (Chaussee du Vexin, France). Folin–Ciocalteau, DPPH, aluminum chloride, sodium carbonate, quercetin, gallic acid, Trolox, FRAP, acetic acid, TPTZ, $FeCl_3$, $FeSO_4.7H_2O$, potassium sorbate, ammonium nitrate (NH_4NO_3), sulfur dioxide, and the other reagents were purchased from Sigma Chemical Co. (St. Louis, MO, USA). Sulfuric acid (H_2SO_4), sodium hydroxide (NaOH), and copper sulfate ($CuSO_4$) were obtained from Merck KGaA (Darmstadt, Germany); mercury sulfate ($HgSO_4$), hydrochloric acid (HCl), and potassium sulfate (K_2SO_4) were obtained from Scharlau (Barcelona, España). Sodium chloride (NaCl) was provided by Pronalab (Lisbon, Portugal). Light petroleum was supplied by Chem-LAB (Zedelgem, Belgium).

2.3. Fermentations

Liquid medium fermentation: A liquid spontaneous control fermentation (LSF) of the leaves was conducted alongside another liquid fermentation inoculated with *Saccharomyces cerevisiae* yeast (LYF), following the methodology reported by Ferreira et al. [20]. In the initial experiment, 10% and 30% glucose/g of leaves were utilized. Based on the results obtained from the first fermentation, the subsequent experiment proceeded with 15% and 20% glucose/g of sample to optimize the conditions of the fermentation process. The liquid medium fermentation involved preparing a medium of 90.00 mL of saline solution (0.85%), 0.60 g/L of commercial nutrients (Enovit—AEB Group), and 10.00 mL of distilled water. To achieve a sterile medium, autoclaving was performed (121 °C for 60 min). After cooling, 3 g of the leaf sample was introduced, followed by immersion in a water bath

(80 °C for 5 min) and a thermal shock in ice. Finally, 125 μL of 6% sulfur dioxide per 100 g of sample and *Saccharomyces cerevisiae* yeast (1×10^6 CFU/g) was added from the collection of microorganisms belonging to ESA. In the case of LSF, neither pasteurization nor inoculum addition was carried out. The experiment was conducted for 14 days at 25 °C. To monitor the fermentations, the following parameters were evaluated every 24 h: flask weight, Brix degree, and assessment of colony number/mL on YPD solid medium. In the latter case, the results were expressed in natural logarithm colony-forming units per milliliter of sample (ln CFU/mL).

Solid medium fermentation: To assess the behavior of *Saccharomyces cerevisiae* in solid-state medium, two fermentations were conducted, one spontaneous (SSF) and one inoculated (SYF), using the methodology described by Dulf et al. [21] with slight adaptations. The medium was prepared using a 0.003% saline solution, 2.0 g of glucose, and 2.0 g of NH_4NO_3. The solution was autoclaved (121 °C for 60 min), and after cooling, 30.0 g of leaf sample was introduced. It underwent pasteurization and thermal shock, and finally, the inoculum was introduced (1×10^6 CFU/g of leaves). Fermentation took place for 9 days at 25 °C. Microbiological control was also assessed by plating on YPD medium at days 1, 3, and 9. From the plate counts on YPD solid medium, the number of natural logarithm colony-forming units per gram of sample (ln CFU/g) was determined. A linear regression analysis was performed using Microsoft Corporation's Excel program (2016) with the natural logarithm values of CFU/g (ln CFU/g) versus time, which are not presented in this work. From the exponential growth phase, the log phase, the equation of the line was obtained to determine the specific growth rate (μ_c) for each of the fermentations. The doubling time (DT) was calculated by dividing ln2 by the specific growth rate (μ_c).

2.4. Extract Procedure

To prepare the extract from liquid medium fermentation, the fermented products were first placed in sterile tubes and centrifuged (5810 R, New York, NY, USA) to separate the solid part. The supernatant was then evaporated to remove part of the alcohol obtained in the fermentation process and subsequently lyophilized (Labconco, Freezone 4.5 model, Kansas City, MO, USA). The extracts were obtained by double hydroalcoholic extraction with 80% ethanol, as described by Paula [22] followed by drying in a rotary evaporator (IKA RV8, Deutschland, Germany) at 40 °C. For sample preparation, 0.1 g of extract (4 mg/mL) was added to a 25 mL volumetric flask with 80% ethanol.

2.5. Yogurt Production

For yogurt production, natural yogurt (Continente) was used as an inoculum at a ratio of 120 g/L of milk (12.0%, *w*/*v*). The natural yogurt used as the inoculum contained a concentration of lactic acid starter cultures (live bacterial cultures) of approximately 10^8 CFU/g. These starter cultures consist of *L. bulgaricus* and *S. thermophilus* bacteria. They are essential for fermenting milk and converting lactose into lactic acid, which gives yogurt its characteristic texture and flavor. In fact, Celik and Temiz [23] report that wild strains are able to produce quantitatively and/or qualitatively different flavor compounds that give yogurt its own desirable characteristics. Initially, 1 L of ultra-high temperature processing (UHT; Continente) whole milk (3.6% fat) was boiled along with sugar (12.6%, *w*/*v*) and powdered milk (1.0% fat; 1.0%, *w*/*v*). We used UHT milk because this type of milk is the most available in our country and has been used in several prior studies focused on yogurt production [24].

To inoculate the yogurt, the sample needed to be cooled to 42 ± 2 °C. The sample was then conditioned in a hermetic container in a thermostat bath for 8 h at 42 ± 2 °C (to reach the pH 4.6, which is required for the product to curdle) [25], followed by placement in a refrigerator (4 ± 2 °C) for 48 h. After this period, the coagulum was broken with a sterile glass rod, and pasteurized strawberry pulp (10.0%, *w*/*v*) was added. The process of obtaining strawberry pulp involved cooking only sanitized and frozen strawberries over medium heat until they reached a gelatinous texture, followed by storage in a sterile container. The extracts that were added to the yogurt were prepared as described in Section 2.4. The yogurts were filled in a bio safety cabinet (Microflow Class 2, Gravesend, UK) packaged in sterile Schott bottles with lids as follows: control yogurt (Cy) with potassium sorbate (0.1%, *w/v*); yogurt SSFy with extract from spontaneous solid-state fermentation (0.1%, *w/v*); yogurt SYFy with extract from yeast solid-state fermentation (0.1%, *w/v*); and yogurt LYFy with extract from liquid-state yeast fermentation (0.1%, *w/v*). A different yogurt was used for each analysis timepoint: one (1 day), second (7 days), and third (12 days). Each sample preparation was performed with 80 g in triplicate and refrigerated (4 ± 2 °C) for subsequent analysis.

2.6. Physicochemical Properties

The physicochemical composition, including the ash and lipid content of grapevine leaves and fermentations, was evaluated. The ash content was determined in triplicate by incinerating samples in a muffle furnace at 550 °C (Lindberg, model 51894, Atlanta, GA, USA; ±1.5 °C). The quantification of mineral residue was calculated by the difference between the mass sample and the mass of the ash. Total lipid content was determined using the continuous extraction method using a Soxhlet apparatus (Behrotest, Labor-Technik, Düsseldorf, Germany). All of the analyses were conducted in accordance with AOAC [26] guidelines.

For yogurt samples, analyses of proteins, ash, lipids, pH, and titratable acidity were performed at days 1, 7, and 12. Ash and lipid analyses were performed as described above. The pH of the yogurt samples was measured using a Potentiometric (Mettler Toledo Model, Mumbai, India) with a combined pH electrode. The percentage of lactic acid in yogurt was quantified using titration with NaOH (0.08 N), with phenolphthalein as the indicator. The Kjeldahl method was used for protein quantification and total nitrogen content [26].

2.7. Bioactive Compounds and Antioxidant Activity

The analysis of bioactive compounds and antioxidant activity was carried out both for the leaves and their fermented products and for the yogurts enriched with the extracts. The leaf extracts and their fermentations were prepared according to Section 2.4. Yogurt samples also underwent the drying and concentration process in a lyophilizer for subsequent extraction, according to Section 2.4.

To determine the total phenolic content, the methodology adapted from Singleton, Orthofer, and Lamuela-Raventos [27] was applied, using gallic acid equivalents (GAEs) as the standard. The results were expressed as the mg of GAE/g of a sample, as obtained from the calibration curve (y = 0.0098x + 0.0081; R^2 = 0.999).

The flavonoid content was determined according to the methodology reported by Savi et al. [28], using quercetin (QE) as the standard. Results were expressed as mg of QE/g of sample, as obtained from the calibration curve (y = 0.027x − 0.082; R^2 = 0.996).

The ferric-reducing ability (FRAP) was quantified following the methodology employed by Santos et al. [29], where ferrous sulfate served as the standard. The results were expressed in mg Fe(II)E/g of sample, as obtained from the calibration curve (y = 0.0026x − 0.05591; R^2 = 0.999).

The DPPH assay was performed based on the work of Paula [22], using Trolox as the standard. Results were expressed as mg TroloxE/g of sample, as obtained from the calibration curve (y = 1.307x − 2.829; R^2 = 0.997). The concentration of DPPH was calculated based on its blocking effect using the following equation:

$$\%\ \text{inhibition} = [(\text{Abs}_{\text{DPPH}} - \text{Abs}_{\text{S}})/\text{Abs}_{\text{DPPH}}] \times 100 \tag{1}$$

where % inhibition = percentage of antioxidant activity; Abs_{DPPH} = solution absorbance with the sample and the free radical DPPH; Abs_{S} = absorbance of the solution with sample without the free radical DPPH.

2.8. Microbiological Analysis

To determine microbiological quality, the resulting products from fermentations were evaluated on days 1, 3, and 9, and yogurt samples were assessed at days 1, 7, and 12. Sample preparation involved adding 1 g of sample with buffered water (0.1%) to a sterile tube, resulting in a concentration of 10^{-1} g/mL. Successive dilutions were then made as necessary to obtain the result for each analysis.

The quantification of molds and yeasts was performed using Rose Bengal CAF Agar medium, Liofilchem, Italy [30], following the manufacturer's instructions. Growth was observed from the second day until day 7 of incubation at 25 °C.

The methodology for lactobacilli counting involved immersing the sample in a double layer of MRS Agar (Man, Rogosa, and Sharpe), as described by Ferreira et al. [20]. Plates were covered with Parafilm and incubated at 30 °C for 72 h.

The analysis for the detection of sulfite-reducing clostridia spores was based on the AOAC manual [26]. Using aseptic conditions, the samples were immersed in a double layer of Modified Iron Sulphite Agar (HIMEDIA) in test tubes to ensure anaerobiosis. These tubes were covered with Parafilm and incubated at 37 °C for 5 days.

The SimPlate kit (Biocontrol®) was used for the determination of coliforms at 37 °C and *Escherichia coli* according to the manufacturer's specifications. Total coliform presence was quantified by color change from blue to pink, and *E. coli* detection and quantification were conducted by counting wells exhibiting fluorescence under exposure to ultraviolet (UV) light at 365 nm.

Following the methodology used by Santos et al. [29], coagulase-positive staphylococci counting was performed on Baird-Parker medium (BP-HIMEDIA) supplemented with potassium tellurite and egg yolk saline solution. Incubation was carried out for 48 h at 37 °C.

Overall, microorganism counts were expressed as colony-forming units (CFUs) per gram or milliliter of sample. Only the result for sulfite-reducing Clostridium was indicated by presence or absence.

2.9. Statistical Analysis

Overall, the assumptions for each ANOVA analysis confirmed that the data showed homogeneity of variances (Levene's test p-value > 0.05) and normality (Shapiro–Wilk test p-value > 0.05). To compare the samples, the results underwent one-way ANOVA (analysis of variance) followed by Tukey's test (5.0%), utilizing SPSS software (version 20.0).

3. Results and Discussion

3.1. Fermentations

3.1.1. Liquid Media Fermentation

In this study, it was found that *Saccharomyces cerevisiae* grown in a culture medium containing grapevine leaves (1×10^6 CFU/g of leaves) and 10% glucose was identical to that observed in a medium in which the glucose concentration was increased to 30%, as can be seen in Figure 1.

It can be seen in Figure 1 that, in both cases, fermentation terminus occurred at 48 h. To increase the duration of the fermentation process, other concentrations of glucose were tested.

This study revealed that no microbial growth occurred during the 48 h of spontaneous fermentation. This suggests that the result obtained was likely due to a lack of nutrients (carbon source). In this circumstance, glucose was added in subsequent experiments. Based on these results, it was decided to use 15% and 20% glucose/g of sample in both yeast-inoculated (*S. cerevisiae*) and spontaneous fermentation.

The weight and °Brix (%) of the samples were also monitored to follow the performance of the fermentations (shown in Figures S1 and S2 of the Supplementary Materials, respectively).

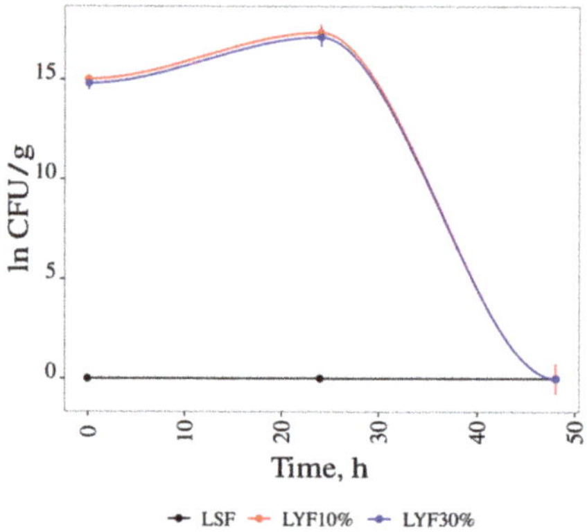

Figure 1. Fermentation with 10% and 30% glucose. LSF = liquid spontaneous fermentation; LYF10% = liquid yeast fermentation with 10% of glucose; LYF30% = liquid yeast fermentation with 30% of glucose.

It was observed that the weight of the flasks decreased over time as a consequence of substrate (glucose) consumption during fermentation. However, spontaneous fermentation with 20% glucose/g sample (LSF20%) exhibited less weight loss, and the °Brix (%) showed no variation, indicating that even with 20% substrate, spontaneous fermentation of the leaves practically did not occur.

From the analysis of the figure, it is noted that the other fermentations (LSF15%, spontaneous fermentation with 15% of glucose, LYF15%, yeast fermentation with 15% of glucose, and LYF20%, yeast fermentation with 20% of glucose) exhibited a similar behavior regarding the reduction in flask weight and °Brix (%) throughout the fermentation process, suggesting substrate consumption by *S. cerevisiae*.

Figure 2 illustrates the variation in the ln of colony-forming units (CFUs) as a function of time in the various experimental conditions over the fermentation time. The results indicate that in fermentations containing 15% glucose (LSF15% and LYF15%), yeast growth was higher than in fermentations containing 20% glucose (LSF20% and LYF20%); therefore, in subsequent studies, only 15% glucose/g sample was used.

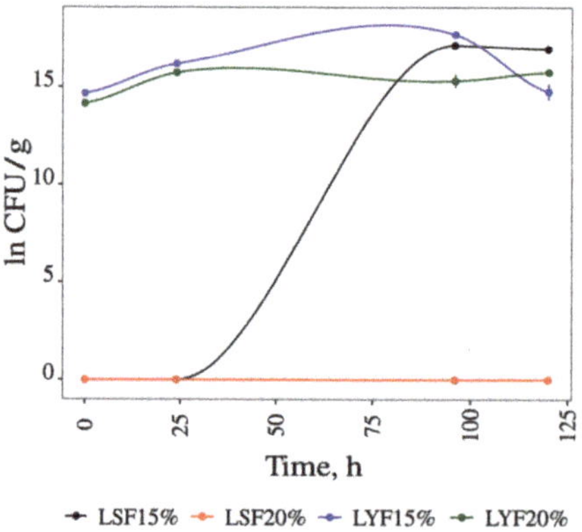

Figure 2. Variation of the natural logarithm of CFUs over time for *Saccharomyces cerevisiae* grown in liquid medium with 15% and 20% glucose. LSF15% = liquid spontaneous fermentation with 15% of glucose; LSF20% = liquid spontaneous fermentation with 20% of glucose; LYF15% = liquid yeast fermentation with 15% of glucose; LYF20% = liquid yeast fermentation with 20% of glucose.

The results obtained in the analysis of °Brix (%) and, consequently, the consumption of sugars using 15% glucose in all the fermentations are shown (see Figure S3, Supplementary Materials). These figures demonstrate that both fermentations started 24 h after inoculation, as there was a reduction in °Brix (%) and an increase in final biomass over time. Figure 3 shows that in spontaneous fermentation (LSF), the final biomass obtained at 216 h of fermentation was 17.69 ln CFU/mL. This growth was due to the microbiota present on the leaf. Meanwhile, in *S. cerevisiae* fermentation, the final biomass was 19.76 ln CFU/mL, and fermentation ended 72 h after inoculation. After, the °Brix (%) remained constant.

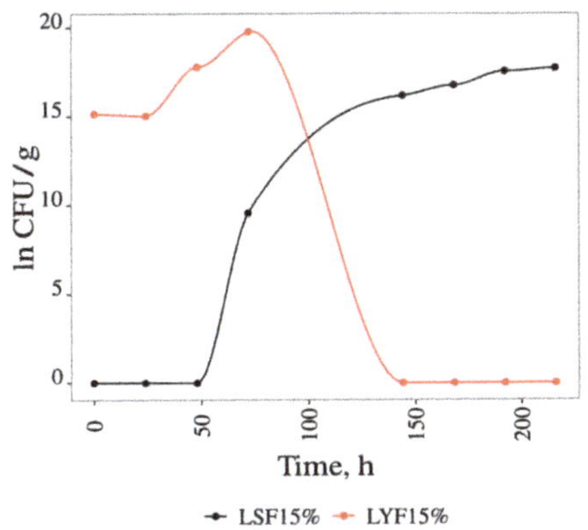

Figure 3. Variation in the ln of the CFU related with time of *S. cerevisiae* grown in liquid media with 15% of glucose. LSF15% = liquid spontaneous fermentation with 15% of glucose; LYF15% = liquid yeast fermentation with 15% of glucose.

From the results obtained for the variation in CFU, the speed (rate) of growth of *S. cerevisiae* was calculated under all the experimental conditions. Table 1 shows the values obtained for the Lag phase (h), the specific growth rate (μc in h^{-1}), the doubling time (DT in h), and the final biomass (FB in CFU/mL).

Table 1. Growth parameters of *S. cerevisiae* across the various study conditions in liquid media.

Liquid Medium Fermentation				
Sample	**Lag Phase (h)**	**μ_c (h^{-1})**	**DT (h)**	**FB (CFU/mL)**
LSF15%	24	0.095	7.27	$4.80 \times 10^7 \pm 0.00$
LYF10%	0	0.097	7.18	$3.40 \times 10^7 \pm 0.04$
LYF15%	0	0.130	5.32	$3.80 \times 10^8 \pm 0.04$
LYF20%	0	0.133	5.21	$3.40 \times 10^7 \pm 0.05$
LYF30%	0	0.096	7.23	$2.70 \times 10^7 \pm 0.32$

LSF15% = Liquid spontaneous fermentation with 15% glucose; LYF10%: = liquid yeast fermentation with 10% glucose; LYF15% = liquid yeast fermentation with 15% glucose; LYF20% = liquid yeast fermentation with 20% glucose; LYF30% = liquid yeast fermentation with 30% glucose; Lag phase = adaptation time (h); μ_c = specific rate of growth (h^{-1}); DT = duplication time (h); FB = final biomass (CFU/mL).

Table 1 shows that the duration of the Lag phase (adaptation phase of the microorganism to the culture medium) in the spontaneous fermentation (LSF15%) conducted in medium with 15% glucose was high (24 h); in all the inoculated fermentations, the duration of this growth parameter was 0 h.

The LSF15% growth rate (0.095 h^{-1}) was identical to that obtained in the medium with 10% glucose (0.097 h^{-1}). In the LYF15% and LYF20% tests, the specific growth rates were similar and higher than the others (0.130 and 0.133 h^{-1}, respectively). The results suggest that these sugar concentrations stimulated growth. However, the use of 30% glucose inhibited the specific growth rate (0.096 h^{-1}). This is in line with reports in the literature that *S. cerevisae* is a Crabtree-negative yeast (repression of respiratory metabolism by glucose) [31]. The doubling times (DT) for LYF15% and LYF20% were shorter (5.32 and 5.21 h), as expected, because DT is inversely proportional to μ_c.

3.1.2. Solid Media Fermentation

To study solid-medium fermentation, ammonium nitrate (NH_4NO_3) was incorporated into the yeast growth medium with the aim of providing better growth conditions for the yeast [21]. Indeed, nitrogen (N) is an essential nutrient for yeast growth and metabolism, and it is fundamental for good fermentative performance [32]. According to Dulf et al. [21], it was possible to increase the antioxidant power and the content of total phenols in fermentations of plum grains in solid medium with the addition of NH_4NO_3.

In Figure 4, the results of microbial growth in ln CFU/g over time (9 days) of solid spontaneous fermentation (SSF) and solid yeast fermentation (SYF) are shown. From the analysis of the results, it can be observed that after 50 h of incubation, the microorganisms entered the stationary phase. The specific growth rate (μ_c in h^{-1}), doubling time (DT in h), and final biomass (FB in CFU/mL) were also determined (shown in Table S1, Supplementary Materials). The results suggest that in the fermentations conducted in solid media, the duration of the latency phase (Lag) was the same in SSF and SYF. The SYF conducted using *S. cerevisiae* obtained a final biomass of 18.32 ln CFU/g, while in the case of solid spontaneous fermentation (SSF), the maximum growth was 7.43 ln CFU/g. The growth rates and, consequently, doubling times were identical in both cases (Figure 4). In fact, as described in the literature, adequate supplementation of nitrogen sources such as ammonia and amino acids can improve the fermentation process, as it is essential for yeast growth and metabolism, influencing the production of bioactive compounds. In addition, nitrogen affects fermentation kinetics and cell productivity, making it essential for the quality and yield of fermentation processes [33,34]. Ruiz et al. [35] reported that nitrogen is an essential nutrient during wine fermentation because its deficiency causes sullage and tarring in fermentations.

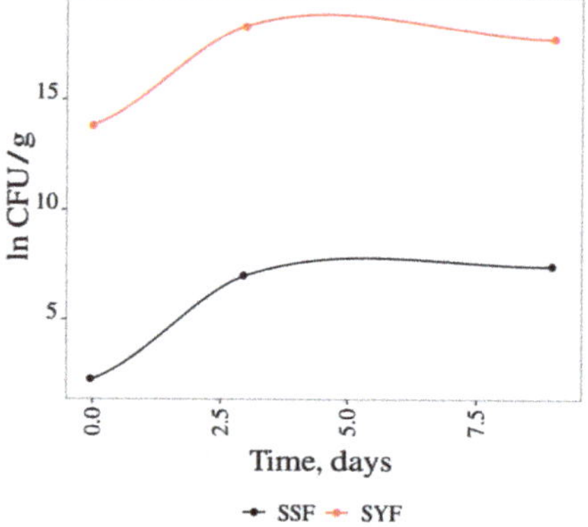

Figure 4. Fermentation results in solid media in ln CFU/g for 9 days. SSF = spontaneous fermentation in solid media; SYF = solid media yeast fermentation.

3.2. Characterization of V. vinifera L. Leaves and Fermentation Products

3.2.1. Physicochemical Properties

The grapevine leaves were dried, resulting in a loss of 69.90% of water and a color change to dark green. Reduced humidity in relation to grapevine leaves leads to increased quality and stability, thus reducing the rate of microbial deterioration [36]. The analysis of ash, resulting from the complete incineration of plant material found in vine leaves, averaged 3.90%, a value higher than that found by Lima [9] in leaves of the Touriga Franca grape varieties in natura (2.13%). The percentage of fat found in the leaves was 3.30%, close to the values found by Lima [9] in leaves of the Malvasia Fina grape varieties in natura (4.59%) and after bleaching (4.54%). The values obtained in the other samples analyzed by this researcher were higher than 5.00%.

Table 2 shows the results obtained from the physicochemical analyses (ash and fat content) of the leaves of *V. vinifera* "in natura" and the products resulting from the various fermentation processes.

Table 2. Results of the physicochemical analysis (ash and fat content) of the leaves of *V. vinifera* "in natura" and after the various fermentation processes.

Assay	Ash (%)	Fat (%)
Leaf	5.86 ± 0.69 [a]	3.30 ± 0.03 [a]
SSF	1.21 ± 0.83 [b]	0.85 ± 0.15 [b]
SYF	2.22 ± 0.33 [b]	0.80 ± 0.26 [b]
LYF	37.19 ±1.01 [c]	0.48 ± 0.28 [b]

Means followed by qual letters are not statistically different among themselves by Tukey test at 5% probability. SSF = solid spontaneous fermentation; SYF = solid yeast fermentation; LYF = liquid yeast fermentation. In each variable, different letters indicate significant mean differences (a–c).

Table 2 shows that the percentage of ash in the solid fermented products, SSF and SYF, was lower than the ash content found in the leaves not subject to fermentation (5.86%). This decrease in the percentage of ash may be related to the use of some substrates during fermentation by the microorganisms that conducted it. However, the content of completely incinerated material of plant origin observed in fermentation in liquid medium (LYF) was higher (37.19%), probably since the speed of growth and the final biomass of the yeasts were higher when growth took place in liquid medium.

The analysis of total lipids showed a high percentage of fats in the vine leaf sample (3.30%) compared to the content found in the fermented leaves (0.48–0.85%) and this difference was significant ($p < 0.05$). However, the fermentations (SSF, SYF, and LYF) showed no significant difference. The decrease in lipid content in the yeast-inoculated trials is probably related to yeast growth; in fact, the lack of long-chain fatty acids and sterols is one of the main causes of difficult fermentations and increased volatile acidity in fermented products.

3.2.2. Chemical Properties

According to the literature, vine leaf extracts contain a promising phenolic content and antioxidant capacity [37]. The extraction yield of *V. vinifera* leaves with 80% ethanol was 24.52%, and the extract was analyzed to determine the total phenols and flavonoid content and the antioxidant activity. The analytical results obtained are presented in Table 3. The results show a total phenol content of 32.32 ± 3.64 mg gallic acid equivalent (GAE)/g of sample. According to the literature, the value of total phenols found in fresh grapevine leaves by Lima [9] was between 139.0 and 170.0 mg GAE/g of extract. The divergence found in this research may have been influenced by the post-harvest processing conditions, namely the drying process to which the leaves were subjected, as well as the time of harvest, climate, and soil, which directly interfere with the composition and quantity of bioactive compounds in the leaves [38].

Table 3. Total phenol and flavonoid contents and antioxidant activity (DPPH and FRAP) of the grapevine leaves.

	Total Phenols mg GAE/g	Total Flavonoids mg QE/g	DPPH mg TroloxE/g	FRAP mg Fe(II)E/g
Leaves of *V. vinifera*	32.32 ± 3.64	17.41 ± 2.68	37.65 ± 2.18	79.34 ± 6.74

Results expressed in mean of analysis ± standard deviation. GAE: gallic acid equivalent; QE: quercetin equivalent; DPPH: 2,2-Diphenyl-1-picrylhydrazyl; FRAP: ferric-reducing antioxidant power.

The flavonoid content observed in this study was 17.41 ± 2.68 mg QE/g, equivalent to that reported by Loizzo et al. [38], which ranged from 2.20–26.2 mg QE/g. This variation can be explained by the fact that environmental factors, genotypes, and post-harvest processing directly influence the content of bioactive compounds present in the vine. This finding corroborates the study by Dhull et al. [16], which demonstrates that solid-state fermentation conducted by *Aspergillus awamori* is a process capable of significantly increasing the total phenolic content and the condensed tannin content.

The result of the analysis of this capacity evaluated by the DPPH method was 37.65 ± 2.18) mg TroloxE/g (51.27%). Our results were identical to those reported by Katalinic et al. [39] and Loizzo et al. [38]. These researchers obtained inhibition values of 34.6–75.4% and 7.19–30.28%, respectively, in different varieties of *V. vinifera*. The result obtained when determining the antioxidant activity using the FRAP method was 79.34 ± 6.74 mg Fe(II)E/g; this value corroborates those found by Loizzo et al. [38], who obtained values of 67.06–100.41 mg Fe(II)E/g.

The results of the total phenolic content of the vine leaves and fermentation products (SSF, SYF, and LYF) are shown in Table 4.

Table 4. Total phenol content of the leaves of *V. vinifera* and of the leaf fermentations.

Assay	Total Phenol Content mg GAE/g
Leaf	32.32 ± 3.64 [a]
SSF	41.62 ± 0.57 [b]
SYF	52.58 ± 0.35 [c]
LYF	39.23 ± 1.85 [b]

Results expressed in mean ± standard deviation. Means followed by the same letters are not statistically different using the Tukey test at 5% probability. GAE: gallic acid equivalent; SSF: spontaneous fermentation in solid media; SYF: yeast fermentation in solid media; LYF: yeast fermentation in liquid media. Different letters indicate significant mean differences (a–c).

Analyzing the evaluation results of total phenols, fermentation caused the phenolic content of the samples to increase. All the fermentations had a higher phenolic content than the leaves as they were. The spontaneous fermentation showed a phenolic content of 41.62 (±0.57) mg GAE/g, while the fermentations inoculated with *S. cerevisiae* showed a content of 52.58 (±0.35) and 39.23 (±1.85) mg GAE/g in the solid and liquid fermentations, respectively. All fermentations showed significant differences compared to the raw leaf sample. However, the total phenolic contents of SSF and SYF were similar ($p > 0.05$). SYF had the highest phenolic content. These results only prove what has already been mentioned by Dhull et al. [16] and Dulf et al. [21], that fungi and yeasts during fermentation produce hydrolytic enzymes such as beta-glucosidase that catalyze and release aglycones, increasing phenolic compounds, anthocyanins, and antioxidant capacity [16,21]. Studies conducted by Dhull et al. [16] have shown that fermentation can induce the hydrolysis of polymers capable of releasing phenolic compounds from the walls of plant matrices, making them soluble, allowing for an increase in their concentration and the antioxidant potential of the extracts. In addition, the species of microorganism also influences the increase in phenolic content produced during fermentation. *S. cerevisiae* strains produce enzymes such as beta-glucosidases, carboxylesterases and feruloyl esterases, which are effective in releasing insoluble bound phenolics.

3.2.3. Microbiological Analysis Results

To determine the microbiological quality of the leaves subjected to different fermentation processes, the following parameters were assessed: mold and yeast count, lactic acid bacteria count, sulfite-reducing clostridial spore count, *E. coli*/coliform count and coagulase-positive staphylococci count. Table 5 summarizes the results of the mold and yeast count of the *V. vinifera* leaves and fermented products at 1, 3, and 9 days.

Table 5. Total enumeration of molds and yeasts in grapevine leaves and throughout the fermentation process.

Days	Yogurt	Mold and Yeast (log CFU/mL)
1	Leaf	1.00 ± 0.07 [a]
	SSF	1.00 ± 0.01 [a]
	SYF	2.16 ± 0.08 [b]
	LYF	3.85 ± 0.17 [c]
3	SSF	2.26 ± 0.07 [b]
	SYF	4.11 ± 0.21 [c]
	LYF	4.87 ± 0.55 [c]
9	SSF	1.18 ± 0.20 [a]
	SYF	6.36 ± 0.38 [d]
	LYF	7.36 ± 0.64 [d]

Means followed by the same letters do not differ statistically from each other according to the Tukey test at a 5% probability level. SSF = solid spontaneous fermentation; SYF = solid yeast fermentation; LYF = liquid yeast fermentation. Different letters indicate significant mean differences (a–d).

Table 5 shows that the mold and yeast counts increased during all the leaf fermentation processes. As expected, the fermentations inoculated with *S. cerevisiae* (SYF and LYF) showed higher mold and yeast counts throughout the fermentation process (1, 3, and 9 days) compared to spontaneous fermentation (SSF). LYF showed marked growth throughout the fermentation process, with colony-forming unit (CFU/g) content varying from 3.85 ± 0.17 log CFU/g to 7.36 ± 0.64 log CFU/g.

Sulfite-reducing clostridial spores were absent in all the samples analyzed. The *E. coli*/Total coliform and coagulase-positive Staphylococci count was 1 log CFU/mL in all samples.

3.3. *Yogurt Analysis*

3.3.1. Physicochemical Analysis Results of the Yogurt Samples

The physicochemical characteristics (ash, fat, protein, pH, and titratable acidity) of the four yogurt samples were determined: the control with the addition of potassium sorbate (Control yogurt, Cy) and the other three with the incorporation of vine leaf fermentation extract (SSFy, SYFy, and LYFy), at storage times of 1, 7, and 12 days at 4 °C. Table 5 shows the results of the physicochemical composition of the four samples. For each variable, the comparison of means showed that there were no significant differences.

The results in Table 6 show that the ash, fat, protein, pH, and titratable acidity content of the four yogurt samples (Cy, SSFy, SYFy, and LYFy) did not differ significantly ($p > 0.05$) from each other. Overall, the carbohydrate content was about 18.6% (calculated value according to AOAC Official Method 986.25.) for each yogurt sample. The ash content was similar to the values observed by Santos et al. [29], who observed 0.71% in terms of Cy, and Afiyah et al. [40], who noted 0.82%. In our study, ash values ranging from 0.77 - 0.69% were observed in terms of Cy, while values ranging from 0.58–0.94% were observed in yogurts with incorporated vine leaf extracts (SSFy, SYFy, and LYFy). Santos et al. [29] observed values of 0.70% for this parameter in yogurt containing red propolis, while Afiyah et al. [40] obtained results ranging from 0.75–0.79% in yogurt containing mango juice (*Mangifera indica* L.). The lipid content ranged from 2.00–2.27%, which is lower than the values found by Santos et al. [29], Afiyah et al. [40] and Haq et al. [41] in yogurt supplemented with lentil flour. The lower fat content observed in this work may be a consequence of the ingredients used to prepare it, such as low-fat milk powder and low-fat pasteurized milk. According to European legislation, our yogurt would be classified as semi-skimmed [42]. The protein content of the yogurts studied ranged from 2.15–2.63%, which was slightly lower than those reported by Santos et al. [29], Haq et al. [41], and Fagnani and Boniatti [42] enriched with grape seed flour. None of the yogurt samples evaluated showed significant differences ($p > 0.05$) in terms of pH values. The pH values found ranged from 4.16–4.34. It should be noted that the control sample had the highest pH values at all the times analyzed. All the samples showed a decrease in pH values over time (12 days), which was to be expected since homofermentative microorganisms, such as *Lactobacillus acidophilus*, continue to ferment lactose and reduce the pH by forming lactic acid (LA) [43]. The titratable acidity of the yogurts ranged from 0.90–1.08 g of LA/100 g. All the samples showed a slight variation in titratable acidity over the storage period. However, it is believed that over the course of the yogurt's shelf life, all the samples would show an increase in acidity, which would be advantageous for the yogurt since acidity helps to control the undesirable growth of pathogenic and spoiling bacteria [44]. It was found that replacing potassium sorbate (Cy) with vine leaf extract (SSFy, SYFy, and LYFy) did not significantly alter the physicochemical composition of yogurt samples.

Table 6. Physicochemical properties of the yogurts at different timepoints (1, 7, and 12 days).

Day	Yogurt	Ash (%)	Fat (%)	Protein (%)	pH	TA g of LA/100 g
1	Cy	0.75 ± 0.04	2.14 ± 0.04	2.63 ± 0.05	4.34 ± 0.04	1.08 ± 0.02
	SSFy	0.69 ± 0.15	2.07 ± 0.17	2.51 ± 0.08	4.18 ± 0.10	0.98 ± 0.05
	SYFy	0.94 ± 0.07	2.00 ± 0.12	2.15 ± 0.04	4.24 ± 0.02	0.93 ± 0.08
	LYFy	0.69 ± 0.01	2.12± 0.07	2.48 ± 0.04	4.26 ± 0.01	1.05 ± 0.07
7	Cy	0.77 ± 0.02	2.09 ± 0.05	2.48 ± 0.11	4.34 ± 0.05	0.95 ± 0.07
	SSFy	0.71 ± 0.05	2.12 ± 0.13	2.48 ± 0.13	4.20 ± 0.03	0.95 ± 0.07
	SYFy	0.58 ± 0.11	2.20 ± 0.06	2.51± 0.07	4.25 ± 0.03	0.98 ± 0.05
	LYFy	0.71 ± 0.02	2.24 ± 0.01	2.43 ± 0.10	4.19 ± 0.07	0.93 ± 0.06
12	Cy	0.69 ± 0.10	2.15 ± 0.03	2.53 ± 0.02	4.28 ± 0.07	0.94 ± 0.00
	SSFy	0.72 ± 0.01	2.27 ± 0.02	2.84 ± 0.06	4.16 ± 0.05	1.00 ± 0.04
	SYFy	0.72 ± 0.12	2.16 ± 0.10	2.93 ± 0.10	4.20 ± 0.00	0.90 ± 0.04
	LYFy	0.75 ± 0.03	2.25 ± 0.14	2.80 ± 0.07	4.19 ± 0.00	0.97 ± 0.04

TA = titratable acidity; LA = lactic acid; Cy = control yogurt; SSFy = yogurt with an extract from solid spontaneous fermentation; SYFy = yogurt with an extract from solid yeast fermentation; LYFy = yogurt with an extract from liquid yeast fermentation.

3.3.2. Chemical Analysis Results of the Yogurt Samples

The results of the analyses of total phenols and antioxidant activity (DPPH and FRAP) of the yogurts at 1 and 12 days of storage are summarized in Table 7.

Table 7. Total phenolic content and antioxidant activity (DPPH and FRAP) of yogurts at 1 and 12 days of storage.

Days	Yogurt	Total Phenols mg GAE/g	DPPH mg TroloxE/g	FRAP mg Fe(II)E/g
1	Cy	1.23 ± 0.05 ^{a}	1.19 ± 0.34 ^{a}	4.22 ± 0.16 ^{a}
	SSFy	1.60 ± 0.13 ^{b}	1.23 ± 0.06 ^{a}	5.06 ± 0.19 ^{b}
	SYFy	1.87 ± 0.01 ^{b}	1.51 ± 0.07 ^{a}	5.28 ± 0.25 ^{b}
	LYFy	1.34 ± 0.13 ^{ab}	1.19 ± 0.13 ^{a}	3.80 ± 0.27 ^{a}
12	Cy	1.44 ± 0.07 ^{ab}	1.01 ± 0.06 ^{a}	4.00 ± 0.36 ^{a}
	SSFy	1.94 ± 0.14 ^{b}	1.33 ± 0.06 ^{a}	5.30 ± 0.12 ^{b}
	SYFy	2.16 ± 0.28 ^{c}	1.34 ± 0.08 ^{a}	5.77 ± 0.21 ^{c}
	LYFy	1.43 ± 0.11 ^{ab}	1.38 ± 0.06 ^{a}	4.32 ± 0.33 ^{a}

Results expressed in mean of the analyses ± standard deviation. Means followed by same letters are not statistically different using the Tukey test at 5% probability. GAE = gallic acid equivalent; DPPH: 2,2-Diphenyl-1-picrylhydrazyl; FRAP = ferric-reducing antioxidant power; Cy = control yogurt; SSFy = yogurt with extract from solid spontaneous fermentation; SYFy = yogurt with extract from solid yeast fermentation; LYFy = yogurt with extract from liquid yeast fermentation. In each variable, different letters indicate significant mean differences (a–c).

The total phenol content and antioxidant activity observed in the yogurts were higher in the samples with added *V. vinifera* leaf extract (SSFy, SYFy, and LYFy) compared to the control (Cy). It was found that in all the samples, the total phenol content was higher at 12 days of storage. The significant increase in total phenolics corroborates the achievement of the objectives outlined in this study. The observed improvement can be directly attributed to the incorporation of fermented extracts into yogurt formulations, highlighting their direct influence on the antioxidant properties of dairy products.

The yogurt sample with the highest phenol content was SYFy (2.16 ± 0.28 mg GAE/g of yogurt). In addition, it was the only sample that showed a significant difference ($p < 0.05$) between 1 and 12 days. The results of the analysis of antioxidant activity using the DPPH method showed that none of the samples presented any statistical difference. The LYFy sample was the only one with an increase in antioxidant content using the DPPH method, from 1.19 ± 0.34–1.38 ± 0.06 mg TroloxE/g of yogurt. The FRAP results showed that only the control (Cy) had a reduced antioxidant content (from 4.22 ± 0.16–4.00 ± 0.36 mg Fe(II)E/g of yogurt), while the other yogurt samples (SSFy, SYFy, and LYFy) had increased antioxidant content, particularly the SYFy sample, which was the only one to show a significant difference (from 5.28 ± 0.25–5.77 ± 0.21 mg Fe(II)E/g of yogurt). In yogurt containing added vine leaf extracts, especially the fermented ones, the phenolic content and the antioxidant activity were higher than in the control. This suggests that these compounds probably originated from leaf extracts, which may increase the bioactive properties of the final product. This shows that the addition of the leaf extract makes it possible to obtain foods with higher nutritional quality and health benefits due to their antioxidant and antimicrobial properties [11].

3.3.3. Microbiological Analysis Results of the Yogurt Samples

Over its shelf life, yogurt is susceptible to microbiological changes. The conditions in which the product is stored interfere with its microbial stability, particularly temperature, pH, oxygen, and the energy source (carbohydrates, proteins, and fats). In addition, the growth of deteriorating microorganisms, such as coliforms, molds and yeasts, can lead to changes in the appearance, aroma, and taste of the yogurt, as well as indicating that sanitary practices were poor during production [35]. Antimicrobial additives used in food products are necessary to inhibit these types of microorganisms [45]. The effect of the extracts added to yogurt corroborates the studies carried out by De Andrade et al. [46] and Katalinic et al. [39], which indicate that the bioactive compounds present in vine leaves have antioxidant and antimicrobial potential since these compounds induce metabolic effects that are beneficial to health and offer possibilities for their application in the prevention of oxidative and/or microbial deterioration of food products. The results obtained for the lactic acid bacteria in yogurts with strawberry pulp on days 1, 7, and 12 of storage are shown in Table 8.

Table 8. Lactic bacteria count of samples at different storage times (1, 7, and 12 days).

Days	Yogurt	Lactic Bacteria log CFU/mL
1	Cy	7.97 ± 0.01 [b]
	SSFy	7.25 ± 0.01 [c]
	SYFy	7.44 ± 0.02 [c]
	LYFy	7.28 ± 0.01 [c]
7	Cy	8.04 ± 0.07 [b]
	SSFy	8.43 ± 0.19 [a]
	SYFy	8.44 ± 0.15 [a]
	LYFy	8.44 ± 0.18 [a]
12	Cy	8.08 ± 0.05 [b]
	SSFy	8.21 ± 0.13 [ab]
	SYFy	8.44 ± 0.05 [a]
	LYFy	7.56 ± 0.01 [c]

Cy = control yogurt; SSFy = yogurt with extract from solid spontaneous fermentation; SYFy = yogurt with extract from solid yeast fermentation; LYFy = yogurt with extract from liquid yeast fermentation. Different letters indicate significant mean differences (a–c).

The results of the microbiological analysis showed an absence of *E. coli*/total coliforms, in accordance with the European parameters for the microbiological quality of yogurt. According to the microbiological criteria of the European Union, all samples showed count values of <10 CFU/mL for molds and yeasts, sulphite-reducing clostridia and coagulase-positive *Staphylococcus* [47]. The absence of these microorganisms is an indication of good practices during the production of this product. The lactic acid bacteria count was stable in all the yogurt samples. Figure 5 illustrates the behavior of lactic acid bacteria during the 12 days of evaluation.

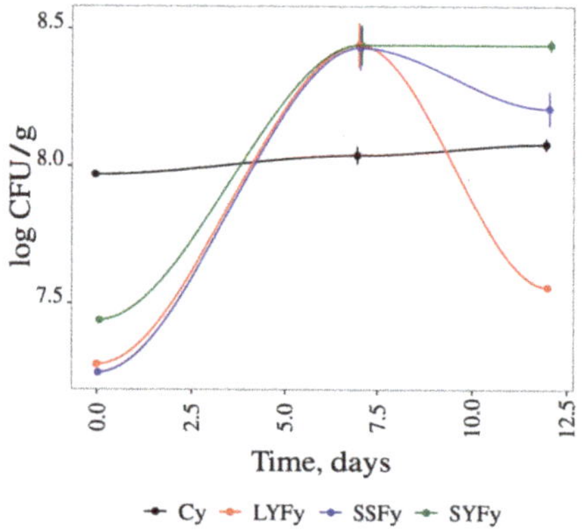

Figure 5. Lactic bacteria count of yogurt samples during 12 days of storage.

The lactic acid bacteria count in the control varied from 7.97 ± 0.01–8.08 ± 0.05 log CFU/g over the 12 days of storage (not significant mean differences). The SSFy, SYFy, and LYFy samples had a maximum count of 8.43 ± 0.19, 8.44 ± 0.15 and 8.44 ± 0.18, respectively, at 7 days. At 12 days of storage, the amount of these bacteria decreased slightly in the SSFy sample (8.21 ± 0.13 log CFU/g) and significantly decreased in the LYFy sample (7.56 ± 0.01 log CFU/g) compared to the results at 7 days. Meanwhile, SYFy maintained a count of 8.44 log CFU/g from 7 to 12 days of storage. In this context, it is considered that introducing or replacing an additive in yogurt helps preserve and stabilize the product, as well as enabling the growth of lactic acid bacteria.

4. Conclusions

In this study, we characterized the fermentative process of *V. vinifera* L. leaves (grapevine). We then evaluated the effects on bioactivity and shelf life by mixing the fermented leaves with yogurt, exploring their potential role as preserving agents.

The replacement of potassium sorbate preservative in yogurts by fermented grapevine leaf extract increased the phenolic content, enhancing its antioxidant activity. Microbiological evaluation indicated that grapevine leaf fermentation extract possesses antimicrobial potential without interfering with the growth of lactic acid bacteria. Among the fermentations, the one conducted in a solid medium demonstrated the best performance and enhanced biological properties. Additionally, yogurts incorporating the extract showed no significant differences in physicochemical properties, pH, or acidity compared to the control. Therefore, we provide evidence that may allow taking advantage of an agricultural by-product to increase the quality and nutritional value of a widely consumed food product.

As the demand for natural additives continues to increase, the products developed could be considered an alternative source to introduce into the food additives market. However, future research should assess the sensory aspect and toxicity of this natural additive, which is currently a highly abundant and wasted residue. Additionally, it will be relevant to evaluate the antimicrobial activity of fermented grapevine leaves against specific spoilage microorganisms (challenge test).

Supplementary Materials: The following supporting information can be downloaded at: https://www.mdpi.com/article/10.3390/foods13132053/s1. Figure S1. Evolution of the weights of the flasks in the fermentations in liquid media with 15% and 20% of glucose. LSF15% = Liquid spontaneous fermentation with 15% of glucose; LSF20%= Liquid spontaneous fermentation with 20% of glucose; LYF15% = Liquid yeast fermentation with 15% of glucose; LYF20% = Liquid yeast fermentation with 20% of glucose; Figure S2. Soluble solids of the fermentations in liquid media with 15% and 20% of glucose. LSF15% = Liquid spontaneous fermentation with 15% of glucose; LSF20% = Liquid spontaneous fermentation with 20% of glucose; LYF15% = Liquid yeast fermentation with 15% of glucose; LYF20%: Yeast fermentation with 20% of glucose; Figure S3. Soluble solids of the fermentations in liquid media with 15% of glucose. LSF15% = Liquid spontaneous fermentation with 15% of glucose; LYF15% = Liquid yeast fermentation with 15% of glucose; Table S1. Growth parameters of S. cerevisiae in solid media.

Author Contributions: Conceptualization, L.M.E.; methodology L.F., V.B.P., L.G.D. and L.M.E.; formal analysis, L.F. and L.M.E.; investigation, L.F. and V.B.P.; data curation, L.F., V.B.P., L.G.D. and L.M.E.; writing—original draft preparation, L.F. and M.S.-D.; writing—review and editing, L.F., M.S.-D., L.G.D. and L.M.E.; supervision, L.M.E. All authors have read and agreed to the published version of the manuscript.

Funding: This work was supported by national funds through FCT/MCTES (PIDDAC): CIMO, UIDB/00690/2020 (DOI: 10.54499/UIDB/00690/2020) and UIDP/00690/2020 (DOI: 10.54499/UIDP/00690/2020); and SusTEC, LA/P/0007/2020 (DOI: 10.54499/LA/P/0007/2020). Grant of L.F. are due to CERTRA—Development of Traditional Cereal Value Chains for Sustainable Food in Portugal, Investment RE-C05-i03—Research and Innovation Agenda for Agriculture, Food, and Agro-Industry Sustainability, No. 12/C05-i03/2021—PRR-C05-i03-I-000161, R&D+I Projects Research and Innovation Projects—Sustainable Food (action line 1.2 Production), source of funding: PRR – Recovery and Resilience Plan Program (funding by the European Union).

Informed Consent Statement: Not applicable.

Data Availability Statement: The original contributions presented in the study are included in the article/Supplementary Material, further inquiries can be directed to the corresponding author.

Acknowledgments: The authors are grateful to CERTRA—Development of Traditional Cereal Value Chains for Sustainable Food in Portugal, Investment RE-C05-i03—Research and Innovation Agenda for Agriculture, Food, and Agro-Industry Sustainability, No. 12/C05-i03/2021—PRR-C05-i03-I-000161, R&D+I Projects Research and Innovation Projects—Sustainable Food (action line 1.2 Production).

Conflicts of Interest: The authors declare no conflicts of interest.

References

1. Mathias, T.R.S.; Carvalho Junior, I.C.; Carvalho, C.W.P.; Sérvulo, E.F.C. Rheological characterization of coffee-flavored yogurt with different types of thickener. *Alim. Nutr.* **2011**, *22*, 521–529.
2. Lopes, R.P.; Mota, M.J.; Pinto, C.A.; Souza, S.; Silva, J.A.L.; Gomes, A.M.; Delgadillo, I.; Saraiva, J.A. Physicochemical and microbial changes in yogurts produced under different pressure and temperature conditions. *LWT* **2019**, *99*, 423–430. [CrossRef]
3. Santos, J.V.R.; Miranda, E.S.M.; Oliveira, A.T.C.; Damaceno, M.N.; Silva, M.S.; Cavalcante, A.B.D. Cinética da fermentação de leite adicionado de farinha de banana verde na produção de iogurte. *Res. Soc. Dev.* **2020**, *9*, e295985316. [CrossRef]
4. Kefi, B.B.; Baccouri, S.; Torkhani, R.; Koumba, S.; Martin, P.; M'hamdi, N. Application of response surface methodology to optimize solid-phase extraction of benzoic acid and sorbic acid from food drinks. *Foods* **2022**, *11*, 1257. [CrossRef]
5. Zhang, W.; Yang, F.; Xu, J.; Wang, L.; Zhou, K. Determination of nine preservatives in food samples by solid phase extraction coupled with capillary electrophoresis. *Int. J. Electrochem. Sci.* **2021**, *16*, 21022. [CrossRef]
6. Dey, S.; Nagababu, B.H. Applications of food color and bio-preservatives in the food and its effect on the human health. *Food Chem.* **2022**, *1*, 100019. [CrossRef]
7. European Commission. Commission Regulation (EU) No 1129/2011 of 11 November 2011 Amending Annexes II and III to Regulation (EC) No 1333/2008 of the European Parliament and of the Council by Establishing a Union List of Food Additives. 2011. Available online: https://eur-lex.europa.eu/legal-content/PT/TXT/PDF/?uri=CELEX:32011R1129&from=es (accessed on 18 May 2024).
8. Sat, I.G.; Sengul, M.; Keles, F. Use of grape leaves in canned food. *Pak. J. Nutr.* **2002**, *1*, 257–262.
9. Lima, A.F. Caracterização da Bioatividade de Folhas de Diferentes Castas de Videira Quando Sujeitas a Processamento Alimentar. Master's Thesis, Instituto Politécnico de Bragança, Bragança, Portugal, 2015.

10. Nzekoue, F.K.; Nguefang, M.L.K.; Alessandroni, L.; Mustafa, A.M.; Vittori, S.; Caprioli, G. Grapevine leaves (*Vitis vinifera*): Chemical characterization of bioactive compounds and antioxidant activity during leave development. *Food Biosci.* **2022**, *50*, 102120. [CrossRef]
11. Sahu, A.; Singh, D.; Shukla, R. Bioactive compounds and reported pharmacological activities of *Vitis vinifera* L.—An overview. *World J. Pharm. Res.* **2023**, *12*, 27–38.
12. Caleja, C.S.G. Incorporação de Ingredientes Naturais em Diferentes Matrizes Alimentares como Potenciadores de Conservação e Promotores de Saúde. Ph.D. Thesis, Porto University, Porto, Portugal, 2018.
13. Liu, N.; Song, M.; Wang, N.; Wang, Y.; Wang, R.; An, X.; Qi, J. The effects of solid-state fermentation on the content, composition and in vitro antioxidant activity of flavonoids from dandelion. *PLoS ONE* **2020**, *15*, e0239076. [CrossRef]
14. Lee, B.H. *Fundamentos de Biotecnología de los Alimentos*; Editorial Acribia: Zaragoza, Spain, 1996; p. 475. ISBN 84-200-0922-9.
15. Rhodes, A.; Fletcher, D.L. *Principles of Industrial Microbiology*; Pergamon Press: Oxford, UK, 1975; p. 320. ISBN 0080119050.
16. Dhull, S.B.; Punia, S.; Kidwai, M.K.; Kaur, M.; Chawla, P.; Purewal, S.S.; Sangwan, M.; Palthania, S. Solid-state fermentation of lentil (*Lens culinaris* L.) with *Aspergillus awamori*: Effect on phenolic compounds, mineral content, and their bioavailability. *Legume Sci.* **2020**, *2*, e37. [CrossRef]
17. Xu, L.; Guo, S.; Zhang, S. Effects of solid-state fermentation with three higher fungi on the total phenol contents and antioxidante properties of diverse cereal grains. *FEMS Microbiol. Lett.* **2018**, *365*, fny163. [CrossRef] [PubMed]
18. García, L.R.P.; Bianchi, V.L.D. Efeito da fermentação fúngica no teor de compostos fenólicos em casca de café robusta. *Semin. Ciências Agrárias* **2015**, *36*, 777–786. [CrossRef]
19. Kosar, M.; Küpeli, E.; Malyer, H.; Uylaser, V.; Türkben, C.; Baser, K.H.C. Effect of brining on biological activity of leaves of *Vitis vinifera L.* (Cv. Sultani Cekirdeksiz) from Turkey. *J. Agric. Food Chem.* **2007**, *55*, 4596–4603. [CrossRef] [PubMed]
20. Ferreira, L.M.M.; Ferreira, A.M.; Benevides, C.M.J.; Melo, D.; Costa, A.S.G.; Faia, A.M.; Oliveira, M.B.P.P. Effect of controlled microbial fermentation on nutritional and functional characteristics of cowpea bean flours. *Foods* **2019**, *8*, 530. [CrossRef] [PubMed]
21. Dulf, F.V.; Vodnar, D.C.; Socaciu, C. Effects of solid-state fermentation with two filamentous fungi on the total phenolic contents, flavonoids, antioxidant activities and lipid fractions of plum fruit (*Prunus domestica* L.) by-products. *Food Chem.* **2016**, *209*, 27–36. [CrossRef]
22. Paula, V.M.B. Caraterização Química e Biológica do Própolis da "Serra de Bornes" por TLC. Master's Thesis, Instituto Politécnico de Bragança, Bragança, Portugal, 2012.
23. Celik, O.F.; Temiz, H. Lactobacilli isolates as potential aroma producer starter cultures: Effects on the chemical, physical, microbial, and sensory properties of yogurt. *Food Biosci.* **2022**, *48*, 101802. [CrossRef]
24. Muncan, J.; Tei, K.; Tsenkova, R. Real-time monitoring of yogurt fermentation process by aquaphotomics near-infrared spectroscopy. *Sensors* **2020**, *21*, 177. [CrossRef]
25. Water, J.V.; Naiyanetr, P. Yoghurt and immunity: The health benefits of fermented milk products that contain lactic acid bacteria. In *Handbook of Fermented Functional Foods*; CRC Press: Boca Raton, FL, USA, 2003; pp. 113–144.
26. AOAC. *Association of Official Analytical Chemistry: Official Methods of Analysis*, 19th ed.; AOAC: Gaithersburg, MD, USA, 2012; pp. 1–3000.
27. Singleton, V.L.; Orthofer, R.; Lamuela-Raventos, R.M. Analysis of total phenols and other oxidation substrates and antioxidants by means of Folin–Ciocalteu reagent. *Methods Enzymol.* **1999**, *299*, 152–178.
28. Savi, P.R.S.; Santos, L.; Gonçalves, A.M.; Biesek, S.; Lima, C.P. Análise de flavonoides totais presentes em algumas frutas e hortaliças convencionais e orgânicas mais consumidas na região sul do Brasil. *Demetra Aliment. Nutr. Saúde* **2017**, *12*, 275–287. [CrossRef]
29. Santos, M.S.; Estevinho, L.M.; Carvalho, C.A.L.; Morais, J.S.; Conceição, A.L.S.; Paula, V.B.; Guedes, K.M.; Almeida, R.C.C. Probiotic yogurt with brazilian red propolis: Physicochemical and bioactive properties, stability, and shelf life. *J. Food Sci.* **2019**, *84*, 3429–3436. [CrossRef] [PubMed]
30. Liofilchem. Rose Bengal CAF Agar. Selective Medium for Detection of Yeasts and Moulds from Food and Environmental Materials. 2015. Available online: http://www.liofilchem.net/login/pd/ifu/10034_IFU.pdf (accessed on 10 October 2023).
31. Dai, Z.; Huang, M.; Chen, Y.; Siewers, V.; Nielsen, J. Global rewiring of cellular metabolism renders *Saccharomyces cerevisiae* Crabtree negative. *Nat. Commun.* **2018**, *9*, 3059. [CrossRef] [PubMed]
32. Mendes-Ferreira, A.; Sampaio-Marques, B.; Barbosa, C.; Rodrigues, F.; Costa, V.; Mendes-Faia, A.; Ludovico, P.; Leao, C. Accumulation of non-superoxide anion reactive oxygen species mediates nitrogen-limited alcoholic fermentation by *Saccharomyces cerevisiae*. *Appl. Environ. Microbiol* **2010**, *76*, 7918–7924. [CrossRef] [PubMed]
33. Vidal, E.E. Influência da Fonte de Nitrogênio no Perfil Fermentativo, Transcriptômico, e na Produção de Álcoois Superiores em *Saccharomyces cerevisiae*. Master's Thesis, Universidade Federal de Pernambuco, Recife, Pernambuco, Brazil, 2012.
34. Pereira, A.F. Suplementação de Nitrogênio Sobre a Fermentação Alcoólica para Produção de Cachaça, Cerveja e Vinho. Master's Thesis, Universidade Federal de Viçosa, Viçosa, Minas Gerais, Brazil, 2007.
35. Ruiz, J.; Celis, M.; Toro, M.; Mendes-Ferreira, A.; Rauhut, D.; Santos, A.; Belda, I. Phenotypic and transcriptional analysis of *Saccharomyces cerevisiae* during wine fermentation in response to nitrogen nutrition and co-inoculation with *Torulaspora delbrueckii*. *Int. Food Res.* **2020**, *137*, 109663. [CrossRef]

36. Câmara, G.B.; Oliveira, T.K.B.; Macedo, C.D.S.; Leite, D.D.D.F.; Soares, T.D.C.; Lima, A.R.N.; Vasconcelos, S.H.; Soares, T.C.; Barbosa, M.L.; Trigueiro, L.S.D.L. Physico-chemical, toxicological and nutritional characterization of dry and in *natura Moringa oleifera Lam* leaves. *Res. Soc. Dev.* **2019**, *8*, e178111450. [CrossRef]
37. Fernandes, F.; Ramalhosa, E.; Pires, P.; Verdial, J.; Valentão, P.; Andrade, P.; Bento, A.; Pereira, J.A. *Vitis vinífera* leaves towards bioactivity. *Ind. Crops Prod.* **2013**, *43*, 434–440. [CrossRef]
38. Loizzo, M.R.; Sicari, V.; Pellicanò, T.; Xiao, J.; Poiana, M.; Tundis, R. Comparative analysis of chemical composition, antioxidant and antiproliferative activities of Italian *Vitis vinifera* by-products for a sustainable agro-industry. *Food Chem. Toxicol.* **2019**, *127*, 127–134. [CrossRef] [PubMed]
39. Katalinic, V.; Mozina, S.S.; Generalic, I.; Skroza, D.; Ljubenkov, I.; Klancnik, A. Phenolic profile, antioxidant capacity, and antimicrobial activity of leaf extracts from six *Vitis vinífera L. varieties. Int. J. Food Prop.* **2013**, *16*, 45–60. [CrossRef]
40. Afiyah, D.N.; Sarbini, R.N.; Huda, M.S. Analysis of the yogurt nutrient content and antioxidant activity by adding Podang Urang Mango juice (*Mangifera indica* L.). *J. Ternak* **2022**, *13*, 47–52. [CrossRef]
41. Haq, F.U.; Sameen, A.; Zaman, Q.U.; Mushtaq, B.S.; Hussain, M.B.; Javed, A.; Plygun, S.; Korneeva, O.; Shariati, M.A. Development and evalutation of yogurt supplemented with lentil flour. *JMBFS.* **2019**, *8*, 1005–1009. [CrossRef]
42. Fagnani, R.; Boniatti, P.M.S. Formulação de iogurte concentrado enriquecido com farinha de semente de uva: Atividade antioxidante e cinética de fermentação. *Ensaios* **2020**, *24*, 189–193. [CrossRef]
43. Lopes, R.P. Effects of high hydrostatic pressure on yogurt production. Master's Thesis, Departamento de Química, Universidade de Aveiro, Aveiro, Portugal, 2013.
44. Duarte, M.C.K.H.; Cortez, N.M.S.; Cortez, M.A.S.; Franco, R.M.; Macedo, N. Ação antagonista de *Lactobacillus acidophilus* frente a estirpes patogênicas inoculadas em leite fermentado. *J. Bioenergy Food Sci.* **2016**, *3*, 1–10. [CrossRef]
45. Anari, H.N.B.; Majdinasab, M.; Shaghaghian, S.; Khalesi, M. Development of a natamycin-based non-migratory antimicrobial active packaging for extending shelf-life of yogurt drink (Doogh). *Food Chem.* **2022**, *366*, 130606. [CrossRef]
46. De Andrade, R.B.; Machado, B.A.S.; Barreto, G.d.A.; Nascimento, R.Q.; Corrêa, L.C.; Leal, I.L.; Tavares, P.P.L.G.; Ferreira, E.d.S.; Umsza-Guez, M.A. Syrah Grape Skin Residues Has Potential as Source of Antioxidant and Anti-Microbial Bioactive Compounds. *Biology* **2021**, *10*, 1262. [CrossRef]
47. European Commission. Commission Regulation (EC) No. 2073/2005 of 15 November 2005 on microbiological criteria for foodstuffs. *Off. J. Eur. Union* **2005**, *50*, 1–26.

foods

MDPI

Article

Evaluation of Antimicrobial and Preservative Effects of Cinnamaldehyde and Clove Oil in Catfish (*Ictalurus punctatus*) Fillets Stored at 4 °C

Rosemary I. Ebirim and Wilbert Long III *

Department of Human Ecology, Delaware State University, 1200 North Dupont Highway, Dover, DE 19901, USA
* Correspondence: wlong@desu.edu; Tel.: +1-(302)-857-6883

Abstract: This study aimed to evaluate cinnamaldehyde (CN) and clove oil (CO) effectiveness in inhibiting growth and killing spoilage and total aerobic bacteria when overlaid with catfish fillet stored at 4 °C. A 1.00 mL concentration of CO inhibited growth by 2.90, 1.96, and 1.96 cm, respectively, for *S. baltica*, *A. hydrophilia*, and total bacteria. Similarly, treatment with 1.00 mL of CN resulted in ZIB of 2.17, 2.10, and 1.10 cm, respectively, for *S. baltica*, *A. hydrophilia*, and total bacteria from catfish exudates. Total bacteria from catfish exudates treated with 0.50 mL CN for 40 min, resulted in a 6.84 log decrease, and treatment with 1.00 mL resulted in a 5.66 log decrease at 40 min. Total bacteria exudates treated with 0.50 mL CO resulted in a 9.69 log reduction at 40 min. Total bacteria treated with 1.00 mL CO resulted in a 7.69 log decrease at 7 days, while untreated pads overlaid with catfish resulted in ≥9.00 CFU/mL. However, treated absorbent pads with catfish at 7 days, using 0.50 mL and 1.00 mL CN, had a bacterial recovery of 5.53 and 1.88 log CFU/mL, respectively. Furthermore, CO at 0.50 mL and 1.00 mL reduced the bacteria count to 5.21 and 1.53 log CFU/mL, respectively, at day 7.

Keywords: storage; spoilage; essential oil; cinnamaldehyde; clove oil; catfish; absorbent food pads

1. Introduction

High protein content, water activity, and pH, amongst other things, limit storage, safety, and consumption, thus the need for different methods of preservation [1,2]. Fish and fish products are rich in protein, vitamins, minerals, and, often, essential omega 3 fatty acids, which have been linked to health benefits including improved longevity, promotion of fetal development, and improvement of cardiovascular function and health [3–5]. Fish have been associated with numerous foodborne outbreaks and foodborne illnesses [6,7]. The rate at which the human population is growing has led to an increased need for sources of protein such as fish. Fish harvesting and consumption have increased at different levels and have resulted in a global industry with increased international exports, providing employment, especially in rural riverine areas where the main occupation is fishing [8–11]. Specific spoilage organisms (SSOs) such as *Shewanella baltica* (NCTC strain 10735) and *Aeromonas hydrophila* (ATCC strain 7966) have been identified as producing metabolites that affect the sensory properties of seafood and impact the economic value chain, consumption, and utilization of fish [12,13]. Additionally, fish spoilage and contamination may make the fish unsafe and unfit for human consumption. The SSOs that affect fish are commonly associated with specific fish species, processing, storage conditions, and microbial interactions [14–16]. One of the most conventional preservation techniques used in the seafood industry is cold treatment. Cold processing and preservation treatments including refrigeration, chilling, super-chilling, and freezing, with or without non-synthetic or synthetic chemical preservatives, have shown effectiveness in limiting or reducing pathogens in seafood [15,17].

The use of several synthetic chemical preservatives such as nitrites, nitrates, benzoates, sulfur dioxide, and many more to destroy or delay the growth of bacteria, yeast, and molds has been applied in food preservation. These chemical preservatives have shown the ability to prevent changes in texture and color, the development of unpleasant flavors and off-odors, and the loss of nutrients in seafood during storage at low temperatures [18–20]. However, their use has been linked to potential hazards that can cause serious health issues, including allergic reactions, asthma, neurological damage, and cancer [19,21]. Biological methods of food preservation include the use of plant materials. Some of these plants may contain antimicrobials, antioxidants, and natural extracts that have been harnessed for food preservation [22–24].

Natural extracts of plant origin are gaining significant interest in the food industry as consumers demand safer and more natural preservatives. Their secondary metabolite content, especially those with phenolic groups, are considered the most effective [25,26]. They are extracted from different plant parts like flowers, bark, herbs, wood, leaves, seeds, buds, twigs, fruit, and roots. These phenolic compounds include eugenol, thymol, carvacrol, vanillin, allicin, cinnamic, aldehyde, and allyl isothiocyanate and are usually found in plants such as cloves, cinnamon, thyme, oregano, orchids, garlic, and mustard. They are being considered as preferable antimicrobial alternatives to synthetic preservatives [25,27,28]. Clove oil and cinnamaldehyde have been generally recognized as safe (GRAS) for their intended use as flavoring agents in food by the Flavor and Extract Manufacturers Association of the USA (FEMA) (21 CFR 182.60) and the Food and Drug Administration of the United States (FDA) [9,28].

Cinnamaldehyde [C6H5CH=CHCHO] is an organic aromatic aldehyde, a natural extract derived from the inner bark and leaves of cinnamon trees of the genus Cinnamomum (e.g., *C. zeylanicum*, *C. cassia*, and *C. camphor*) [26]. It is a yellow essential oil with a cinnamon odor and sweet taste. It has a wide application in the food industry, herbal remedies, and home care products. Due to its diverse applications, CN is a potential biomarker that can be used for tracing and authentication of various products [29,30]. It is commonly used in washing solutions as a decontaminant, in active food packaging as a hurdle, or incorporated in food packages as an antimicrobial agent [27,31,32]. Various studies have shown the antimicrobial activity of cinnamaldehyde against different species of microorganisms [31,33–35].

Clove oil has traditionally been used as a seasoning and as an antimicrobial agent in food and food packages. It has also been used as an antiseptic for oral infections. Clove oil is an essential oil derived from clove trees known as *Syzygium aromaticum* [36,37]. The main active compounds in clove oil are eugenol, eugenyl acetate, and caryophyllene. It is a yellow or colorless essential oil with a spicy, pungent taste [38–40]. It has a wide range of applications in the food industry and the oral hygiene industry as well. Studies have shown its antimicrobial actions against a good number of pathogenic organisms and also spoilage organisms. It has both antioxidant and scavenging activity, as reported by [41].

Catfish refers to any fish of the order *Siluriformes*, which is predominantly composed of freshwater stout-bodied and scaleless bony fish, known for their long barbels that are present near the mouth of the fish and resemble a cat's whiskers [42]. They are bottom dwellers that mostly scavenge and feed on almost any kind of animal or vegetable matter, which exposes them to a wide range of pollutants that are toxic to humans and may result in significant levels of potential contamination [43]. Catfish are of considerable commercial and economic importance, and many of the larger species are farmed. The highly diverse nature of catfish can be seen by their wide distribution in tropical South America, Asia, and Africa; however, only one family is native to North America and one family to Europe. In the United States, the three primary species of catfish, are blue catfish, channel catfish, and flathead catfish, which are found in most rivers, lakes, and reservoirs of the United States and are readily available for food consumption [42,44–46].

Absorbent food pads amended with natural antimicrobial agents have been applied in the food industry as a means of actively packaging food products to reduce microbial contamination, thereby extending shelf life and promoting food safety and the sanitary conditions of refrigerated food products to improve consumer acceptability [47–49]. Various studies have been carried out where antimicrobial substances or materials have been incorporated into absorbent food pads [50–52]. For example, in one such study absorbent food pads were treated by spraying oregano essential oil of 1.5% distillate water on meat exudate absorbent pads used to extend the shelf life of overwrap packed fresh chicken drumsticks stored at 4 °C by approximately 2 days, according to a study carried out by Oral et al. [52].

When evaluating antimicrobials to be used in food packaging for shelf life capabilities, it is important to understand their ability to inhibit bacterial growth and minimal inhibitory concentration (MIC) as well as to reduce bacterial concentrations. Therefore, the purpose of this study was to investigate if essential oils of CN and CO might be used in catfish packaging as a preservative. Specifically, would CN and CO independently reduce microbial population recovery from absorbent food pads in direct contact with catfish fillets stored at 4 °C; inhibit the growth of SSOs and aerobic bacteria from catfish; and reduce the recovery of viable aerobic bacteria from catfish exudates, thereby extending the shelf life of catfish fillet in cold storage.

2. Materials and Methods

2.1. Antimicrobials Formulation

The essential oils used in this study were of commercial grade. Extracts of CN were purchased from Sigma-Aldrich (St. Louis, MO, USA). Samples were 100% pure and were extracted by steam distillation; similarly, CO was purchased from Piping Rock Health Products LLC (Ronkonkoma, NY, USA) and was 100% pure and extracted by steam distillation. The preparation of the essential oil treatment concentrations was carried out by diluting and mixing tap water in a laboratory vortex at room temperature directly before use.

2.2. Bacterial Strains, Growth, Inoculum Preparation and Zone of Inhibition

Specific spoilage organisms (SSOs), *Shewanella baltica* (NCTC strain 10735), and *Aeromonas hydrophila* (ATCC strain 7966), stored at −80 °C were obtained from the Food Biotechnology Laboratory in the Department of Human Ecology, College of Agriculture, Science, and Technology at Delaware State University. A loop of the SSO stock cultures was inoculated into separate 10 mL tryptic soy broth (TSB; Carolina Biological Supply Co., Burlington, NC, USA) and allowed to grow for 24 h at 27 and 37 °C, respectively. Thereafter, 1.00 mL was taken and plated on tryptic soy agar (TSA; MP Biomedicals, LLC (Solon, OH, USA). After that, treatment concentrations using 5 mL of tap water with concentrations of 0.125, 0.25, 0.50, 0.75, and 1.00 (mL/mL) CN or CO were prepared independently. Then, 20 μL of each treatment concentration was spot inoculated independently on the TSA plates containing the SSOs to determine the zone of inhibition. Commercial catfish were purchased from seafood markets within Dover, DE, USA. Samples were aseptically dived into 5 g pieces, which was stomached using 5.00 mL of TSB for 30 s. Then, 1.00 mL of the exudates was plated onto TSA. Twenty microliters of the essential oil treatment concentrations were spot inoculated on each TSA plate containing exudates from the catfish fish samples and incubated at an optimal growth temperature of 27 °C for 24 h to check for the zone of inhibition.

2.3. Bacterial Death Curve

Five grams of catfish fillet sample was stomached using 5 mL TSB, and 1 mL of stomached exudate from catfish fillets was plated on TSA and incubated for 24 h at 27 °C. Afterwards, loops of bacteria colonies isolated from the incubated stomached exudates were inoculated in 10 mL TSB tubes for use for each of the different treatment concentrations and incubated for 24 h at 27 °C. Thereafter, 20 mL of the incubated samples were mixed homogeneously in 50 mL conical tubes for each of the different treatment concentrations. The essential oil treatment concentrations of 0.50 and 1.00 mL CN or CO were then added to the 20 mL sample tubes at a 1:10 mL concentration and vortexed for 5 s. At 5, 10, 20, 30, and 40 min, 1.00 mL of the sample mixtures was pipetted and serially diluted in TSB. Then, 1.00 mL was removed from serially diluted tubes and plated on TSA. The plates were incubated for 24 h at 27 °C and bacterial death was enumerated.

2.4. Absorbent Food Pad Inoculation

Absorbent food pads were inoculated using a modified technique by Ren et al. [53]. Dri-loc absorbent food pads were donated by Novipax (Oak Brook, IL, USA). Treatment of the absorbent food pads and packaging materials was conducted by immersing the absorbent food pads into essential oil treatment concentrations of 0.50 and 1.00 mL of CN or CO, respectively, at a 1:10 ratio (mL/mL) and then allowing them to sit for 30 min at room temperature. Approximately, 50 g of untreated catfish fillets were aseptically cut with a sterilized knife and placed on both treated and untreated absorbent food pads and then stored for 7 days at 4 °C. The effect of the treatments over time was evaluated on days 1, 3, 5, and 7, respectively, by stomaching each treated absorbent food pad and each untreated control absorbent food pad in 5.00 mL of TSB. One milliliter was taken from each stomached absorbent food pad sample and then serially diluted in TSB tubes, and 1.00 mL was plated on TSA to enumerate the total aerobic bacteria/mL on the absorbent food pads.

2.5. Statistical Analysis

Using the SPSS statistical software program 26 (SPSS Inc., Chicago, IL, USA), one-way analysis of variance (ANOVA) was applied in this study. Tukey's HSD test was used for the comparison of logarithmic values to determine statistical significance between mean values of various treatments. Standard deviation was calculated for the data generated using Microsoft Excel 2016 (Microsoft Inc., Redmond, WA, USA). All experiments were carried out in triplicate. Significance was defined as $p \leq 0.05$.

3. Results

3.1. Zone of Inhibition

The inhibitory effects of different concentrations (0.125, 0.25, 0.50, 0.75, and 1.00 mL) of CN or CO are seen in Table 1. Cinnamaldehyde exhibited ZIB ranging from 1.00 to 2.10 cm. At 0.125 mL, CN to *A. hydrophila* plates exhibited a ZIB of 1.00 ± 0.00 cm. The zone of inhibition significantly increased to 1.40 ± 0.07 cm when the treatment concentration increased to 0.25 mL. At a concentration of 0.50 mL CN, the ZIB significantly increased even further to 1.70 ± 0.01 cm from the previous CN concentration. However, there was no significant difference ($p \leq 0.05$) in ZIB observed on *A. hydrophila* plates when treated with 0.50 and 0.75 mL CN. At 1.00 mL, CN treatment further increased the ZIB to 2.10 ± 0.17 cm, with no significant difference from the previous concentration. Furthermore, *A. hydrophila* plates treated with 0.125, 0.25, 0.50, 0.75, or 1.00 mL CO exhibited ZIB ranging from 1.03–1.96 cm. At all levels, with increasing concentrations of CO treatment ZIB increased significantly. At 0.125 mL CO, *A. hydrophila* treated plates revealed a ZIB of 1.03 ± 0.05 cm, which slightly increased to 1.23 ± 0.11 cm when treated with 0.25 mL CO. Upon increasing the concentration to 0.50 mL CO, the ZIB increased to 1.50 ± 0.00 cm with significance difference from the two previous treatment concentrations. Upon further increase of CO concentration to 0.75 mL, there was a further increment in the ZIB to 1.90 ± 0.17 cm. Further increasing the concentration of CO to 1.00 mL resulted in a maximum ZIB of 0.46 cm in diameter.

Both CN and CO revealed the ability to prevent *A. hydrophila* growth. However, CN and CO treatments at 0.75 mL and 1.00 mL concentrations had no significant difference ($p > 0.05$) in their bacteriostatic effects on *A. hydrophila*. Table 2 shows *Shewanella baltica* plates treated with different concentrations of CN or CO (0.125, 0.25, 0.50, 0.75, and 1.00 mL), where CN showed ZIB ranging from 1.03 to 2.17 cm. At 0.125 mL concentration of CN, a ZIB of 1.03 ± 0.06 cm was observed. No significant difference ($p \leq 0.05$) was observed when CN concentration increased to 0.25 mL. When the concentration of CN was further increased to 0.05 mL, a ZIB of 1.90 ± 0.01 cm resulted. No further significant increase ($p > 0.05$) was observed from 0.50 to 1.00 mL.

Table 1. Zone of Inhibition of SSO (*Aeromonas hydrophila*) treated with CN and CO in sterilized tap water (mL/mL).

Treatment Conc. (mL/mL)	ATCC *Aeromonas hydrophila* and CN			
	Trial 1 (cm)	Trial 2 (cm)	Trial 3 (cm)	Average
0.125	1.00	1.00	1.00	1.00 ± 0.00 ^{a}
0.25	1.40	1.50	1.30	1.40 ± 0.10 ^{c}
0.50	1.70	1.80	1.60	1.70 ± 0.01 ^{d}
0.75	1.90	1.90	1.80	1.85 ± 0.07 ^{de}
1.00	2.00	2.30	2.00	2.10 ± 0.17 ^{e}
	ATCC *Aeromonas hydrophila* and CO			
	Trial 1 (cm)	Trial 2 (cm)	Trial 3 (cm)	Average
0.125	1.00	1.10	1.00	1.03 ± 0.05 ^{a}
0.25	1.10	1.30	1.30	1.23 ± 0.11 ^{b}
0.50	1.50	1.50	1.50	1.50 ± 0.00 ^{c}
0.75	1.70	2.00	2.00	1.90 ± 0.17 ^{e}
1.00	2.00	2.00	1.90	1.96 ± 0.05 ^{e}

Means of averages are expressed in cm ± standard deviation. Different letters (a, b, c, d, and e) superscripted on the mean indicate a significant difference in the two treatments (CN and CO). Significance was defined as $p \leq 0.05$. Data were analyzed by one-way ANOVA.

Table 2. Zone of Inhibition of SSO (*Shewanella baltica*) treated with CN and CO in sterilized tap water (mL/mL).

Treatment Conc. (mL/mL)	NTCC 10735, *Shewanella baltica* and CN			
	Trial 1 (cm)	Trial 2 (cm)	Trial 3 (cm)	Average
0.125	1.00	1.10	1.00	1.03 ± 0.06 ^{a}
0.25	1.20	1.10	1.00	1.10 ± 0.10 ^{a}
0.50	2.00	1.90	1.80	1.90 ± 0.01 ^{b}
0.75	2.20	2.00	2.00	2.07 ± 0.12 ^{b}
1.00	2.30	2.20	2.00	2.17 ± 0.15 ^{b}
	NTCC 10735 *Shewanella baltica* and CO			
	Trial 1 (cm)	Trial 2 (cm)	Trial 3 (cm)	Average
0.125	0.80	0.80	0.80	0.80 ± 0.00 ^{a}
0.25	1.80	1.70	1.75	1.75 ± 0.07 ^{b}
0.50	2.50	2.70	2.60	2.60 ± 0.14 ^{c}
0.75	2.80	2.60	2.70	2.70 ± 0.14 ^{c}
1.00	2.90	3.00	2.95	2.90 ± 0.07 ^{c}

Means of averages are expressed in cm ± standard deviation. Different letters (a, b, and c) superscripted on the mean indicate a significant difference between the two treatments (CN and CO). Significance was defined as $p \leq 0.05$. Data were analyzed by one-way ANOVA.

Clove oil treatment concentrations on *S. baltica* inoculated plates resulted in ZIB ranging from 0.80 ± 0.00 to 2.95 ± 0.07 cm in diameter. Control, untreated inoculated plates resulted in no ZIB. At 0.125 mL concentration of CO in *S. baltica*, a ZIB of 0.80 ± 0.00 cm was seen. This value was not statistically different ($p > 0.05$) with 0.125 mL and 0.25 mL of CN treatment of *S. baltica*. The ZIB was increased significantly ($p \leq 0.05$) by 0.95 cm when the concentration of CO was increased to 0.25 mL. Increasing the concentration to 0.50 mL, 0.75 mL, and 1.00 mL did not result in any additional significant difference ($p > 0.05$) in the ZIB.

The inhibitory effects of CN or CO on catfish total aerobic bacteria are shown in Table 3. Catfish plates treated with 0.125, 0.25, 0.50, 0.75, and 1.00 mL of CN exhibited

ZIB ranging from 0.80 to 1.10 cm. At 0.125 mL CN treatment, catfish plates exhibited a ZIB of 0.80 ± 0.00 cm, and no significant difference ($p > 0.05$) was observed when CN concentration increased up to1.00 mL.

Table 3. Zone of inhibition of catfish total aerobic bacteria exudate treated with CN and CO in sterilized tap water (mL/mL).

Treatment Conc. (mL/mL)	Catfish and CN			
	Trial 1 (cm)	Trial 2 (cm)	Trial 3 (cm)	Average
0.125	0.80	0.80	0.80	0.80 ± 0.00 ^{a}
0.25	0.90	0.90	0.90	0.90 ± 0.00 ^{ab}
0.50	1.00	1.00	1.00	1.00 ± 0.00 ^{ab}
0.75	1.00	1.00	1.10	1.03 ± 0.05 ^{bc}
1.00	1.10	1.20	1.00	1.10 ± 0.10 ^{bc}
	Catfish and CO			
	Trial 1 (cm)	Trial 2 (cm)	Trial 3 (cm)	Average
0.125	1.00	1.10	1.00	1.03 ± 0.05 ^{bc}
0.25	1.10	1.30	1.30	1.23 ± 0.11 ^{c}
0.50	1.50	1.50	1.50	1.50 ± 0.00 ^{d}
0.75	1.70	2.00	2.00	1.90 ± 0.17 ^{e}
1.00	2.00	2.00	1.90	1.96 ± 0.05 ^{e}

Means of averages are expressed in cm ± standard deviation. Different letters (a, b, c, d, and e) superscripted on the mean indicate a significant difference between the two treatments (CN and CO). Significance was defined as $p \leq 0.05$. Data were analyzed by one-way ANOVA.

Catfish plates treated with 0.125, 0.25, 0.50, 0.75, and 1.00 mL concentrations of CO exhibited ZIB ranging from 1.03 ± 0.05 to 1.96 ± 0.05 cm. At 0.125 mL CO, catfish treated plates revealed a ZIB of 1.03 ± 0.05 cm. No significant difference ($p > 0.05$) was observed with 0.25 mL CO treatment of catfish plates. Upon increasing the concentration to 0.50 mL CO, the ZIB increased to 1.50 ± 0.00 cm. Further increase of CO concentration to 0.75 mL and 1.00 mL, there was a significant ($p \leq 0.05$) increase in the ZIB from the previous CO concentrations 1.90 ± 0.17cm and. 1.96 ± 0.05 cm, respectively. However, there was no significant difference ($p > 0.05$) between both treatments.

3.2. Bacterial Death Curve

Figure 1 shows that catfish exudates in TSB at time zero and treated with 0.50 mL and 1.00 mL CN had 9.00 ± 0.06 and 8.98 ± 0.03 log CFU/mL, respectively. At 5 min after 0.50 mL and 1.00 mL addition, bacterial concentrations for 0.50 mL did not significantly ($p > 0.05$) decrease. However, for 1.00 mL bacterial counts decreased on average by 0.86 log to 8.12 log CFU/mL. When increasing the length of treatment to 10 min, bacterial concentrations for 0.50 mL and 1.00 mL decreased on average by 1.35 log to 7.53 ± 0.01 log CFU/mL and 0.46 log to 7.66 ± 0.01 log CFU/mL, respectively. Bacterial concentration was further significantly ($p \leq 0.05$) reduced at 20 min for both treatment concentrations, on average by 1.01 log to 6.52 ± 0.01 log CFU/mL and 0.50 log and 7.15 ± 0.06 log CFU/mL, respectively. Thirty minutes after treatment with 0.50 mL of CN, bacterial counts decreased on average by 2.03 log and dropped to 4.49 ± 0.03 log CFU/mL, and for 1.00 mL concentration, the bacterial concentration decreased by 1.56 log and dropped to 5.59 ± 0.11 log CFU/mL. At 40 min after treatment, 0.50 mL bacterial counts significantly ($p \leq 0.05$) reduced from the previous concentration by 2.33 log to 2.16 ± 0.08 log CFU/mL. Similarly, at 40 min 1.00 mL CN reduced the average bacterial counts by 1.71 log to 3.88 ± 0.03 log CFU/mL.

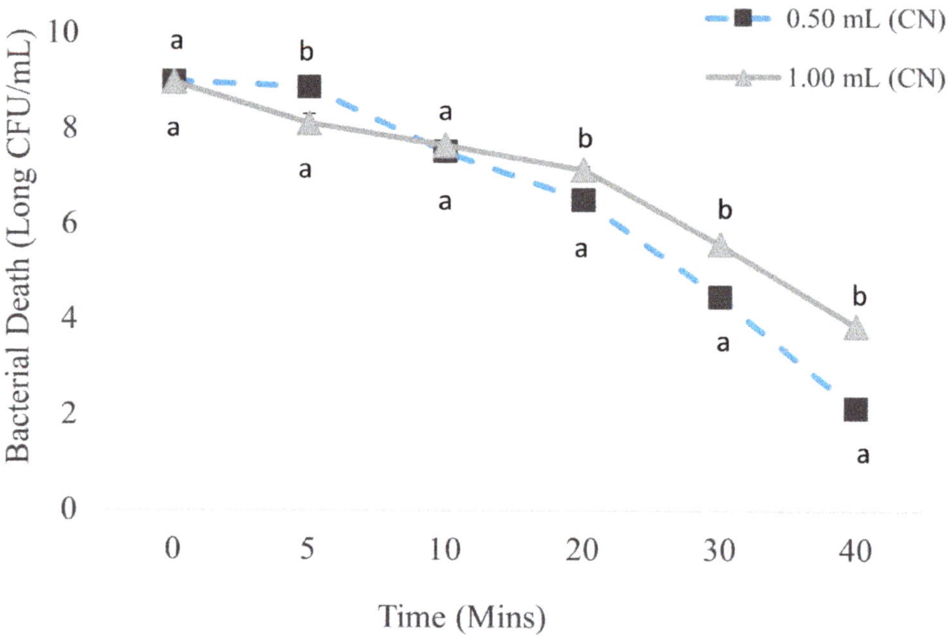

Figure 1. Reduction of bacterial count in store-bought catfish treated with CN. n = 3 for each concentration on each day. Concentration of treatment to water was at 1:10 mL per each treatment. Significance was defined as $p \leq 0.05$. Differences within each day are represented by a and b. Data were analyzed by one-way ANOVA.

Figure 2 shows the bactericidal effects of CO on catfish exudates in TSB. At time zero min, when treated with 0.50 and 1.00 mL CO the catfish exudates had 9.69 ± 0.02 and 9.77 ± 0.03 log CFU/mL, respectively. At 5 min after 0.50 and 1.00 mL addition of CO, bacterial concentrations significantly ($p \leq 0.05$) decreased, on average by 1.48 log to 8.21 ± 0.06 and 2.77 to 7.00 ± 00 log CFU/mL, respectively. When increasing the treatment time to 10 min, the bacterial counts for 0.50 mL further decreased significantly ($p \leq 0.05$) by 0.35 log to 7.86 ± 0.15 and 4.67 to 2.33 ± 10 log CFU/mL. However, there was no significant ($p > 0.05$) decrease in bacterial concentration for CO 1.00 mL from 20 up to 40 min. At 0.50 mL concentrations at 20 and 30 min, the bacterial concentration further significantly ($p \leq 0.05$) reduced, on average by 4.52 log to 3.34 ± 0.02, and 0.87 log to 2.47 ± 0.05 and 0.32 log, respectively. When further increasing the time of treatment to 40 min for 0.50 mL concentrations, there were no counts of viable bacteria at the lowest recovery level.

In Figure 3, untreated absorbent food pads from catfish packaging stayed at ≥9 log CFU/mL for 7 days of study. Absorbent food pads treated with 0.50 mL CN resulted in observably lower bacterial counts on days 1 3, 5, and 7, with results of 5.74 ± 0.04 log CFU/mL, 5.92 ± 0.02 log CFU/mL, 4.19 ± 0.05 log CFU/mL, and 5.53 ± 0.03 log CFU/mL, respectively. Of these treatments, the lowest significant ($p \leq 0.05$) reduction was observed on day 5. Absorbent pads treated with 1.00mL on days 1, 3, 5, and 7, showed results of 3.59 ± 0.011 log CFU/mL, 3.39 ± 0.02 log CFU/mL, 3.60 ± 0.01 log CFU/mL, and 1.31 ± 0.11 log CFU/mL, respectively. Observable reductions were seen each day between CN treatments of 0.50 mL and 1.00 mL of 2.51, 2.53, 0.59, 4.22, and 2.6 log CFU/mL, respectively.

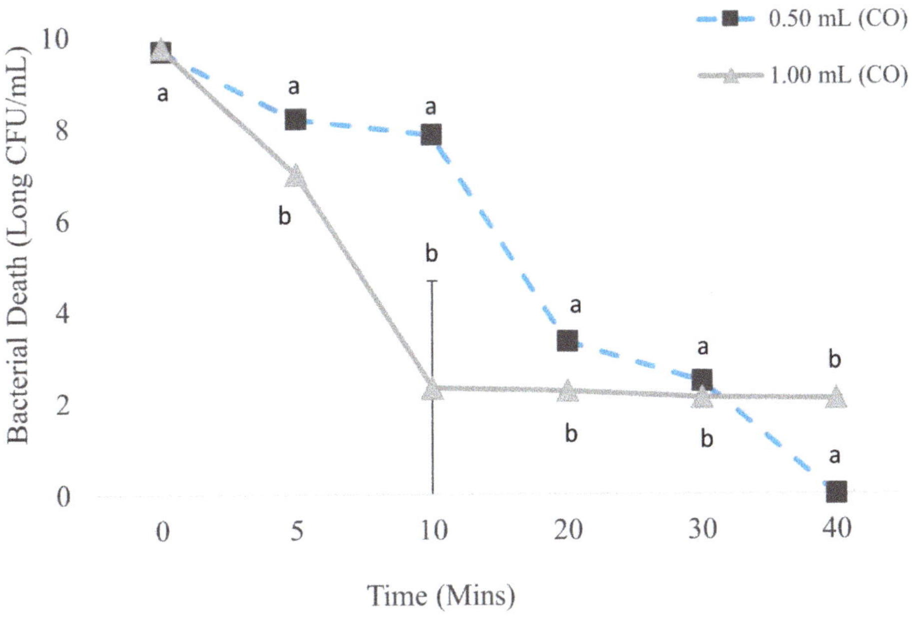

Figure 2. Reduction of bacterial count in store-bought catfish treated with CO. n = 3 for each concentration on each day. Concentration of treatment to water was at 1:10 mL per each treatment. Significance was defined as $p \leq 0.05$. Differences within each day are represented by a and b. Data were analyzed by one-way ANOVA.

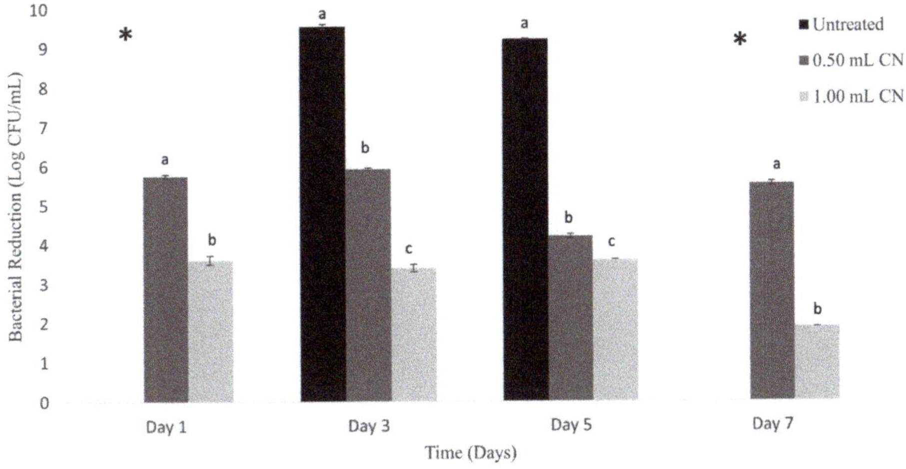

Figure 3. Absorbent food pads treated with CN for storage of catfish at 4 °C. * represents plates that were above the highest level of bacteria count, too numerous to count. n = 3 for each concentration and trial. One mL of each prepared concentration was added to 5 mL of bacteria. The different letters (a, b, c) indicate significant differences between treatments on each day. Significance was defined as $p \leq 0.05$. Statistical analysis for this study was conducted by means of one-way ANOVAs.

Our results suggest that the application of CN to absorbent food pads overlaid with catfish fillet can reduce bacteria in absorbent food pads over 7 days of storage at 4 °C.

Increasing the concentration of CN from 0.50 mL to 1.00 mL also increased the reduction of bacteria concentration in the absorbent pads.

We found that, as shown in Figure 4, untreated absorbent food pads with packaged catfish fillet had ≥9.59 log CFU/mL of total aerobic bacteria for all seven days. Absorbent food pads treated with 0.5 mL CO, resulted in observably lower bacterial counts on days 1, 3, 5, and 7 with results of 5.17 ± 0.15 log CFU/mL, 6.63 ± 0.02 log CFU/mL, 7.90 ± 0.04 log CFU/mL, and 3.12 ± 0.17 log CFU/mL respectively. Of these treatments, the lowest significant ($p \leq 0.05$) reduction was observed on day 7. Absorbent pads treated with 1.00 mL on days 1, 3, 5, and 7 resulted in a bacterial reduction of 3.00 ± 0.00 log CFU/mL, 5.30 ± 0.24 log CFU/mL, 5.90 ± 0.11 log CFU/mL, respectively, and zero counts of total bacteria on day 7. Results suggest that CO embedment on absorbent food pads for storage of catfish can significantly ($p \leq 0.05$) reduce bacteria counts in absorbent food pads over 7 days of storage at 4 °C.

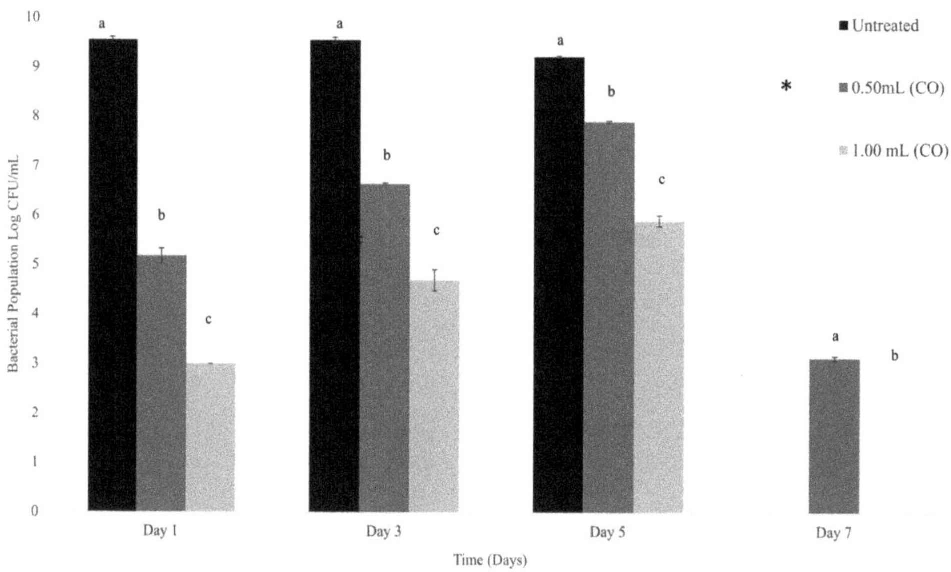

Figure 4. Absorbent food pads treated with CO for storage of catfish at 4 °C. * represents plates that were above the highest level of bacteria count, too numerous to count. Treatment with 1.00 mL CO resulted in no growth. n = 3 for each concentration and trial. One ml of each prepared concentration was added to 5 mL of bacteria. The different letters (a, b, c) indicate significant differences between treatments on each day. Significance was defined as $p \leq 0.05$. Statistical analysis for this study was conducted by means of one-way ANOVAs.

In addition, sensory observations were made on some quality attributes of the catfish sample. The sensory observation and quality attributes of color, aroma, and texture were observed on days 1, 3, 5, and 7 of storage. Catfish fillet samples with untreated and treated absorbent food pads stored at 4 °C were evaluated for texture, smell, and color. Fish fillet samples with the untreated absorbent food pads had a more off-odor and slimy texture by the end of the 7 days at 4 °C of storage. Catfish samples on treated absorbent food pads had less visual degradation, especially in locations in which the fish was in direct contact with the pad. Fish fillet samples on absorbent food pads with CN had a slightly yellow color and CN aroma, while CO treated absorbent food pads had a pale-yellow color and CO aroma. The color change may potentially be due to the uptake of coloration from the natural antimicrobial preservatives CN and CO.

4. Discussions

Natural extracts from plant origins have been in use as antimicrobials for food safety and preservation. Various studies have shown CN and CO to be effective against some pathogenic and food spoilage organisms. Our results establish that different concentrations of CN and CO at different levels of investigation show antimicrobial effectiveness against SSOs and total aerobic bacteria from catfish. The results of this study are in line with studies carried out by [54], which used cinnamaldehyde-incorporated and eugenol-incorporated methylcellulose films as antimicrobial packaging materials to investigate antimicrobial activity against target microorganisms using both an agar-disc diffusion technique and a vapor diffusion technique. At a concentration of 50 μL/mL, cinnamaldehyde and eugenol revealed antimicrobial activity against *Aeromonas hydrophila*, *Enterococcus faecalis*, and some other test strains where ZIB ranged from 0.87 to 3.01 cm, which is in the range of the results from this study, where CN and CO showed ZIB against *Aeromonas hydrophila* ranging from 1.0 to 2.10 cm and from 1.17 to 2.40 cm, respectively. Furthermore, the average range for the ZIB of CN and CO from this study against *Shewanella baltica* was 0.87 to 2.17 cm, which is in line with the findings from [54] against some SSOs. Abdel et al. [55] reported the antimicrobial and inhibitory characteristics of clove oil seen against eight microorganisms, with CO showing a ZIB range of 2.5 and 3.6 cm at low and high concentrations against *Escherichia coli*, which is also within the range of the results obtained from this study, where CO showed a ZIB range against total aerobic bacteria from catfish of 1.03 and 1.96 cm at low and high concentrations. The broad spectrum of inhibitory effects against microorganisms by cinnamaldehyde and eugenol has been reviewed, and their activity and highlighted potential use in antimicrobial packaging agents have been reported [30,34]. This is in line with the findings of this study that showed the antimicrobial activity of CN and CO used in active packaging and absorbent food pads against total aerobic bacteria from catfish exudates.

At the highest concentrations of CN and CO used to investigate the bacteriostatic effects of the treatments, CO exhibited a larger ZIB on the SSOs and the total aerobic bacteria from catfish. Clove oil, furthermore, showed a greater bactericidal effect on the total aerobic bacteria from catfish exudates. The findings from this study also, show that treating absorbent food pads with either 0.50 mL or 1.00 mL concentrations of CN or CO, could limit the increase in bacterial count on absorbent pads used for packaging and storage of catfish fillet at 4 °C and thus extend the shelf life of fish. Furthermore, catfish fillet samples overlaid on the absorbent food pads had the aroma of either CN or CO, with slight yellowish discoloration.

5. Conclusions

Our results establish that the present study has revealed the effectiveness of CN and CO for controlling microbial populations for channel catfish (*Ictalurus punctatus*) fillet packaging using absorbent food pads. These natural extracts used at different concentrations and different phases in this study showed both bactericidal and bacteriostatic abilities in controlling the microbial population of specific spoilage organisms and total aerobic bacteria in absorbent food pad packaging of catfish fillet. Therefore, incorporation of CN and CO into absorbent food pads may be applied successfully in fish packaging to prevent and control bacterial spoilage of fish, especially in developing countries where there is poor infrastructural development in the area of post-harvest handling of fish to delay and reduce microbial proliferation in harvested fish before proper handling and storage. These findings indicate that natural extracts such as CN and CO can potentially extend the shelf life of fish such as catfish in effective active packaging using absorbent food pads. Some sensory characteristics of the catfish were altered in terms of color and aroma, which can be attributed to the color and aroma of the CN and CO extracts used. Therefore, their effect on sensory quality and acceptability by consumers needs to be further studied.

Author Contributions: Conceptualization, Methodology, Investigation, and Writing were conducted by R.I.E. and W.L.III. All authors have read and agreed to the published version of the manuscript.

Funding: This research received no external funding.

Institutional Review Board Statement: Not applicable.

Informed Consent Statement: Not applicable.

Data Availability Statement: The original contributions presented in the study are included in the article, further inquiries can be directed to the corresponding author.

Acknowledgments: The authors would like to acknowledge Gulnihal Ozbay and her laboratory for their indispensable assistance, equipment, and use of laboratory space. Novipax (Oak Brook, IL, USA) is acknowledged for material donated.

Conflicts of Interest: The authors declare no conflicts of interest.

References

1. Mishra, P.; Sharma, M.; Kewat, A. Spoilage in Fish and Shellfish Products. *Limnology* **2022**, *2*, 59.
2. Odeyemi, O.A.; Alegbeleye, O.O.; Strateva, M.; Stratev, D. Understanding spoilage microbial community and spoilage mechanisms in foods of animal origin. *Compr. Rev. Food Sci. Food Saf.* **2020**, *19*, 311–331. [CrossRef] [PubMed]
3. Mishra, S.P.; Pradesh, U. Significance of fish nutrients for human health. *Int. J. Fish. Aquat. Res.* **2020**, *5*, 47–49.
4. Venegas-Calerón, M.; Napier, J.A. New alternative sources of omega-3 fish oil. In *Advances in Food and Nutrition Research*; Elsevier: Amsterdam, The Netherlands, 2023; pp. 343–398.
5. Krittanawong, C.; Isath, A.; Hahn, J.; Wang, Z.; Narasimhan, B.; Kaplin, S.L.; Jneid, H.; Virani, S.S.; Tang, W.W. Fish consumption and cardiovascular health: A systematic review. *Am. J. Med.* **2021**, *134*, 713–720. [PubMed]
6. Qiu, Q.; Dewey-Mattia, D.; Subramhanya, S.; Cui, Z.; Griffin, P.M.; Lance, S.; Lanier, W.; Wise, M.E.; Crowe, S.J. Food recalls associated with foodborne disease outbreaks, United States, 2006–2016. *Epidemiol. Infect.* **2021**, *149*, e190. [CrossRef] [PubMed]
7. Ali, A.; Parisi, A.; Conversano, M.C.; Iannacci, A.; D'Emilio, F.; Mercurio, V.; Normanno, G. Food-borne bacteria associated with seafoods: A brief review. *J. Food Qual. Hazards Control* **2020**.
8. Garcia, S.M.; Rosenberg, A.A. Food security and marine capture fisheries: Characteristics, trends, drivers and future perspectives. *Philos. Trans. R. Soc. B Biol. Sci.* **2010**, *365*, 2869–2880. [CrossRef] [PubMed]
9. Guillen, J.; Natale, F.; Carvalho, N.; Casey, J.; Hofherr, J.; Druon, J.-N.; Fiore, G.; Gibin, M.; Zanzi, A.; Martinsohn, J.T. Global seafood consumption footprint. *Ambio* **2019**, *48*, 111–122. [CrossRef] [PubMed]
10. Andrews, N.; Bennett, N.J.; Le Billon, P.; Green, S.J.; Cisneros-Montemayor, A.M.; Amongin, S.; Gray, N.J.; Sumaila, U.R. Oil, fisheries and coastal communities: A review of impacts on the environment, livelihoods, space and governance. *Energy Res. Soc. Sci.* **2021**, *75*, 102009. [CrossRef]
11. Gephart, J.A.; Golden, C.D.; Asche, F.; Belton, B.; Brugere, C.; Froehlich, H.E.; Fry, J.P.; Halpern, B.S.; Hicks, C.C.; Jones, R.C.; et al. Scenarios for global aquaculture and its role in human nutrition. *Rev. Fish. Sci. Aquac.* **2020**, *29*, 122–138. [CrossRef]
12. Lou, X.; Zhai, D.; Yang, H. Changes of metabolite profiles of fish models inoculated with Shewanella baltica during spoilage. *Food Control* **2021**, *123*, 107697. [CrossRef]
13. Ekonomou, S.; Parlapani, F.; Kyritsi, M.; Hadjichristodoulou, C.; Boziaris, I. Preservation status and microbial communities of vacuum-packed hot smoked rainbow trout fillets. *Food Microbiol.* **2022**, *103*, 103959. [CrossRef] [PubMed]
14. Zhuang, S.; Hong, H.; Zhang, L.; Luo, Y. Spoilage-related microbiota in fish and crustaceans during storage: Research progress and future trends. *Compr. Rev. Food Sci. Food Saf.* **2021**, *20*, 252–288. [CrossRef]
15. Tavares, J.; Martins, A.; Fidalgo, L.G.; Lima, V.; Amaral, R.A.; Pinto, C.A.; Silva, A.M.; Saraiva, J.A. Fresh fish degradation and advances in preservation using physical emerging technologies. *Foods* **2021**, *10*, 780. [CrossRef]
16. Sikorski, Z.E.; Kołakowska, A.; Burt, J.R. Postharvest biochemical and microbial changes. In *Seafood*; CRC Press: Boca Raton, FL, USA, 2020; pp. 55–75.
17. Kumar, K. *Food Processing and Preservation*; Scientific Publisher: Hackensack, NJ, USA, 2020; p. 13.
18. Wu, T.; Wang, M.; Wang, P.; Tian, H.; Zhan, P. Advances in the formation and control methods of undesirable flavors in fish. *Foods* **2022**, *11*, 2504. [CrossRef]
19. Ebirim, R.I. Spoilage and Sensory Observation of Cinnamaldehyde and Clove Oil Application to Control Microbial Populations in Catfish (*Ictalurus punctatus*) and Trout (*Oncorhynchus mykiss*) Fillet Packaging. Master's Thesis, Delaware State University ProQuest Dissertations Publishing, Delaware State University, Dover, DE, USA, 2020.
20. Huang, Y.; Zhang, H.; Lv, B.; Tang, C.; Du, J.; Jin, H. Sulfur dioxide: Endogenous generation, biological effects, detection, and therapeutic potential. *Antioxid. Redox Signal.* **2022**, *36*, 256–274. [CrossRef] [PubMed]
21. Brender, J.D. Human health effects of exposure to nitrate, nitrite, and nitrogen dioxide. In *Just Enough Nitrogen: Perspectives on How to Get There for Regions with too Much and too Little Nitrogen*; The Institute of Food Technologist: Chicago, IL, USA, 2020; pp. 283–294.
22. Singh, V.P. Recent approaches in food bio-preservation-a review. *Open Vet. J.* **2018**, *8*, 104–111. [CrossRef]

23. Ranathunga, N.S.; Wijayasekara, K.N.; Abeyrathne, E.D.N.S. Application of bio-preservation to enhance food safety: A review. *Food Sci. Preserv.* **2023**, *30*, 179–189. [CrossRef]
24. Hussein, A.R. Foods bio-preservation: A review. *Int. J. Res. Appl. Sci. Biotechnol.* **2022**, *9*, 212–217.
25. Kirchner, M.T.; Bläser, D.; Boese, R.; Thakur, T.S.; Desiraju, G.R. Weak C—H··· O hydrogen bonds in anisaldehyde, salicylaldehyde and cinnamaldehyde. *Acta Crystallogr. Sect. C Cryst. Struct. Commun.* **2011**, *67*, o387–o390. [CrossRef]
26. Ashakirin, S.N.; Tripathy, M.; Patil, U.K.; Majeed, A.B.A. Chemistry and bioactivity of cinnamaldehyde: A natural molecule of medicinal importance. *Int. J. Pharm. Sci. Res.* **2017**, *8*, 2333–2340.
27. Bakry, A.M.; Abbas, S.; Ali, B.; Majeed, H.; Abouelwafa, M.Y.; Mousa, A.; Liang, L. Microencapsulation of oils: A comprehensive review of benefits, techniques, and applications. *Compr. Rev. Food Sci. Food Saf.* **2016**, *15*, 143–182. [CrossRef]
28. Gooderham, N.J.; Cohen, S.M.; Eisenbrand, G.; Fukushima, S.; Guengerich, F.P.; Hecht, S.S.; Rietjens, I.M.; Rosol, T.J.; Davidsen, J.M.; Harman, C.L.; et al. FEMA GRAS assessment of natural flavor complexes: Clove, cinnamon leaf and West Indian bay leaf-derived flavoring ingredients. *Food Chem. Toxicol.* **2020**, *145*, 111585. [CrossRef] [PubMed]
29. Medina, S.; Pereira, J.A.; Silva, P.; Perestrelo, R.; Câmara, J.S. Food fingerprints–A valuable tool to monitor food authenticity and safety. *Food Chem.* **2019**, *278*, 144–162. [CrossRef] [PubMed]
30. Suppakul, P. Cinnamaldehyde and eugenol: Use in antimicrobial packaging. In *Antimicrobial Food Packaging*; Elsevier: Amsterdam, The Netherlands, 2016; pp. 479–490.
31. Ribeiro-Santos, R.; Andrade, M.; de Melo, N.R.; Sanches-Silva, A. Use of essential oils in active food packaging: Recent advances and future trends. *Trends Food Sci. Technol.* **2017**, *61*, 132–140. [CrossRef]
32. Long, W., III; Sarker, M.I.; Liu, C.K. Cinnamaldehyde/Lactic Acid Spray Wash Treatment for Meat Safety and Byproduct Quality Assurance. *Am. J. Food Sci. Technol.* **2018**, *6*, 280–289.
33. Shen, S.; Zhang, T.; Yuan, Y.; Lin, S.; Xu, J.; Ye, H. Effects of cinnamaldehyde on Escherichia coli and Staphylococcus aureus membrane. *Food Control* **2015**, *47*, 196–202. [CrossRef]
34. Siddiqua, S.; Anusha, B.A.; Ashwini, L.S.; Negi, P.S. Antibacterial activity of cinnamaldehyde and clove oil: Effect on selected foodborne pathogens in model food systems and watermelon juice. *J. Food Sci. Technol.* **2015**, *52*, 5834–5841. [PubMed]
35. Zamuner, C.F.; Dilarri, G.; Bonci, L.C.; Saldanha, L.L.; Behlau, F.; Marin, T.G.; Sass, D.C.; Bacci, M.; Ferreira, H. A cinnamaldehyde-based formulation as an alternative to sodium hypochlorite for post-harvest decontamination of citrus fruit. *Trop. Plant Pathol.* **2020**, *45*, 701–709. [CrossRef]
36. Sharma, S.; Barkauskaite, S.; Duffy, B.; Jaiswal, A.K.; Jaiswal, S. Characterization and antimicrobial activity of biodegradable active packaging enriched with clove and thyme essential oil for food packaging application. *Foods* **2020**, *9*, 1117. [CrossRef]
37. Gengatharan, A.; Rahim, M.H.A. The application of clove extracts as a potential functional component in active food packaging materials and model food systems: A mini-review. *Appl. Food Res.* **2023**, *3*, 100283. [CrossRef]
38. Hammer, K.A.; Carson, C.F.; Riley, T.V. Antimicrobial activity of essential oils and other plant extracts. *J. Appl. Microbiol.* **1999**, *86*, 985–990. [CrossRef]
39. Hossen Jenia, S. Study on Nutritional Composition, Bioactive Compounds, Antioxidant and Antimicrobial Activity of the Clove (*Syzygium Aromaticum*). Master's Thesis, Chattogram Veterinary and Animal Sciences University, Chattogram, Bangladesh, 2019.
40. Carson, C.F.; Hammer, K.A. Chemistry and bioactivity of essential oils. *Lipids Essent. Oils Antimicrob. Agents* **2011**, 203–238.
41. Jirovetz, L.; Buchbauer, G.; Stoilova, I.; Stoyanova, A.; Krastanov, A.; Schmidt, E. Chemical composition and antioxidant properties of clove leaf essential oil. *J. Agric. Food Chem.* **2006**, *54*, 6303–6307. [CrossRef]
42. Diogo, R.; Oliveira, C.; Chardon, M. The origin and transformation of the palatine-maxillary system of catfish (Teleostei: Siluriformes): An example of macroevolution. *Neth. J. Zool.* **2000**, *50*, 373–388. [CrossRef]
43. Gormaz, J.G.; Fry, J.P.; Erazo, M.; Love, D.C. Public health perspectives on aquaculture. *Curr. Environ. Health Rep.* **2014**, *1*, 227–238. [CrossRef]
44. Triantafyllidis, A.; Krieg, F.; Cottin, C.; Abatzopoulos, T.J.; Triantaphyllidis, C.; Guyomard, R. Genetic structure and phylogeography of European catfish (*Silurus glanis*) populations. *Mol. Ecol.* **2002**, *11*, 1039–1055. [CrossRef]
45. Day, J.J.; Wilkinson, M. On the origin of the Synodontis catfish species flock from Lake Tanganyika. *Biol. Lett.* **2006**, *2*, 548–552. [CrossRef] [PubMed]
46. Dauda, A.B.; Natrah, I.; Karim, M.; Kamarudin, M.S.; Bichi, A.U.H. African catfish aquaculture in Malaysia and Nigeria: Status, trends and prospects. *Fish. Aquac. J.* **2018**, *9*, 1–5. [CrossRef]
47. Castrica, M.; Miraglia, D.; Menchetti, L.; Branciari, R.; Ranucci, D.; Balzaretti, C.M. Antibacterial effect of an active absorbent pad on fresh beef meat during the shelf-life: Preliminary results. *Appl. Sci.* **2020**, *10*, 7904. [CrossRef]
48. Pettersen, M.K.; Nilsen-Nygaard, J.; Hansen, A.; Carlehög, M.; Liland, K.H. Effect of liquid absorbent pads and packaging parameters on drip loss and quality of chicken breast fillets. *Foods* **2021**, *10*, 1340. [CrossRef] [PubMed]
49. Ren, T.; Hayden, M.; Qiao, M.; Huang, T.-S.; Ren, X.; Weese, J. Absorbent pads containing N-halamine compound for potential antimicrobial use for chicken breast and ground chicken. *J. Agric. Food Chem.* **2018**, *66*, 1941–1948. [CrossRef]
50. Gouvêa, D.M.; Mendonça, R.C.S.; Lopez, M.E.S.; Batalha, L.S. Absorbent food pads containing bacteriophages for potential antimicrobial use in refrigerated food products. *LWT-Food Sci. Technol.* **2016**, *67*, 159–166. [CrossRef]
51. Kilinc, B.; Altas, S. Effect of absorbent pads containing black seed or rosemary oils on the shelf life of sardine [*Sardina pilchardus* (Walbaum, 1792)] fillets. *J. Appl. Ichthyol.* **2016**, *32*, 552–558. [CrossRef]

52. Oral, N.; Vatansever, L.; Sezer, Ç.; Aydın, B.; Güven, A.; Gülmez, M.; Başer, K.H.C.; Kürkçüoğlu, M. Effect of absorbent pads containing oregano essential oil on the shelf life extension of overwrap packed chicken drumsticks stored at four degrees Celsius. *Poult. Sci.* **2009**, *88*, 1459–1465. [CrossRef]
53. Ren, T.; Qiao, M.; Huang, T.-S.; Weese, J.; Ren, X. Efficacy of N-halamine compound on reduction of microorganisms in absorbent food pads of raw beef. *Food Control* **2018**, *84*, 255–262. [CrossRef]
54. Sanla-Ead, N.; Jangchud, A.; Chonhenchob, V.; Suppakul, P. Antimicrobial Activity of cinnamaldehyde and eugenol and their activity after incorporation into cellulose-based packaging films. *Packag. Technol. Sci.* **2012**, *25*, 7–17. [CrossRef]
55. Sulieman, A.M.E.; El-Boshra, I.M.; El-Khalifa, E.A. Nutritive value of clove (*Syzygium aromaticum*) and detection of antimicrobial effect of its bud oil. *Res. J. Microbiol.* **2007**, *2*, 266–271.

foods

MDPI

Communication

Antimicrobial Activity of Grapefruit Seed Extract on Edible Mushrooms Contaminations: Efficacy in Preventing *Pseudomonas* spp. in *Pleurotus eryngii*

Marcello Murgia [1], Sara Maria Pani [1,*], Adriana Sanna [1], Luisa Marras [2], Cristina Manis [3], Alessandro Banchiero [1] and Valentina Coroneo [1]

[1] Department of Medical Sciences and Public Health, University of Cagliari, Cittadella Universitaria Monserrato-S.P. Monserrato-Sestu Km 0.700, 09042 Monserrato, CA, Italy; coroneo@unica.it (V.C.)
[2] Analysis Laboratory, ASL Cagliari, Via Piero della Francesca, 1, 09047 Su Planu, CA, Italy
[3] Department of Life and Environmental Sciences, Blocco A, Room 13, University of Cagliari, Cittadella Universitaria Monserrato-S.P. Monserrato-Sestu Km 0.700, 09042 Monserrato, CA, Italy
* Correspondence: saramariapani@gmail.com or s.pani4@studenti.unica.it

Abstract: *Pleurotus eryngii* is an edible mushroom that suffers significant losses due to fungal contamination and bacteriosis. The Pseudomonadaceae family represents one of the most frequent etiologic agents. Grapefruit seed extract (GSE) is a plant extract that contains different bioactive components, such as naringin, and exhibits a strong antibacterial and antioxidant activity. Over the last decade, GSE use as an alternative to chemical treatments in the food sector has been tested. However, to our knowledge, its application on mushroom crops has never been investigated. This study focuses on evaluating GSE efficacy in preventing *P. eryngii* yellowing. GSE antibiotic activity, inhibitory and bactericidal concentrations, and antibiofilm activity against several microorganisms were tested with the Kirby–Bauer disk diffusion assay, the broth microdilution susceptibility test, and the Crystal violet assay, respectively. In vitro, the extract exhibited antimicrobial and antibiofilm activity against *Staphylococcus aureus* 6538 and MRSA (wild type), *Escherichia coli* ATCC 8739, and *Pseudomonas* spp. (*Pseudomonas aeruginosa* 9027, *P. fluorescens* (wild type)). GSE application in vivo, in pre- and post-sprouting stages, effectively prevented bacterial infections and subsequent degradation in the mushroom crops: none of the *P. eryngii* treated manifested bacteriosis. Our findings support the use of GSE as an eco-friendly and sustainable alternative to chemical treatments for protecting *P. eryngii* crops from bacterial contamination, consequently ensuring food safety and preventing financial losses due to spoilage. Furthermore, GSE's potential health benefits due to its content in naringin and other bioactive components present new possibilities for its use as a nutraceutical in food fortification and supplementation.

Keywords: plant extracts; eco-friendly sustainable antimicrobials; spoilage microorganisms; edible mushrooms; yellowing

1. Introduction

Pleurotus eryngii is a basidiomycete mushroom known in Italy as "cardoncello" [1,2]. It is also known as "royal trumpet" or "royal oyster" and is one of the most valuable edible mushrooms, with several varieties. Typically found in southern Europe, Northern Africa, the Middle East, and Central Asia, it is considered one of the most widely spread species of *Pleurotus* and has been known since ancient times for its excellent medicinal and nutritional characteristics [3–5]. This mushroom possesses a low percentage of calories and has a good concentration of major nutrients, such as protein, peptides, minerals, terpenoids, traces of various elements, fiber, and polysaccharides (*P. eryngii* polysaccharides—PEPs). These characteristics have aroused particular interest because of the anti-cancer, hepatoprotective, anti-lipidemia, immune system strengthening, and other activities shown in vitro and in animal models [6–10]. For this reason, *P. eryngii* is classified among functional foods [11–13] and used to create "healthy snacks" [14], namely, foods with high nutritional and biological value (rich in fiber and protein, low in salt, sugar, fat, and calories) [11]. Although this edible mushroom has several beneficial properties for our body, some peculiarities, such as a susceptibility to contaminations by mold and bacterial diseases [15], may compromise its quality. Generally, fungus-induced contaminations are attributed to *Cladobotryum mycophilum*, responsible for "spider's web disease" [16–18], *Gliocladium roseum*, responsible for "brown spot" [19], and *Trichoderma* spp., responsible for "green mold" [20]. The bacterial blight culprits belong to several species, such as *Pantoea* spp. [14], *Erwinia beijingensis*, *Ewingella americana* [21,22], *Enterobacter amnigenus*, and *Staphylococcus* spp. However, Pseudomonadaceae, such as *Pseudomonas tolaasii* and *P. fluorescens*, are the most relevant and responsible for the stem's yellowing [15,23,24]. It has indeed been noted that the initial colonization of the cap induces a loss in production yield; the size of the fruiting body is affected by the bacterial populations in the pre-harvest stage, consequently jeopardizing the quality of the harvested products [15,25]. The yellowing of *P. eryngii* is a bacteriosis that manifests with small yellow or light brown spots on the pileum accompanied by water-rich elongated and coalescing areas on the stem. *P. eryngii* affected by yellowing show a setback in the growth process, turn reddish-brown, and reach the state of rot (Figure 1), which manifests in the final stages with an unpleasant and nauseating odor [23,26]. The occurrence of the infection, as well as its intensity, are influenced by particular environmental conditions, such as high humidity in the growing chambers and hot muggy winds, such as sirocco [15], which are typical of the Mediterranean climate.

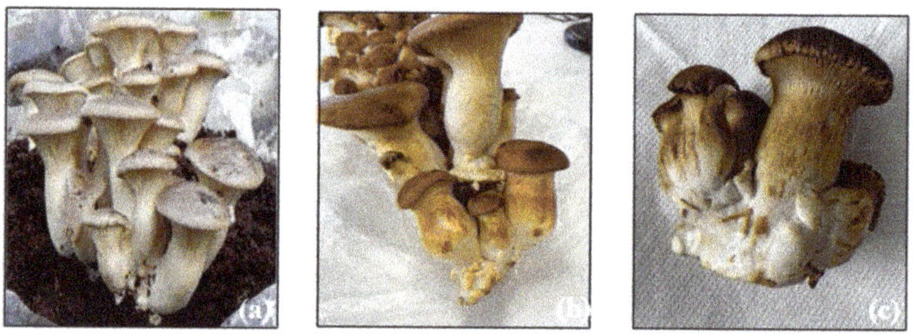

Figure 1. (**a**) *Pleurotus eryngii* without contamination; note the white stem without spots; (**b**) *P. eryngii* contaminated after a few days from first signs: appearance of rusty-red spots on the entire stem; (**c**) *P. eryngii* with bacteriosis after one week.

Innovative experiments and alternatives to the use of chemicals, potentially polluting substances, are currently being evaluated to prevent such diseases, which, with a drastic decrease in *P. eryngii* sporophores, cause considerable financial losses to producers. For example, repeated applications of white wine vinegar in fungal cultures at different concentrations have been evaluated; the acetic acid with 3% concentration has, in fact, an antimicrobial activity on *P. aeruginosa* and other bacteria [23,27]. The scientific community is focused on finding effective molecules propelling biotechnology in an eco-friendly and sustainable direction. Along this line, the present study evaluates grapefruit seed extract's antimicrobial and antibiofilm activity on several microorganisms in vitro and in vivo, focusing on preventing *Pseudomonas* spp., in particular *P. fluorescens.*, in *Pleurotus eryngii*.

Grapefruit seed extract (GSE) is a well-known plant extract with strong antibacterial and antioxidant activity [28,29]. GSE applications span from use in the food sector as a food preservative and infusion into packaging matrices [30,31] to pharmaceutics (e.g., diet supplements, wound healing, glucose and lipid blood level management, etc.) [29]. GSE contains different bioactive components, such as flavonoids, polyphenols, organic acids, and others, that are considered responsible for the antimicrobial and antioxidant activity. Different studies have investigated the mechanism of action of GSE against a wide range of microorganisms, including *Pseudomonas* spp. GSE antimicrobic activity has been attributed to the disruption of the bacterial membrane and liberation of the cytoplasmatic content [28,32–34]. In the literature, the difference in efficacy is reported to depend on the concentration of polyphenols, especially citrus flavonoids, such as naringin [28,35,36]. Naringin (5,7,40-trihydroxyflavanone-7-O-neohexperidoside) is a flavanone glycoside, soluble in water and metabolized by intestinal flora into its aglycone derivative, naringenin [37]. It is a molecule found in several fruits, such as grapes and tomatoes, and especially in citrus fruits, to which it attributes a characteristic bitter taste [38–40]. Naringin, being biologically active, expresses several beneficial proprieties in vitro and in vivo, such as anti-cancer and antioxidant activity [41–43]. In addition, several models show its role in decreasing the concentration of blood lipids, impacting hypertension, hyperlipidemia, and obesity conditions [39,42,44–46]. The possible role of naringin and naringenin as nutraceuticals against several conditions affecting human health is currently under study [47–50], opening a promising line of research relating not only to food supplements but also to food fortification.

Due to GSE's functional properties, there has been a growing interest in using GSE in the food sector as an alternative to chemical treatments over the last decade. Several studies have investigated the efficacy and safety of GSE application on foods, whether directly, in combination with coating materials, or incorporated into edible films [29,51]. However, to our knowledge, none of these studies have evaluated the effectiveness of GSE in preventing *P. eryngii* yellowing.

2. Materials and Methods

From September 2022 to June 2023, specialized technical staff from the Hygiene Laboratory of Cagliari University (accredited according to UNI EN ISO IEC 17025:2017 [52]) carried out several inspections on a mushroom farm growing *P. eryngii* var. *eryngii* in the south of Sardinia (Italy). During the inspections, the technical staff evaluated mushroom contamination while sampling *P. eryngii* specimens (according to UNI EN ISO 7218:2013 [53]); eight basidiomata presenting signs of yellowing and ten without signs of yellowing were collected. Additionally, the producers assisted the technical staff in evaluating the quality of mushrooms based on their appearance, size, color, and texture. The sampled basidiomata were placed in refrigerators at 8 °C, transported to the Hygiene Laboratory of Cagliari University, and analyzed for *Pseudomonas aeruginosa*. Then, the antimicrobial and antibiofilm activity of the GSE was evaluated on *Pseudomonas* spp. and other microorganisms. The field experiment took place after in vitro testing, and GSE was atomized on the mushroom growth substrate surface before and after sprouting (see below for a detailed description of each step).

2.1. Grapefruit Seed Extract

The grapefruit seed extract (100 mg; DSLD (Dietary Supplement Label Database): 296039) used for this experiment was a dietary supplement produced by the certified company Solaray, est. 1973 (Park City, UT, USA) and falls under the FDA regulations for production, marketing, and sale. The company policy includes testing at three different stages during the manufacturing process (suppliers: raw materials; factory: at intake and before bottling) and up to six different quality tests of the product, including microbial testing and contaminant testing to guarantee the absence of potentially harmful chemicals and pesticides. The company facility is 455-2 GMP (Good Manufacturing Practices) certified, and its laboratory is ISO 17025:2017 [52] certified. The GSE was formulated with substances from natural origin only. The other ingredients declared by the producer were vegetable glycerin and natural grapefruit flavor, as stated on the label.

Below is a description of the methods we used to analyze the GSE to ensure the absence of synthetic compounds and measure flavonoid compounds: ultra-high performance liquid chromatography–quadrupole time-of-flight mass spectrometry analysis (UHPLC–Qtof-MS) and gas chromatography–mass spectrometry (GC–MS).

UHPLC–Qtof-MS analysis: To 10 μL of extract, 990 μL of methanol was added. The diluted samples were analysed with a 6560 Q-TOF/MS coupled with an Agilent 1290 Infinity II LC system (Agilent Technologies, Palo Alto, CA, USA). An aliquot of 2.0 μL from each sample was injected in a BEH Amide, 1.7 μm, 150 mm × 2.1 mm chromatographic column (Waters Corporation, Milford, MA, USA). The mobile phase consisted of water containing 0.1% formic acid (A) and a mixture of acetonitrile:methanol (9:1) with 0.1% formic acid (B), flowing at a rate of 0.150 mL/min. This phase was applied using the following linear gradient elution, starting with 85% A and 15% B for 0 min, followed by a gradual increase to 21% B over the next 8 min, then an increase to 40% B over the next 4 min, further to 60% B over the next 7 min, and finally to 90% B over the final 2 min. The mass spectrometric analysis was performed with a QToF-MS equipped with an ESI source with Jet Stream technology using the following parameters: drying gas (N2) flow rate, 11.0 L/min; drying gas temperature, 250 °C; nebulizer, 35 psig; sheath gas temperature, 325 °C; sheath gas flow, 10 L/min; capillary, 3500 V; skimmer, 65 V; Oct RF V, 800 V; fragmentor voltage, 100 V. Each sample was analysed in the mass range of m/z 100–1500. During the HPLC–Qtof-MS analysis, the standard compounds (Rutin (Sigma-Aldrich, Milan, Italy, CAS 153-18-4); Naringin (Sigma-Aldrich, Milan, Italy, CAS 10236-47-2; Hesperidin (Sigma-Aldrich, Milan, Italy, CAS 520-26-3; Naringenin (Sigma-Aldrich, Milan, Italy, CAS 67604-48-2)) were co-chromatographed with the samples under the same analytical conditions for identification purposes.

GC–MS analysis: To 10 μL of extract was added 90 μL of BSTFA (N,O-Bis(trimethylsilyl)-trifluoroacetamide), and the mixture was placed in an oven for 15 min. After derivatization, each sample was diluted in a 1:2 ratio with hexane. A Trace 1300 gas chromatograph coupled with a TSQ 9000 triple quadrupol mass spectrometer (Thermo Fisher Scientific Inc., Waltham, MA, USA) was used for the sample analysis. The volume injection was of 1 μL in the splitless mode. The injector temperature was set at 200 °C. The gas flow rate was 1 mL/min. The column was a DB5-MS (0.25 μm, 30 m × 0.25 mm) (J&W scientific, Folsom, CA, USA). Initially, the oven temperature was set at 50 °C and held for 10 min. Then, it was increased to 300 at 10 °C/min and held at 300 °C for 10 min. Ions were recorded at 1.6 scan/s in the mass range m/z 50–550. Confirmation of sample components was performed by (a) comparison of their relative retention times and mass fragmentation with those of pure standards and (b) computer matching against NIST, as well as retention indices as calculated according to Kovats for C7–C40 n-alkane standard mixtures in dichloromethane (Sigma-Aldrich, Milan, Italy; product ID: 49452-U, Lot. LRAC3116).

GSE preparation for analysis: The commercial GSE had a density of 0.0033 g/mL (3333.33 μg/mL). GSE was diluted with TBS (tryptic soy broth), and different concentrations were evaluated empirically during the preparation of the laboratory tests. The target concentration identified was 52,000 μg/mL.

2.2. Culture Investigations

The presence and concentration of *Pseudomonas aeruginosa* were measured (UNI EN ISO 16266:2006 [54]) in the samples (n = 8) presenting symptoms of yellowing disease and in the samples (n = 10) without yellowing. The strains ATCC *P. aeruginosa* ATCC 9027 and *P. fluorescens* (wild type) were used as reference microorganisms.

2.3. Antimicrobial Activity—Preliminary Assay

The Kirby–Bauer disk diffusion assay was used as a preliminary assay for evaluating the antimicrobial activity of grapefruit seed extract against several microorganisms: *P. aeruginosa* ATCC 9027, *P. fluorescens* (wild type), *S. aureus* ATCC 6538, *S. aureus* MRSA wild type, and *C. albicans* ATCC 2091. In the case of *Pseudomonas* spp., the microbial suspension was prepared with *P. fluorescens* and *P. aeruginosa* previously isolated from the specimens with bacterial disease. Each suspension presented a corresponding concentration of 1 McF (OD600). In addition, Muller–Hinton medium (agar 17.0 g/L, beef infusion solids 2.0 g/L, casein hydrolysate 17.5 g/L, starch 1.5 g/L) was used in standard-diameter

Petri dishes (90 mm), with a medium thickness of 4–5 mm. Then, a sterile swab dipped into the suspension was used for surface seeding, repeating this operation four times by rotating the plate 90 degrees each time. After the inoculum absorption, three 6 mm paper discs impregnated with the grapefruit seed extract were positioned on the growth medium. The plates were incubated at 37 °C for 24 h ± 2. Readings were taken the following day by measuring the diameter (mm) of the inhibition halos and obtaining values comparable to standard values per microbial strain, indicating them as sensitive, intermediate, or resistant.

2.4. Inhibitory and Bactericidal Concentration–Broth Microdilution Susceptibility Test

P. aeruginosa ATCC 9027 and *P. fluorescens* (wild type) inocula were placed in 96-well microplates with a concentration of 100,000 CFU/mL in nutrient broth with the test substance at different concentrations. A set of positive and negative controls was run in triplicate with the samples to ensure reliability. Negative controls consisted of culture broth and GSE to test sterility; positive controls were placed in the microwells with the culture broth and without any treatment to verify inoculum vitality. The same technique was used for *S. aureus* ATCC 6538, *S. aureus* MRSA wild type, *Escherichia coli* 25922, and *C. albicans* ATCC 2091. The MIC (minimum inhibitory concentration) and MBC (minimum bactericidal concentration) were determined [55]. After 24 h of incubation, the results were read by observing the formation of a pellet; MIC was read (first microwell with no growth), and MBC was determined (first well with no growth after the transfer of a given volume into universal agarized medium).

2.5. Antibiofilm Activity

A crystal violet assay [56] was applied to evaluate the antibiofilm activity of grapefruit seed extract on the microorganisms targeted. The microbial strains were revitalized in TSB (tryptic soy broth) for 24 h at 37 °C. Then, 100 μL of the microbial suspension, equal to 0.5 McF with OD600, was transferred to three wells of the microplate with supplementation of 10 μL of 1% glucose solution and incubated at 37 °C for 24 h without agitation. Positive and negative controls and the samples were run in triplicate. For positive controls, 100 μL of microbial suspension was prepared by adding 10 μL of the glucosate solution and 100 μL of 1% DMSO. For negative controls, 100 μL of TBS, 100 μL of grapefruit seed extract at the target concentration, and 10 μL of 1% gluconate solution were added. In the next 24 h of incubation, TBS was withdrawn with a micropipette and replaced with TBS-containing grapefruit seed extract at the target concentration. The controls did not undergo any treatment. After 24 h of incubation at 37 °C, the suspended medium was removed from the treated and positive controls in all microwells, and the formed biofilms were subjected to three washes with 300 μL of 0.01 mol phosphate-buffered saline (PBS, pH = 7.4) to remove weakly bound cells. It was then allowed to dry for one hour inside a thermostat at 37 °C. In the second step, the cells, bound on the surface, were fixed with 200 μL of methanol for 20 min. Excess methanol was removed and allowed to dry for 24 h. Next, staining with 200 μL of 2% Hucker's Crystal Violet for 15 min was performed. The biofilm thus impregnated was washed three times with 300 μL sterile deionized water to remove the unbound dye and then dried at room temperature for 30 min. Then, 200 μL of the biofilm-bound crystal violet was dissolved in 33% glacial acetic acid, and absorbance was measured at 570 nm (Cary 60 UV–Vis Spectrophotometer). The percentage of biofilm eradication was calculated using the following formula (*O.D.*: optical density):

$$\%\ of\ eredication = \frac{O.D.\ positive\ control - O.D.\ treated}{O.D.\ positive\ control} \times 100$$

2.6. Field Experiment

The field experiment was conducted at the mushroom farm from September–October 2023. The external climatic conditions were characterized by temperatures ranging from 19 to 29 °C in September and 16 to 27 °C in October; the average humidity was 67% in

September and 72% in October. An optimal microenvironment for the growth of cardoncello was guaranteed by a breathable cloth cover and a wooden support one meter above the ground on which the pre-inoculated mushroom substrate blocks were placed. In addition, to prevent fungal diseases caused by vectors, netting and several traps were placed on the upper arch. The extract was sprayed on mushroom specimens treated (n = 50) at a concentration of 52,000 μg/mL; a control group was sprayed with sterile water (n = 90). The treatment lasted about 20 days for each mushroom block from before the sprouting phase to harvesting. A total of 100 mL of extract diluted in sterile saline solution was sprayed (distance between 20 and 30 cm) on the soil and treated mushrooms (about 5 mL per day diluted in 15 mL of sterile saline solution). The culture was monitored from the early stages of development to the adult stage, recording the number of specimens with bacterial disease both in treated and control groups. The spent substrate from mushroom cultivation was disposed of after the harvesting cycle.

3. Results

3.1. Preliminary Inspections

Among the basidiomata sampled during the inspections on the mushroom farm that took place from September 2022 to June 2023, *Pseudomonas aeruginosa* was found only in the ones affected by yellowing. The features of the healthy mushrooms defined by the producers during routine screenings, namely, appearance, size, and texture, were displayed to the technical staff and considered the standard for the final evaluation of treated *P. eryngii*.

3.2. GSE Analysis

The flavonoid compounds measured in the methanolic phase of GSE through UHPLC–Qtof-MS analysis are reported in Table 1.

Table 1. Quantitative analysis and accurate mass for flavonoids determined using UHPLC–Qtof-MS. RT retention time; m/z ratio of mass to charge; Δ (ppm) mass error of an assignment when comparing a theoretical m/z and the experimentally observed m/z.

Flavonoids	RT (min)	Formula	*m/z* *Experimental*	*m/z* *Theoretical*	Δ (ppm)	Major Fragmentaion	mg/L
Rutin	6.49	$C_{27}H_{30}O_{16}$	611.1606	611.1607	−0.16	303.0496	99.03
Naringin	9.75	$C_{27}H_{32}O_{14}$	581.1864	581.1865	−0.17	273.0752	46.57
Hesperidin	9.25	$C_{28}H_{34}O_{15}$	611.1967	611.1970	−0.49	303.0857	45.76
Neohesperidin	16.11	$C_{28}H_{34}O_{15}$	611.1972	611.1970	0.33	303.0874	168.29
Naringenin	19.23	$C_{15}H_{12}O_5$	273.0758	273.0757	0.36	287.0904	3515.05

The polar metabolites detected in the GSE by GC–MS analysis are shown in Table 2. No synthetic compounds were detected. The Kovats indexes calculated for the different compounds are reported in Table 2 in comparison with the Kovats indexes reported in the NIST (National Institute of Standards and Technology) database.

For HPLC–Qtof-MS and GC–MS chromatograms of GSE see Supplementary Figures S1 and S2.

3.3. In Vitro Analysis

The Kirby–Bauer disk diffusion assay demonstrated the antimicrobial activity of GSE against *P. aeruginosa* ATCC 9027, *P. fluorescens* wild type, *S. aureus* ATCC 6538, *S. aureus* MRSA wild type, *E. coli* ATCC 8739, and *C. albicans* ATCC 2091. The inhibition halos for *P. aeruginosa* ATCC 9027 and *P. fluorescens* wild type were 8 mm and 22 mm, respectively. The inhibition halo diameters for each of the microorganisms tested are shown in Figure 2.

Table 2. Percentage composition of polar metabolites detected in the GSE by GC–MS analysis. RT retention time. NA, not available.

RT	Compounds	%	Calculated Kováts Retention Indexes	Theoretical Kováts Retention Indexes
18.944	Lactic Acid, 2TMS derivative	14.69208	1055	1057
21.066	Diacetin, TMS	3.489502	1105	NA
22.808	Glycerol, 3TMS derivative	56.27466	1279	1282
26.651	Diacetin, TMS	5.599487	1105	NA
26.952	1,2,3-Butanetriol-3TMS	10.27547	1285	1286
27.853	Butane, 1,2,3-tris(trimethylsiloxy)-TMS	0.131453	1285	1285
28.013	Monocaproin, 2TMS	0.13052	1886	1886
28.613	Diglycerol, 4TMS derivative	0.3968	1902	NA
30.696	Ascorbic acid, 4TMS derivative	5.920048	1968	1971
31.777	9-Octadecenenitrile	0.23135	2315	NA
33.338	Citric acid, 4TMS derivative	1.752973	2618	2622
34.64	Oleamide, TMS derivative	1.105656	2763	2765

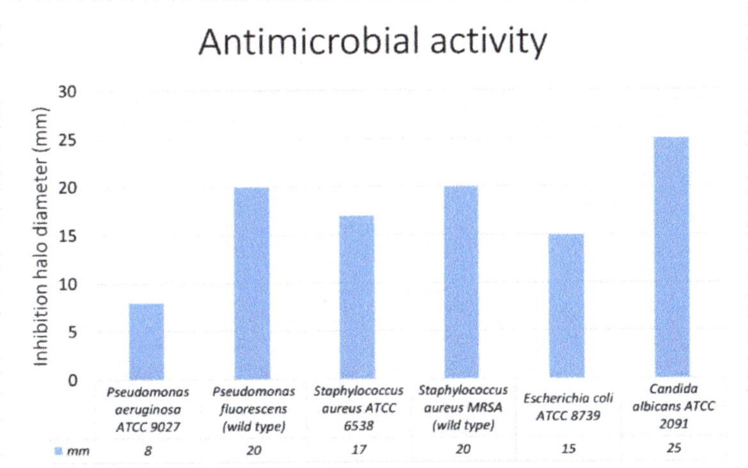

Figure 2. Antimicrobial activity-inhibition halo histograms. Diameters of inhibition halos (Kirby–Bauer disk diffusion assay) related to the grapefruit seed extract action are expressed in mm.

The MIC and MBC values established through the broth microdilution susceptibility test for *P. aeruginosa* ATCC 9027, *P. fluorescens* wild type, *S. aureus* ATCC 6538, *S. aureus* MRSA (wild type), and *E. coli* ATCC 8739 are shown in Table 3.

The percentages of inhibition established through the crystal violet assay for *P. aeruginosa* ATCC 9027 and *P. fluorescens* wild type were 1.15% and 0.15%, respectively. The percentages of inhibition for each of the microorganisms tested are shown in Figure 3.

There was no evidence of bacteriostatic and antibiofilm activity on *C. albicans* ATCC 2091.

Table 3. Minimum inhibitory concentration (MIC) and minimum bactericidal concentration (MBC) values established through the broth microdilution susceptibility test for each of the microorganisms tested.

Target Microorganisms	MIC	MBC
Staphylococcus aureus ATCC 6538	162,5 μg/mL	650 μg/mL
Staphylococcus aureus MRSA wild type	325 μg/mL	325 μg/mL
Pseudomonas aeruginosa ATCC 9027 *Pseudomonas fluorescens wild type*	650 μg/mL	1300 μg/mL
Escherichia coli 8739	650 μg/mL	1300 μg/mL
Candida albicans ATCC 2091	-	-

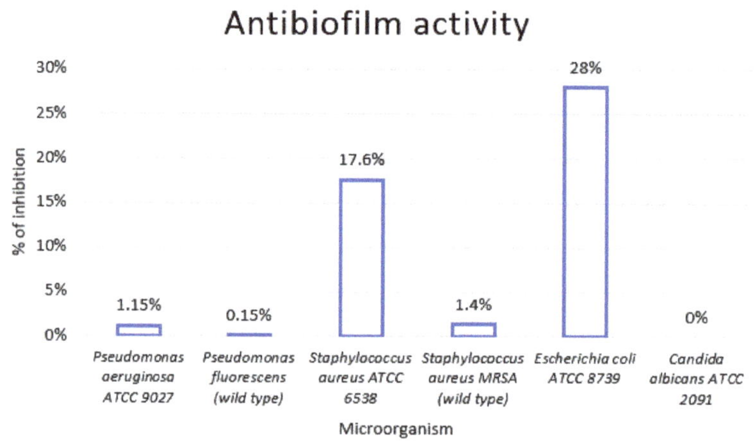

Figure 3. Antibiofilm activity—% of inhibition histograms. Percentages obtained with the crystal violet assay.

3.4. In Vivo Analysis

During the field experiment, none of the *P. eryngii* treated with the grapefruit seed extract manifested bacteriosis, while three cases were observed in the controls. Field testing of the substance demonstrated protective efficacy in preventing contamination and subsequent bacterial debasement. Furthermore, according to manufacturers, the treated mushrooms were of the same quality as the healthy untreated ones, presenting the same appearance, size, color, and a slightly softer stem.

4. Discussion

The bacteriosis of cardoncello manifests as the appearance of reddish-brown cankers extending from the cap to the stem, inducing a change in the color and organoleptic characteristics of the product [23,57,58]. The Pseudomonadaceae family has been identified as a major culprit in the etiology of such bacterial diseases [58,59], particularly the species *aeruginosa*, pathogenic to humans, and *fluorescens*, which may cause acute opportunistic clinical manifestations of bacteremia in individuals with compromised immune systems [60]. Bacterial contamination by Pseudomonadaceae may both directly and indirectly harm the consumer since the presence of lesions on the fungus' surface promotes contamination by other species of microorganisms [61,62]. Furthermore, a decrease in the number and quality of the fungi grown represents an important risk of economic loss for producers. In this context, our study is the first to evaluate the efficacy of the GSE against *P. eryngii* bacterial blight and to suggest its possible use for preventive purposes. Nowadays, great efforts are being devoted to finding innovative and natural alternatives to chemicals to prevent alterations in food products and guarantee food safety and maximum productivity. Pure GSE is among the several bioactive compounds originating from natural sources, and it is widely accepted and recognized as safe for direct or indirect use in food. GSE in its pure form is non-toxic and "chemical-free" (marketing term), namely, safe and environmentally friendly, containing natural ingredients only. Therefore, the use of pure GSE on food or food matrices is not expected to harm consumers and the environment. However, some commercial GSEs contain synthetic compounds [63,64], such as benzethonium chloride and benzalkonium chloride, that may derive from the conversion of unstable polyphenols during GSE extraction and purification. These compounds exhibit potent antimicrobial activity and some toxicity at high concentrations [51]. For these reasons, in the present study, we used a GSE free from these synthetic compounds. This approach enabled a reliable evaluation of GSE efficacy in protecting *P. eryngii* crops from bacterial contamination while considering safety aspects and potential environmental impacts. Therefore, we can state that the antimicrobial activity exhibited by the GSE used in this study is attributable to its natural content in polyphenols, especially flavonoids such as naringin.

The Kirby–Bauer disk diffusion assay demonstrated that the investigated GSE was active against all Gram-positive and Gram-negative bacteria, as well as *C. albicans*. GSE exhibited the largest zones of inhibition for *C. albicans*, *P. fluorescens*, and *Staphylococcus MRSA*. The crystal violet assay showed that GSE exerted antibiofilm activity on all the microorganisms tested except for *C. albicans*. In terms of percentage of inhibition, a certain variability was observed, reflecting, in our opinion, the complexity of the biofilm simulated in vitro. These findings are consistent with the literature concerning *P. aeruginosa* spp., *S. aureus* MRSA, and *E. coli* [29,32,65–67]. In the literature, the mechanism of GSE antimicrobial activity has been attributed to the disruption of the bacterial membrane and liberation of the cytoplasmatic content [33]. The antibiofilm effect of GSE on *S. aureus* and *E. coli* has been attributed to changes in the exopolysaccharide production rate and mobility, as well as changes in hydrophobicity in *E. coli* only [68].

The field experiment demonstrated that spraying GSE twice a day from before the sprouting phase to harvesting can prevent the growth of *P. fluorescens* and *P. aeruginosa* during the cultivation of cardoncello, which is particularly critical under several environmental circumstances. The extract antimicrobial effects were not affected by the 20–30 cm distance required for the application, suggesting that GSE is suitable as a spray. The application of GSE as a measure of prevention of bacterial blight occurrence is worth further investigation not only on *P. eryngii* but also on other foodstuffs. Furthermore, given the current attention to the nutraceutical use of naringin and other flavonoids, the potential added value of foods supplemented with GSE deserves consideration.

Limitations and Future Directions

Since this work is the first to evaluate the effectiveness of GSE in preventing *P. eryngii* yellowing, it should be considered a pilot study. It has several limitations, such as the small number of basidiomata we were able to treat due to economic constraints of the producers. We did not conduct a challenge test with *Pseudomonas* spp. and analyzed the quality of treated mushrooms in terms of appearance, shape, color, and texture only.

The next steps to validate this natural control strategy involve (i) larger scale experimentation with at least three to five biological replicates (including 25–50 mushrooms each), which is also essential to confirm further that the slight changes in the texture due to reiterate nebulization do not affect the final quality of the products; (ii) deep evaluation of treated mushroom quality, including polyphenol oxidase (PPO) activity, sensory analysis, and chemical characterization; and (iii) conducting a challenge test with *Pseudomonas* spp. Furthermore, in the following stages, we plan to compare GSE with chemical compounds and study its mechanism of action on *Pseudomonas* spp. to gain a deeper understanding of GSE characteristics and levels of effectiveness. Moreover, considering that naringin's antioxidant activity is affected by light and high temperatures (>100 °C) [69,70], we plan to investigate the naringin concentration and bioavailability in the final product treated with GSE.

5. Conclusions

P. eryngii yellowing is a disease that can occur in all the basiodiomata development phases, from sprouting to commercial maturation. It can bring huge economic damage to producers due to its rapid spread in *P. eryngii* cultivations and the current lack of standardized control measures. The present work contributes to the knowledge of the antimicrobial efficacy of natural GSE and provides valuable input to the branch of research aimed at preventing and controlling *P. eryngii* yellowing. Our findings support the use of GSE to protect *P. eryngii* crops from bacterial contamination, particularly from *Pseudomonas* spp., which have often been identified as responsible for the yellowing. Atomizing GSE pre- and post-sprouting represents a promising eco-friendly and sustainable control strategy alternative to chemical treatments to ensure food safety and prevent financial losses due to *P. eryngii* spoilage. In addition, due to its content in naringin and other bioactive components, GSE opens new horizons regarding its use as a nutraceutical in food fortification and supplementation. Due to the limitations mentioned above, our preliminary findings, although encouraging, require larger and deeper studies to be further validated.

Supplementary Materials: The following supporting information can be downloaded at: https://www.mdpi.com/article/10.3390/foods13081161/s1. Figure S1: HPLC-Qtof/MS chromatogram of GSE; Figure S2: GC-MS chromatogram of GSE.

Author Contributions: Conceptualization, M.M., S.M.P. and V.C.; data curation, M.M., S.M.P. and V.C.; formal analysis, A.S., L.M., C.M., A.B. and V.C.; investigation, M.M., A.S., C.M. and V.C.; methodology, L.M., A.S., C.M., A.B. and V.C.; resources, V.C.; validation, V.C.; writing—original draft, M.M., S.M.P. and V.C.; writing—review and editing, S.M.P., C.M. and V.C. All authors have read and agreed to the published version of the manuscript.

Funding: This research received no external funding.

Institutional Review Board Statement: Not applicable.

Informed Consent Statement: Not applicable.

Data Availability Statement: The original contributions presented in the study are included in the article, further inquiries can be directed to the corresponding author.

Conflicts of Interest: The authors declare no conflicts of interest.

References

1. Venturella, G.; Palazzolo, E.; Saiano, F.; Gargano, M.L. Notes on a New Productive Strain of King Oyster Mushroom, Pleurotus Eryngii (Higher Basidiomycetes), a Prized Italian Culinary-Medicinal Mushroom. *Int. J. Med. Mushrooms* **2015**, *17*, 199–206. [CrossRef]
2. Carlavilla, J.R.; Manjón, J.L. The King Oyster Mushroom Pleurotus Eryngii Behaves as a Necrotrophic Pathogen of Eryngium Campestre. *Ital. J. Mycol.* **2023**, *52*, 22–31. [CrossRef]
3. Lv, S.; Zhu, X.; Liu, Z.; Hu, L.; Xu, D.; Chitrakar, B.; Mo, H.; Li, H. Edible Pleurotus Eryngii Papery Food Prepared by Papermaking Process. *Foods* **2022**, *11*, 3514. [CrossRef] [PubMed]
4. Teniou, S.; Bensegueni, A.; Hybertson, B.M.; Gao, B.; Bose, S.K.; McCord, J.M.; Chovelon, B.; Bensouici, C.; Boumendjel, A.; Hininger-Favier, I. Biodriven Investigation of the Wild Edible Mushroom Pleurotus Eryngii Revealing Unique Properties as Functional Food. *J. Funct. Foods* **2022**, *89*, 104965. [CrossRef]
5. Stajić, M.; Vukojević, J.; Duletić-Laušević, S. Biology of Pleurotus Eryngii and Role in Biotechnological Processes: A Review. *Crit. Rev. Biotechnol.* **2009**, *29*, 55–66. [CrossRef]
6. Yu, A.; Ji, Y.; Ma, G.; Xu, J.; Hu, Q. Identification and Preparation of Selenium-Containing Peptides from Selenium-Enriched Pleurotus Eryngii and Their Protective Effect on Lead-Induced Oxidative Damage in NCTC1469 Hepatocytes. *J. Sci. Food Agric.* **2023**, *103*, 4522–4534. [CrossRef] [PubMed]
7. Yuan, B.; Ma, N.; Zhao, L.; Zhao, E.; Gao, Z.; Wang, W.; Song, M.; Zhang, G.; Hu, Q.; Xiao, H. In Vitro and in Vivo Inhibitory Effects of a Pleurotus Eryngii Protein on Colon Cancer Cells. *Food Funct.* **2017**, *8*, 3553–3562. [CrossRef] [PubMed]
8. Zhang, B.; Li, Y.; Zhang, F.; Linhardt, R.J.; Zeng, G.; Zhang, A. Extraction, Structure and Bioactivities of the Polysaccharides from Pleurotus Eryngii: A Review. *Int. J. Biol. Macromol.* **2020**, *150*, 1342–1347. [CrossRef] [PubMed]

9. Chen, J.; Yong, Y.; Xing, M.; Gu, Y.; Zhang, Z.; Zhang, S.; Lu, L. Characterization of Polysaccharides with Marked Inhibitory Effect on Lipid Accumulation in Pleurotus Eryngii. *Carbohydr. Polym.* **2013**, *97*, 604–613. [CrossRef] [PubMed]
10. Alam, N.; Yoon, K.N.; Lee, J.S.; Cho, H.J.; Shim, M.J.; Lee, T.S. Dietary Effect of Pleurotus Eryngii on Biochemical Function and Histology in Hypercholesterolemic Rats. *Saudi J. Biol. Sci.* **2011**, *18*, 403–409. [CrossRef]
11. Amerikanou, C.; Tagkouli, D.; Tsiaka, T.; Lantzouraki, D.Z.; Karavoltsos, S.; Sakellari, A.; Kleftaki, S.A.; Koutrotsios, G.; Giannou, V.; Zervakis, G.I.; et al. Pleurotus Eryngii Chips—Chemical Characterization and Nutritional Value of an Innovative Healthy Snack. *Foods* **2023**, *12*, 353. [CrossRef] [PubMed]
12. Gong, P.; Long, H.; Guo, Y.; Wang, S.; Chen, F.; Chen, X. Isolation, Structural Characterization, and Hypoglycemic Activities In Vitro of Polysaccharides from Pleurotus Eryngii. *Molecules* **2022**, *27*, 7140. [CrossRef] [PubMed]
13. Manzi, P.; Marconi, S.; Aguzzi, A.; Pizzoferrato, L. Commercial Mushrooms: Nutritional Quality and Effect of Cooking. *Food Chem.* **2004**, *84*, 201–206. [CrossRef]
14. Hess, J.M.; Jonnalagadda, S.S.; Slavin, J.L. What Is a Snack, Why Do We Snack, and How Can We Choose Better Snacks? A Review of the Definitions of Snacking, Motivations to Snack, Contributions to Dietary Intake, and Recommendations for Improvement. *Adv. Nutr.* **2016**, *7*, 466–475. [CrossRef] [PubMed]
15. Bruno, G.L.; Rana, G.L.; Sermani, S.; Scarola, L.; Cariddi, C. Control of Bacterial Yellowing of Cardoncello Mushroom Pleurotus Eryngii Using Acetic or Hydrochloric Acid Solutions. *Crop Prot.* **2013**, *50*, 24–29. [CrossRef]
16. Gea, F.J.; Carrasco, J.; Suz, L.M.; Navarro, M.J. Characterization and Pathogenicity of Cladobotryum Mycophilum in Spanish Pleurotus Eryngii Mushroom Crops and Its Sensitivity to Fungicides. *Eur. J. Plant Pathol.* **2017**, *147*, 129–139. [CrossRef]
17. Kim, M.K.; Seuk, S.W.; Lee, Y.H.; Kim, H.R.; Cho, K.M. Fungicide Sensitivity and Characterization of Cobweb Disease on a Pleurotus Eryngii Mushroom Crop Caused by Cladobotryum Mycophilum. *Plant Pathol. J.* **2014**, *30*, 82. [CrossRef] [PubMed]
18. Carrasco, J.; Navarro, M.J.; Santos, M.; Diánez, F.; Gea, F.J. Incidence, Identification and Pathogenicity of Cladobotryum Mycophilum, Causal Agent of Cobweb Disease on Agaricus Bisporus Mushroom Crops in Spain. *Ann. Appl. Biol.* **2016**, *168*, 214–224. [CrossRef]
19. Chen, J.T.; Huang, J.W. A Semiselective Medium for Detecting Gliocladium Roseum, the Causal Agent of King Oyster Mushroom Brown Spot. *Plant Pathol. Bull.* **2004**, *13*, 107–116.
20. Kredics, L.; Kocsubé, S.; Nagy, L.; Komoń-Zelazowska, M.; Manczinger, L.; Sajben, E.; Nagy, A.; Vágvölgyi, C.; Kubicek, C.P.; Druzhinina, I.S.; et al. Molecular Identification of Trichoderma Species Associated with Pleurotus Ostreatus and Natural Substrates of the Oyster Mushroom. *FEMS Microbiol. Lett.* **2009**, *300*, 58–67. [CrossRef]
21. González, A.J.; Gea, F.J.; Navarro, M.J.; Fernández, A.M. Identification and RAPD-Typing of Ewingella Americana on Cultivated Mushrooms in Castilla-La Mancha, Spain. *Eur. J. Plant Pathol.* **2012**, *133*, 517–522. [CrossRef]
22. Reyes, J.E.; Venturini, M.E.; Oria, R.; Blanco, D. Prevalence of Ewingella Americana in Retail Fresh Cultivated Mushrooms (Agaricus Bisporus, Lentinula Edodes and Pleurotus Ostreatus) in Zaragoza (Spain). *FEMS Microbiol. Ecol.* **2004**, *47*, 291–296. [CrossRef] [PubMed]
23. Bruno, G.L.; De Corato, U.; Rana, G.L.; De Luca, P.; Pipoli, V.; Lops, R.; Scarola, L.; Mannerucci, F.; Piscitelli, L.; Cariddi, C. Suppressiveness of White Vinegar and Steam-Exploded Liquid Waste against the Causal Agents of Pleurotus Eryngii Yellowing. *Crop Prot.* **2015**, *70*, 61–69. [CrossRef]
24. Iacobellis, N.S.; Lo Cantore, P. Pseudomonas "Reactans" a New Pathogen of Cultivated Mushrooms. In *Pseudomonas Syringae and Related Pathogens*; Springer: Dordrecht, The Netherlands, 2003.
25. Soler-Rivas, C.; Jolivet, S.; Arpin, N.; Olivier, J.M.; Wichers, H.J. Biochemical and Physiological Aspects of Brown Blotch Disease of Agaricus Bisporus. *FEMS Microbiol. Rev.* **1999**, *23*, 591–614. [CrossRef] [PubMed]
26. Bellettini, M.B.; Bellettini, S.; Fiorda, F.A.; Pedro, A.C.; Bach, F.; Fabela-Morón, M.F.; Hoffmann-Ribani, R. Diseases and Pests Noxious to Pleurotus Spp. Mushroom Crops. *Rev. Argent. Microbiol.* **2018**, *50*, 216–226. [CrossRef] [PubMed]
27. Ryssel, H.; Kloeters, O.; Germann, G.; Schäfer, T.; Wiedemann, G.; Oehlbauer, M. The Antimicrobial Effect of Acetic Acid-An Alternative to Common Local Antiseptics? *Burns* **2009**, *35*, 695–700. [CrossRef] [PubMed]
28. Cvetnić, Z.; Vladimir-Knežević, S. Antimicrobial Activity of Grapefruit Seed and Pulp Ethanolic Extract. *Acta Pharm.* **2004**, *54*, 243–250.
29. Kim, T.; Kim, J.H.; Oh, S.W. Grapefruit Seed Extract as a Natural Food Antimicrobial: A Review. *Food Bioprocess Technol.* **2021**, *14*, 626–633. [CrossRef]
30. Yun, D.; Liu, J. Recent Advances on the Development of Food Packaging Films Based on Citrus Processing Wastes: A Review. *J. Agric. Food Res.* **2022**, *9*, 100316. [CrossRef]
31. Zayed, A.; Badawy, M.T.; Farag, M.A. Valorization and Extraction Optimization of Citrus Seeds for Food and Functional Food Applications. *Food Chem.* **2021**, *355*, 129609. [CrossRef]
32. Reagor, L.; Gusman, J.; McCoy, L.; Carino, E.; Heggers, J.P. The Effectiveness of Processed Grapefruit-Seed Extract as an Antibacterial Agent: I. An in Vitro Agar Assay. *J. Altern. Complement. Med.* **2002**, *8*, 325–332. [CrossRef]
33. Heggers, J.P.; Cottingham, J.; Gusman, J.; Reagor, L.; McCoy, L.; Carino, E.; Cox, R.; Zhao, J.G. The Effectiveness of Processed Grapefruit-Seed Extract as an Antibacterial Agent: II. Mechanism of Action and in Vitro Toxicity. *J. Altern. Complement. Med.* **2002**, *8*, 333–340. [CrossRef] [PubMed]

34. Barawi, S.; Hamzah, H.; Hamasalih, R.; Mohammed, A.; Abdalrahman, B.; Abdalaziz, S. Antibacterial Mode of Action of Grapefruit Seed Extract against Local Isolates of Beta-Lactamases-Resistant Klebsiella Pneumoniae and Its Potential Application. *Int. J. Agric. Biol.* **2021**, *26*, 499–508. [CrossRef]
35. Çiçek Polat, D.; Eryilmaz, M.; Akalin, K.; Coşkun, M. Antimicrobial Activity of Grapefruit Seed. *Hacettepe Univ. J. Fac. Pharm.* **2018**, *38*, 1–3.
36. Céliz, G.; Daz, M.; Audisio, M.C. Antibacterial Activity of Naringin Derivatives against Pathogenic Strains. *J. Appl. Microbiol.* **2011**, *111*, 731–738. [CrossRef]
37. Zeng, X.; Zheng, Y.; He, Y.; Zhang, J.; Peng, W.; Su, W. Microbial Metabolism of Naringin and the Impact on Antioxidant Capacity. *Nutrients* **2022**, *14*, 3765. [CrossRef]
38. Bugianesi, R.; Catasta, G.; Spigno, P.; D'Uva, A.; Maiani, G. Naringenin from Cooked Tomato Paste Is Bioavailable in Men. *J. Nutr.* **2002**, *132*, 3349–3352. [CrossRef] [PubMed]
39. Alam, M.A.; Subhan, N.; Rahman, M.M.; Uddin, S.J.; Reza, H.M.; Sarker, S.D. Effect of Citrus Flavonoids, Naringin and Naringenin, on Metabolic Syndrome and Their Mechanisms of Action. *Adv. Nutr.* **2014**, *5*, 404–417. [CrossRef]
40. Jourdan, P.S.; McIntosh, C.A.; Mansell, R.L. Naringin Levels in Citrus Tissues: II. Quantitative Distribution of Naringin in Citrus Paradisi MacFad. *Plant Physiol.* **1985**, *77*, 903–908. [CrossRef] [PubMed]
41. Chen, R.; Qi, Q.L.; Wang, M.T.; Li, Q.Y. Therapeutic Potential of Naringin: An Overview. *Pharm. Biol.* **2016**, *54*, 3203–3210. [CrossRef] [PubMed]
42. Gorinstein, S.; Leontowicz, H.; Leontowicz, M.; Krzeminski, R.; Gralak, M.; Delgado-Licon, E.; Ayala, A.L.M.; Katrich, E.; Trakhtenberg, S. Changes in Plasma Lipid and Antioxidant Activity in Rats as a Result of Naringin and Red Grapefruit Supplementation. *J. Agric. Food Chem.* **2005**, *53*, 3223–3228. [CrossRef] [PubMed]
43. Stabrauskiene, J.; Kopustinskiene, D.M.; Lazauskas, R.; Bernatoniene, J. Naringin and Naringenin: Their Mechanisms of Action and the Potential Anticancer Activities. *Biomedicines* **2022**, *10*, 1686. [CrossRef] [PubMed]
44. Raja Kumar, S.; Mohd Ramli, E.S.; Abdul Nasir, N.A.; Ismail, N.H.M.; Mohd Fahami, N.A. Preventive Effect of Naringin on Metabolic Syndrome and Its Mechanism of Action: A Systematic Review. *Evid. Based Complement. Altern. Med.* **2019**, *2019*, 9752826. [CrossRef] [PubMed]
45. Termkwancharoen, C.; Malakul, W.; Phetrungnapha, A.; Tunsophon, S. Naringin Ameliorates Skeletal Muscle Atrophy and Improves Insulin Resistance in High-Fat-Diet-Induced Insulin Resistance in Obese Rats. *Nutrients* **2022**, *14*, 4120. [CrossRef]
46. Mir, I.A.; Tiku, A.B. Chemopreventive and Therapeutic Potential of "Naringenin," a Flavanone Present in Citrus Fruits. *Nutr. Cancer* **2015**, *67*, 27–42. [CrossRef] [PubMed]
47. Massaro, L.; Raguzzini, A.; Aiello, P.; Valencia, D.V. The Potential Role of Naringin and Naringenin as Nutraceuticals Against Metabolic Syndrome. *Endocr. Metab. Immune Disord. Drug Targets* **2022**, *23*, 428–445. [CrossRef] [PubMed]
48. Murugesan, N.; Woodard, K.; Ramaraju, R.; Greenway, F.L.; Coulter, A.A.; Rebello, C.J. Naringenin Increases Insulin Sensitivity and Metabolic Rate: A Case Study. *J. Med. Food* **2020**, *23*, 343–348. [CrossRef] [PubMed]
49. Nguyen, M.A.; Staubach, P.; Tamai, I.; Langguth, P. High-Dose Short-Term Administration of Naringin Did Not Alter Talinolol Pharmacokinetics in Humans. *Eur. J. Pharm. Sci.* **2015**, *68*, 36–42. [CrossRef]
50. Singh, S.; Sharma, A.; Monga, V.; Bhatia, R. Compendium of Naringenin: Potential Sources, Analytical Aspects, Chemistry, Nutraceutical Potentials and Pharmacological Profile. *Crit. Rev. Food Sci. Nutr.* **2022**, *63*, 8868–8899. [CrossRef] [PubMed]
51. Roy, S.; Zhang, W.; Biswas, D.; Ramakrishnan, R.; Rhim, J.W. Grapefruit Seed Extract-Added Functional Films and Coating for Active Packaging Applications: A Review. *Molecules* **2023**, *28*, 730. [CrossRef] [PubMed]
52. *ISO/IEC 17025:2017*; General Requirements for the Competence of Testing and Calibration Laboratories. International Standard Organization: Geneva, Switzerland, 2018.
53. *ISO 7218:2007/Amd 1:2013*; Microbiology of Food and Animal Feeding Stuffs General Requirements and Guidance for Microbiological Examinations Amendment 1. International Standard Organization: Geneva, Switzerland, 2014.
54. *ISO 16266:2006*; Water Quality Detection and Enumeration of Pseudomonas Aeruginosa Method by Membrane Filtration. International Standard Organization: Geneva, Switzerland, 2008.
55. Angioni, A.; Barra, A.; Cereti, E.; Barile, D.; Coïsson, J.D.; Arlorio, M.; Dessi, S.; Coroneo, V.; Cabras, P. Chemical Composition, Plant Genetic Differences, Antimicrobial and Antifungal Activity Investigation of the Essential Oil of *Rosmarinus officinalis* L. *J. Agric. Food Chem.* **2004**, *52*, 3530–3535. [CrossRef] [PubMed]
56. Stepanović, S.; Vuković, D.; Hola, V.; Di Bonaventura, G.; Djukić, S.; Ćirković, I.; Ruzicka, F. Quantification of Biofilm in Microtiter Plates: Overview of Testing Conditions and Practical Recommendations for Assessment of Biofilm Production by Staphylococci. *APMIS* **2007**, *115*, 891–899. [CrossRef] [PubMed]
57. Osdaghi, E.; Martins, S.J.; Ramos-Sepulveda, L.; Vieira, F.R.; Pecchia, J.A.; Beyer, D.M.; Bell, T.H.; Yang, Y.; Hockett, K.L.; Bull, C.T. 100 Years since Tolaas: Bacterial Blotch of Mushrooms in the 21st Century. *Plant Dis.* **2019**, *103*, 2714–2732. [CrossRef] [PubMed]
58. Gao, Q.; Liu, Y.; Xie, J.; Zhao, S.; Qin, W.; Song, Q.; Wang, S.; Rong, C. Bacterial Infection Induces Ultrastructural and Transcriptional Changes in the King Oyster Mushroom (*Pleurotus Eryngii*). *Microbiol. Spectr.* **2022**, *10*, e01445-22. [CrossRef]
59. Sajben, E.; Manczinger, L.; Nagy, A.; Kredics, L.; Vágvölgyi, C. Characterization of Pseudomonads Isolated from Decaying Sporocarps of Oyster Mushroom. *Microbiol. Res.* **2011**, *166*, 255–267. [CrossRef]

60. Scales, B.S.; Dickson, R.P.; Lipuma, J.J.; Huffnagle, G.B. Microbiology, Genomics, and Clinical Significance of the Pseudomonas Fluorescens Species Complex, an Unappreciated Colonizer of Humans. *Clin. Microbiol. Rev.* **2014**, *27*, 927–948. [CrossRef] [PubMed]
61. Lee, H.I.; Jeong, K.S.; Cha, J.S. PCR Assays for Specific and Sensitive Detection of Pseudomonas Tolaasii, the Cause of Brown Blotch Disease of Mushrooms. *Lett. Appl. Microbiol.* **2002**, *35*, 276–280. [CrossRef] [PubMed]
62. Venturini, M.E.; Reyes, J.E.; Rivera, C.S.; Oria, R.; Blanco, D. Microbiological Quality and Safety of Fresh Cultivated and Wild Mushrooms Commercialized in Spain. *Food Microbiol.* **2011**, *28*, 1492–1498. [CrossRef]
63. Takeoka, G.; Dao, L.; Wong, R.Y.; Lundin, R.; Mahoney, N. Identification of Benzethonium Chloride in Commercial Grapefruit Seed Extracts. *J. Agric. Food Chem.* **2001**, *49*, 3316–3320. [CrossRef] [PubMed]
64. Takeoka, G.R.; Dao, L.T.; Wong, R.Y.; Harden, L.A. Identification of Benzalkonium Chloride in Commercial Grapefruit Seed Extracts. *J. Agric. Food Chem.* **2005**, *53*, 7630–7636. [CrossRef]
65. Han, H.W.; Kwak, J.H.; Jang, T.S.; Knowles, J.C.; Kim, H.W.; Lee, H.H.; Lee, J.H. Grapefruit Seed Extract as a Natural Derived Antibacterial Substance against Multidrug-Resistant Bacteria. *Antibiotics* **2021**, *10*, 85. [CrossRef] [PubMed]
66. Choi, J.S.; Lee, Y.R.; Ha, Y.M.; Seo, H.J.; Kim, Y.H.; Park, S.M.; Sohn, J.H. Antibacterial Effect of Grapefruit Seed Extract (GSE) on Makgeolli-Brewing Microorganisms and Its Application in the Preservation of Fresh Makgeolli. *J. Food Sci.* **2014**, *79*, M1159–M1167. [CrossRef] [PubMed]
67. Wang, K.; Lim, P.N.; Tong, S.Y.; Thian, E.S. Development of Grapefruit Seed Extract-Loaded Poly(ε-Caprolactone)/Chitosan Films for Antimicrobial Food Packaging. *Food Packag. Shelf Life* **2019**, *22*, 100396. [CrossRef]
68. Song, Y.J.; Yu, H.H.; Kim, Y.J.; Lee, N.K.; Paik, H.D. Anti-Biofilm Activity of Grapefruit Seed Extract against Staphylococcus Aureus and Escherichia Coli. *J. Microbiol. Biotechnol.* **2019**, *29*, 1177–1183. [CrossRef]
69. Chaaban, H.; Ioannou, I.; Chebil, L.; Slimane, M.; Gérardin, C.; Paris, C.; Charbonnel, C.; Chekir, L.; Ghoul, M. Effect of Heat Processing on Thermal Stability and Antioxidant Activity of Six Flavonoids. *J. Food Process. Preserv.* **2017**, *41*, e13203. [CrossRef]
70. Ioannou, I.; M'hiri, N.; Chaaban, H.; Boudhrioua, N.M.; Ghoul, M. Effect of the Process, Temperature, Light and Oxygen on Naringin Extraction and the Evolution of Its Antioxidant Activity. *Int. J. Food Sci. Technol.* **2018**, *53*, 2754–2760. [CrossRef]

foods

MDPI

Article

Inhibition of Food-Borne Pathogen Growth and Biogenic Amine Synthesis by Spice Extracts

Ferhat Kuley [1], Nikheel Bhojraj Rathod [2], Esmeray Kuley [1], Mustafa Tahsin Yilmaz [3,*] and Fatih Ozogul [1,4,*]

[1] Department of Seafood Processing Technology, Faculty of Fisheries, University of Cukurova, Balcali, 01330 Adana, Turkey; ekuley@cu.edu.tr (E.K.)
[2] Department of Post Harvest Management of Meat, Poultry and Fish, PG Institute of Post Harvest Technology and Management, Dr. Balasaheb Sawant Konkan Krishi Vidyapeeth, Killa-Roha 402116, Maharashtra State, India; nikheelrathod310587@gmail.com
[3] Department of Industrial Engineering, Faculty of Engineering, King Abdulaziz University, 21589 Jeddah, Turkey
[4] Biotechnology Research and Application Center, Cukurova University, 01330 Adana, Turkey
* Correspondence: myilmaz@kau.edu.sa (M.T.Y.); fozogul@cu.edu.tr (F.O.)

Abstract: Food-borne pathogens and their toxins cause significant health problems in humans. Formation of biogenic amines (BAs) produced by microbial decarboxylation of amino acids in food is undesirable because it can induce toxic effects in consumers. Therefore, it is crucial to investigate the effects of natural additives with high bioactivity like spice extracts to inhibit the growth of these bacteria and the formation of BAs in food. In the present study, the antibacterial effects of diethyl ether spice (sumac, cumin, black pepper, and red pepper) extracts at doses of 1% (*w*/*v*) on Gram-positive (*Staphylococcus aureus* and *Enterococcus faecalis*) and Gram-negative (*Klebsiella pneumoniae*, *Pseudomonas aeruginosa*, *Campylobacter jejuni*, *Aeromonas hydrophila*, *Salmonella* Paratyphi A, and *Yersinia enterocolitica*) food-borne pathogen bacterial strains (FBP) were established. In addition, the accumulation of ammonia (AMN), trimethylamine (TMA), and biogenic amines (BAs) in tyrosine decarboxylase broth (TDB) was investigated by using high performance liquid chromatography (HPLC). Sumac extract exhibited the highest antibacterial potential against all FBPs, followed by cumin and peppers. AMN (570.71 mg/L) and TMA (53.66 mg/L) production were strongly inhibited by sumac extract in the levels of 55.10 mg/L for *Y. enterocolitica* and 2.76 mg/L for *A. hydrophila*, respectively. With the exception of *S. aureus*, black pepper dramatically reduced the synthesis of putrescine, serotonin, dopamine, and agmatine by FBP especially for Gram-negative ones. Furthermore, sumac extracts inhibited histamine and tyramine production by the majority of FBP. This research suggests the application of sumac extracts as natural preservatives for inhibiting the growth of FBPs and limiting the production of AMN, TMA, and BAs.

Keywords: biogenic amine; food pathogenic bacteria; inhibition; spice; food safety

1. Introduction

Due to a rise in the cases of food poisoning-related mortality, there has been a growing global demand for safe food. Food-borne pathogens (FBPs) are known to spread disease through infection, and they can also produce toxins that result in food poisoning. Gram-positive and Gram-negative bacteria are the primary cause of most illnesses and fatalities [1]. Most food-borne illness outbreaks that have been documented are linked to well-known organisms, including *Salmonella*, *Campylobacter*, Norovirus, *Listeria monocytogenes*, and *Escherichia coli* that produces Shiga toxin. *Staphylococcus aureus*, *Clostridium* species, *Bacillus cereus*, *Yersinia enterocolitica*, parasites, and other pathogens have also been shown to cause diseases on occasion [2]. Biogenic amine-related toxins have become a significant concern due to their potential to be poisonous and carcinogenic, as well as to trigger headaches, dizziness, and heart palpitations [3]. As a result of the activities of microbes during processes of decarboxylation, transamination, reducing amination, and compound degradation, poisonous nitrogenous chemicals called biological amines are produced [4]. Consequently, biogenic amines (BAs) are frequently used as a sign of the quality and safety of food.

Histamine and tyramine, the two main BAs found in foods, are among the most hazardous and extensively studied amines [4]. A strict monitoring system and regulatory limits have been established at various levels for histamine and tyramine BAs, taking into account the variations in foods and processing methods. It is known that the other BAs have a synergistic effect on raising the harmfulness of histamine and tyramine [2]. Therefore, different approaches to prevent or manage the concentration of BAs are required to enhance food safety and quality as well as human health [5]. Many practices have been approved for lower BAs, including the following: using food additives or bioactive compounds (phenolic or terpenoids), using multiple starter cultures during the fermentation process, gamma irradiation, cold storage temperatures, high-hydrostatic pressure processing (HHP), food packaging procedures, and so on [6,7]. There has been a growing interest among consumers in clean label foods, which are the foods preserved using natural antimicrobials [8]. Due to their high antioxidative and antibacterial activity, spices are among the most commonly used natural antimicrobials for food preservation.

Spices are made from various plant parts, including roots, rhizomes, stem bark, leaves, fruits, flowers, and seeds [9,10]. Foods are often flavored and colored with spices [11]. In most cases, spices are sold powdered, making them vulnerable to food fraud [12]. It has been reported that many unbranded spices readily available in the markets contain synthetic dyes; therefore, they pose a health risk to humans [13]. As an alternative to their powdered form, extracts of spices have often been used to formulate foods. In this respect, various extraction solvents are used; however, they have an impact on the bioactivity of the extracts. Spice extracts have gained a lot of attention due to their wide range of bioactivities and are generally recognized as safe (GRAS). However, there is currently little research on how spice extracts affect the synthesis of bacterial biogenic amines in various mediums. The presence of biogenic amines can be detected by using a variety of media. Using different types of media for detecting biogenic amines has its advantages and disadvantages. For example, agar plates provide a simple and cost-effective method, but they may have limited sensitivity. On the other hand, liquid media offer higher sensitivity, but they can be more time-consuming and require specialized equipment for analysis. In this respect, different broths have been used to count bacteria that produce amines to determine the presence of BAs. Furthermore, an accumulation of amines is found to be greater in Tyrosine decarboxylase broth (TDB) [14–16]. Therefore, the present study aims at assessing the effects of diethyl ether-extracted spice extract on the growth and generation of biogenic amines by Gram-positive (*Staphylococcus aureus* ATCC 29213 and *Enterococcus faecalis* ATCC 29212) and Gram-negative (*Klebsiella pneumoniae* ATCC 700603, *Pseudomonas aeruginosa* ATCC 27853, *Campylobacter jejuni* ATCC 33560, *Aeromonas hydrophila* NCIMB1135, *Salmonella* Paratyphi A NCTC13, and *Yersinia enterocolitica* NCTC 11175) food-borne bacteria.

2. Materials and Methods

2.1. Spices, Chemicals, and Cultural Media

A total of four different dried spices identified based on their botanical names were used in this study: sumac (*Rhus coriaria* L.), cumin (*Cuminum cyminum* L.), black pepper (*Piper nigrum*), and red pepper (*Capsicum annuum*). All these spices were acquired from a local market in Adana, Turkey. The spices were ground and dried. The compounds of diethyl ether, active carbon, tyrosine, peptone, Lab-Lemco powder, NaCl, pyridoxal–HCl, trimethylamine hydrochloride, ammonium chloride were obtained from Merck (Darmstadt, Germany) and Sigma-Aldrich (Seelze, Germany). Biogenic amines standards, e.g., histamine dihydrochloride, tyramine hydrochloride, tryptamine hydrochloride, putrescine dihydrochloride, 2-phenylethylamine hydrochloride, cadaverine dihydrochloride, spermidine trihydrochloride, spermine tetrahydrochloride, 5-hydroxytryptamine (serotonin), 3-hydroxytyramine hydrochloride (dopamine), agmatine sulphate, trichloroacetic acid, benzoyl chloride, acetonitrile NaOH, were acquired from Merck (Darmstadt, Germany) and Sigma-Aldrich (Seelze, Germany). All of them were of analytical reagent quality. Culture media, e.g., nutrient broth and plate count agar (PCA), were purchased from Merck (Darmstadt, Germany), and Biokar (Beauvais, France) Difco, respectively.

2.2. Bacterial Strains

Reference bacterial strains, e.g., *Staphylococcus aureus* (ATCC 29213), *Klebsiella pneumoniae* (ATCC 700603), *Enterococcus faecalis* (ATCC 29212), *Pseudomonas aeruginosa* (ATCC 27853), and *Campylobacter jejuni* (ATCC 33560) used in this study were obtained from the American Type Culture Collection (Rockville, MD, USA). *Aeromonas hydrophila* (NCIMB1135), *Salmonella* Paratyphi A (NCTC13), and *Yersinia enterocolitica* (NCTC 11175) were obtained from the National Collection of Industrial Food and Marine Bacteria (Aberdeen, UK) and the National Collection of Type Cultures (London, UK).

2.3. Spice Extraction

The solvent extraction technique was used to extract the spices. An extraction thimble (30 × 80 mm, Whatman 2810-338, UK) made from cellulose was used to combine 200 g of powdered spice with 1 L of diethyl ether and carried out in a reflux extractor. The mixture was then extracted for 4 h at 60 °C. Extraction process was carried out twice for each spice. To remove the color of the extracts, 40 g of activated carbon (Merck, Darmstadt, Germany) was used to bleach them for 30 min at 60 °C after extraction. After the extracts had been filtered through Whatman No. 1 filter paper (Maidstone, UK), the impurities were eliminated from the extracts. A rotary evaporator (Heidolph WB 2000, Heidolph Instruments, Schwabach, Germany) was used to extract the organic solvent. Before further use, the dried extracts were stored at −20 °C and protected from light. In order to carry out the antibacterial and biogenic amine analyses, the spice extracts were sterilized for 15 min at room temperature (22 °C) in a Telstar Bio IIA biological cabinet (Telstar, Madrid, Spain) using UV radiation (30 W, 253.7 nm wavelength, 50 cm away from the light source).

2.4. Culture Media and Biogenic Amines (BAs) Extraction

The method outlined by Klausen and Huss [17] was used to measure the synthesis of ammonia (AMN), trimethylamine (TMA), and BAs by reference to FBP staining, in tyrosine decarboxylase broth (TDB). Food-borne pathogens were cultured for two or three days at their ideal growth temperature in nutrient broth. After that, 0.5 mL of each bacterial culture was added to the TDB for tyrosine decarboxylation over the course of 72 h, yielding 106 colony-forming units per mL (106 cfu/mL) as measured by the McFarland cell densitometer (Biosan DEN 1, Riga, Latvia). Spice extracts were added to the TDB at a concentration of 1% (w/v), following bacterial inoculation. All extracts were tested in triplicate on the same day for all groups. As a part of the extraction process, five milliliters of TDB containing food-borne pathogens were divided into separate bottles and then added with two milliliters of trichloroacetic acid (6%, w/v) in order to extract biogenic amines. A filter paper with a pore size of 11 m (Schleicher and Schuell, Dassel, Germany) was then used to filter the extracts; then, they were centrifuged for 10 min at 3000× g. A total of four milliliters of each bacterial supernatant was collected for the analysis, and the procedure was carried out in three duplicates.

2.5. Analysis of BA by HPLC after Derivatization

The method outlined by Özogul [18] was followed in order to prepare a standard amine-mixed aqueous solution containing ammonium chloride, trimethylamine hydrochloride, and twelve amines. Derivatization of a 100 µL standard amine solution containing 10 mg of each amine per microliter was accomplished by adding 40 mL of 2% (v/v) benzoyl chloride in acetonitrile and 1 mL of aqueous 2 M NaOH solution. After shaking the solution for one minute in a vortex mixer, it was allowed to stand at room temperature and shielded from light for 20 min. Afterwards, the derivatization was stopped by adding 2 mL of saturated aqueous NaCl solution. The resultant solution was extracted twice with two milliliters of diethyl ether. The top layer was then separated, put into sterile sample tubes, dried with a nitrogen stream, and combined with one milliliter of acetonitrile. The BAs were separated and quantified by performing triplicate injections of 10 μL of the produced solution into Shimadzu HPLC equipment (Kyoto, Japan), following the HPLC approach previously described by Özogul [19]. The samples of extracted bacterial cultures were prepared in the same manner as those of the standard mixed amine solution, with the exception that 4 mL of each extracted bacterial culture was replaced with 100 mL of the standard mixed amine solution during the derivatization procedure.

The method outlined by Özogul [19] was used to determine the concentrations of BAs, TMA, and AMN. The results were expressed as milligrams of BAs (or TMA and ammonia) per litter of TDB (mg/L). There was an HPLC apparatus used in this study, which was a Shimadzu Prominence HPLC unit (Shimadzu, Kyoto, Japan), equipped with an HPLC ODS Hypersil column, 5 μm (250 × 4.6) mm (Phenomenex, Macclesfield, Cheshire, UK), an autosampler (SIL 20AC), a column oven (CTO-20AC), a communication bus module (CBM-20A) featuring a valve unit FCV-11AL, and two binary gradient pumps (Shimadzu LC-10AT).

2.6. Chromatographic Separation

In order to conduct the chromatographic separation, gradient elutions were performed using acetonitrile (eluant A) and HPLC grade water (eluant B) at a flow rate of 1.2 mL/min. The injection volume was 10 μL, and the overall separation time was less than 20 min. Detection was monitored at 254 nm. Standard curves were created for each amine ranging from 0 to 50 mg/mL. A correlation coefficient of peak area versus amine standard concentrations was computed for each compound following the injection of five duplicates of each standard solution of amine. The curves for each benzoylated amine showed a correlation coefficient (r) greater than 0.99.

2.7. Determination of Different Bacterial Growths in Tyrosine Decarboxylase Broth (TDB)

After appropriate dilutions (10^{-10} CFU/mL) were made of each bacterial culture in the TDB, 0.1 mL was inoculated in triplicate onto plate count agar (PCA, Merck, Darmstadt, Germany) plates using a spread plate approach. Following 72 h of incubation at 30 °C, the results were obtained as the logarithm of total viable colony-forming units per milliliter of broth, log (average standard deviation), and log (CFU/mL).

2.8. Statistical Analysis

The results were calculated using triplicate samples for each spice (per treatment). An analysis of variance (ANOVA) was performed and Duncan's multiple range tests were run on the data when there were significant differences at $p < 0.05$. Statistical differences between the control and spice extracts were determined based on pathogen concentrations and BA contents. All statistical analyses were conducted using SPSS version 19 for Windows (SPSS Inc., Chicago, IL, USA).

3. Results and Discussions

3.1. Bacterial Growth in Tyrosine Decarboxylase Broth

The results of different food-borne pathogen bacterial growths in TDB are shown in Figure 1. Due to sumac's higher antimicrobial activity, sumac extract demonstrated significant inhibition of both Gram-positive (*S. aureus* and *E. faecalis*) and Gram-negative (*K. pneumoniae*, *P. aeruginosa*, *C. jejuni*, *A. hydrophila*, *S.* Paratyphi A, and *Y. enterocolitica*) bacteria. The highest inhibition levels were observed for *Y. enterocolitica*, ranging from 8.61 to 5.05 log (CFU/mL). *P. aureginosa* was inhibited at similar levels by cumin extract and sumac extract. Cumin, red pepper, and black pepper extracts also inhibited *K. pneumoniae* and *E. faecalis* at similar levels. For all microorganisms tested, cumin, black pepper, and red pepper spice extracts inhibited bacteria below 1.0 log (CFU/mL). Sumac extract was shown to be effective against several Gram-positive and Gram-negative bacteria in previous studies [20,21]. The presence of several polyphenolic compounds in sumac was linked to antibacterial activity [22]. On the other hand, a previous study reported that cumin extract exhibited lower bactericidal activities [23]. Impacts of drying technique and extraction solvents on antibacterial activity were earlier discussed, highlighting the role of different drying techniques (degrading the bioactive compound) and solvents (poor solubility of the bioactive compound) on the extraction of bioactive compounds responsible for activity [24, 25].

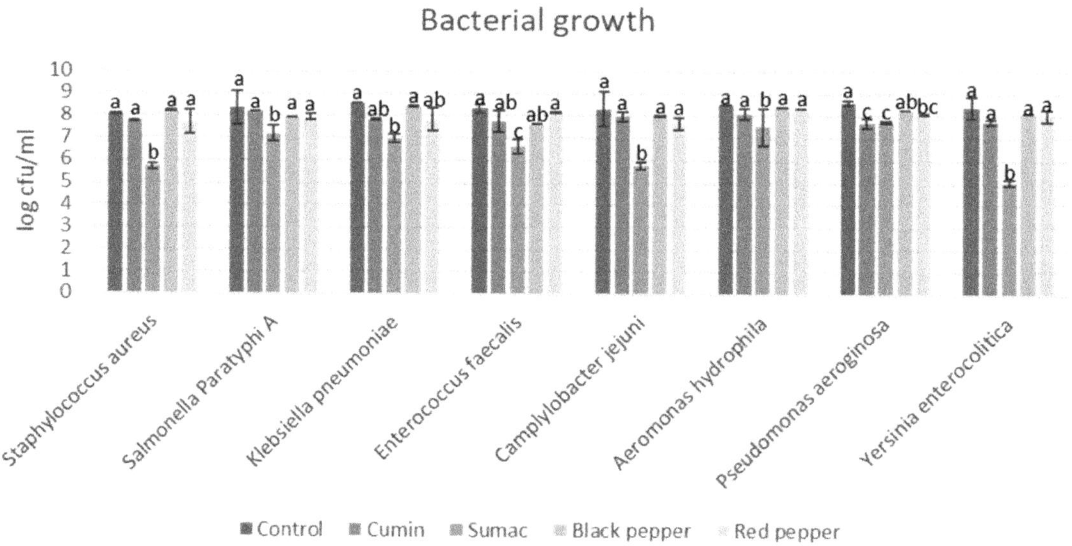

Figure 1. Food-borne pathogen growth in tyrosine decarboxylase broth. a–c indicate significant differences ($p < 0.05$) among groups. Spice extracts were added in tyrosine decarboxylase broth at a concentration of 1% (w/v).

3.2. *Ammonia, Trimethylamine and BAs production in Tyrosine Decarboxylase Broth*

There are three types of BAs that are present in food. Heterocyclic BAs (histamine and tryptamine), aliphatic BAs (putrescine and cadaverine), and aromatic BAs (tyramine and phenylethylamine). A further categorization is based on the quantity of amine groups, which include polyamines (spermidine and spermine), diamines (histamine, putrescine, and cadaverine), and monoamines (tyramine and phenylethylamine) [26]. BAs such as diamines, polyamines, and TMA are detected to monitor the freshness or spoilage rate of food. The most dangerous amines are histamine and tyramine, which are the two primary BAs present in food [4]. Inhibitory effects of four spice extracts, e.g., sumac, black pepper, red pepper, and cumin on the production of ammonia (AMN), trimethylamine (TMA), and the formed BAs (putrescine, cadaverine, spermidine, tryptamine, phenylethylamine, spermine, serotonin, dopamine, and agmatine) produced by eight food-borne bacteria using TDB are presented in Table 1. AMN production was between 543 mg/L by *A. hydrophila* and 844 mg/L by *K. pneumoniae*. A significant inhibition of ammonia production was observed with all spice extracts (>75%), particularly sumac, which inhibited five microbial strains, e.g., *S. aureus* (90%), *S.* Paratyphi A (91%), *K. pneumoniae* (80%), *E. faecalis* (87%), and *Y. enterocolitica* (92%). There was an 80% inhibition of *P. aeruginosa* and a 75% inhibition of *C. jejuni* by black pepper whereas 89% of inhibition of *A. hydrophila* by cumin was observed. The maximum production of cadaverine and putrescin was recorded by *S.* Paratyphi A (4.39

mg/L) and *C. jejuni* (35.49 mg/L). The control sample (without spice extract) had a generally high level of BAs formation (except for tryptamine and phenylethylamine). Extracts of cumin exhibited higher values than control samples. A similar pattern was observed with putrescine, where sumac extract inhibited five species (>50%) other than *K. pneumoniae* (50%), whereas black pepper extract inhibited stronger inhibition for *A. hydrophila* (47%) and *Y. enterocolitica* (90%). The production of cadaverine in all evaluated microorganisms was most resistant to spice extracts, with the exception of *S.* Paratyphi A, for which an inhibition of 60% was observed in the presence of sumac extract. However, among extracts tested, there was a higher increase in cadaverine production with cumin extract over control. Sumac extract was the most effective inhibitor of the four spice extracts evaluated, followed by black pepper and red pepper extracts. Spermidine production peaked at 90.22 mg/L, primarily generated by *E. faecalis*. Black pepper extract was found to be the most effective against spermidine production, inhibiting over 70% of all evaluated microorganisms. Similarly, sumac inhibited spermine production in all samples. In comparison to other extracts, sumac extract were found to promote serotonin generation in bacteria (*S. aureus*, *S.* Paratyphi A, *K. pneumoniae*, *E. faecalis*, and *P. aeruginosa*). Additionally, cumin extract promoted serotonin production for all three remaining microorganisms. The results showed that red pepper extract was effective on inhibition the production of trimethylamine by *A. hydrophila*, while black pepper inhibited the formation of trimethylamine by *P. aeruginosa*, *E. faecalis*, and *S.* Paratyphi A. Among all tested microorganisms, pepper-based extracts significantly retarded dopamine production. Sumac extract inhibited the production of agmatine by all tested microorganisms with the exception of *S. aureus*.

Microorganisms secrete endogenous enzymes (amino acid decarboxylase) and exogenous enzymes for decarboxylation of proteins and amino acids [4]. Cumin extract intensifies the production of tryptamine, phenylethylamine, and spermidine BAs due to its synergistic effect with TDB broth in decarboxylating phenylalanine and tryptophan, thereby producing phenylethylamine and tryptamine. This is the first study to suggest that spice extract increases BAs production, which could be explained by the abundance of alkaloid in cumin, because alkaloid content has been associated with increasing BAs production. [4,27–29]. The sumac extract was the most effective inhibitor, followed by the black pepper, red pepper, and the cumin extracts. In this study, sumac extract was observed to suppress BAS production, which could be ascribed to the fact that sumac extract contained 211 different kinds of phytochemicals, such as polyphenols, organic acids (mallic and tannic acid), and flavonoids [30–34]. Peppers (black and red) were reported to be the sources of bioactive capsaicin, piperine, flavonoid, amide, and organic acid constituents, confirming their ability to inhibit biogenic amine production. Several studies show that the bioactive components in spice extracts are antibacterial, inhibiting the actions of endogenous enzymes, and also targeting Gram-positive and Gram-negative bacteria [8,32,35–39]. However, the antibacterial action led to the inactivation of microorganisms that caused BA production. In addition, bioactive constituents present in spices were reported to inhibit enzymatic activities due to their high antioxidant potential, which is mainly responsible for decarboxylation of amino acids [40–42].

Table 1. Ammonia and biogenic amine production by food-borne pathogens spice extracts in tyrosine decarboxylase broth (mg/L).

	AMN	PUT	CAD	SPD	TRP	PHEN	SPM	SER	TMA	DOP	AGM	Group
Staphylococcus aureus (ATCC29213)	808.05 ± 68.56 ^{a}	15.85 ± 0.54 ^{b}	3.42 ± 0.21 ^{cd}	16.70 ± 1.36 ^{b}	0.28 ± 0.01 ^{c}	0.25 ± 0.01 ^{b}	16.16 ± 0.37 ^{c}	7.69 ± 0.25 ^{d}	3.21 ± 0.03 ^{d}	95.24 ± 1.63 ^{c}	39.18 ± 0.27 ^{c}	C
	84.60 ± 5.66 ^{c}	5.64 ± 0.22 ^{c}	2.01 ± 0.32 ^{d}	2.10 ± 0.20 ^{c}	0.00 ± 0.00 ^{c}	0.00 ± 0.00 ^{b}	5.50 ± 0.13 ^{d}	95.31 ± 1.66 ^{a}	3.96 ± 0.21 ^{d}	190.39 ± 13.94 ^{b}	42.49 ± 3.77 ^{c}	SUM
	323.35 ± 17.46 ^{b}	81.09 ± 4.21 ^{a}	27.99 ± 2.27 ^{a}	90.42 ± 8.11 ^{a}	17.80 ± 0.27 ^{a}	12.77 ± 0.51 ^{a}	42.13 ± 1.91 ^{b}	34.03 ± 2.62 ^{b}	22.34 ± 0.09 ^{b}	503.00 ± 0.34 ^{a}	71.83 ± 1.28 ^{b}	CUM
	344.10 ± 34.97 ^{b}	15.77 ± 1.13 ^{b}	16.87 ± 0.97 ^{b}	1.01 ± 0.01 ^{c}	0.28 ± 0.03 ^{c}	0.00 ± 0.00 ^{b}	9.32 ± 0.07 ^{d}	14.27 ± 1.37 ^{c}	12.79 ± 0.34 ^{c}	494.54 ± 16.05 ^{a}	111.65 ± 8.84 ^{a}	BP
	419.86 ± 30.89 ^{b}	8.34 ± 0.48 ^{c}	6.22 ± 0.40 ^{c}	15.25 ± 1.27 ^{b}	5.80 ± 0.18 ^{b}	0.41 ± 0.01 ^{b}	69.93 ± 3.29 ^{a}	8.99 ± 0.60 ^{d}	30.01 ± 2.15 ^{a}	24.37 ± 1.38 ^{d}	47.11 ± 1.62 ^{c}	RP
Salmonella Paratyphi A (NCTC13)	836.47 ± 62.76 ^{a}	21.49 ± 1.04 ^{b}	4.39 ± 0.40 ^{c}	22.24 ± 2.08 ^{b}	0.00 ± 0.00 ^{b}	0.00 ± 0.00 ^{b}	17.81 ± 1.53 ^{c}	31.40 ± 2.52 ^{b}	18.65 ± 0.92 ^{a}	468.63 ± 41.47 ^{a}	70.90 ± 5.82 ^{a}	C
	71.51 ± 5.94 ^{c}	1.44 ± 0.02 ^{c}	1.73 ± 0.04 ^{d}	5.34 ± 0.48 ^{c}	0.00 ± 0.00 ^{b}	0.00 ± 0.00 ^{b}	2.96 ± 0.16 ^{e}	82.42 ± 2.93 ^{a}	4.24 ± 0.09 ^{c}	355.11 ± 30.40 ^{b}	26.40 ± 1.56 ^{c}	SUM
	286.12 ± 5.54 ^{b}	57.39 ± 4.55 ^{a}	27.41 ± 1.19 ^{a}	60.55 ± 2.34 ^{a}	16.54 ± 0.67 ^{a}	12.81 ± 0.85 ^{a}	58.75 ± 0.63 ^{b}	23.96 ± 0.51 ^{c}	17.17 ± 1.29 ^{a}	482.23 ± 37.95 ^{a}	80.87 ± 6.91 ^{a}	CUM
	78.34 ± 0.79 ^{c}	5.89 ± 0.39 ^{c}	3.07 ± 0.13 ^{cd}	0.00 ± 0.00 ^{d}	0.16 ± 0.02 ^{b}	0.00 ± 0.00 ^{b}	7.28 ± 0.74 ^{d}	8.56 ± 0.35 ^{d}	3.29 ± 0.16 ^{c}	45.24 ± 4.30 ^{c}	32.90 ± 1.59 ^{bc}	BP
	287.30 ± 18.56 ^{b}	4.23 ± 0.07 ^{c}	7.86 ± 0.42 ^{b}	21.25 ± 2.47 ^{b}	0.77 ± 0.01 ^{b}	0.97 ± 0.10 ^{b}	112.37 ± 2.42 ^{a}	19.71 ± 1.09 ^{c}	14.36 ± 1.09 ^{b}	322.84 ± 21.54 ^{b}	43.26 ± 1.82 ^{b}	RP
Klebsiella pneumoniae (ATCC700603)	699.88 ± 45.21 ^{a}	26.36 ± 1.53 ^{a}	3.03 ± 0.04 ^{c}	16.66 ± 0.93 ^{b}	0.00 ± 0.00 ^{b}	0.00 ± 0.00 ^{b}	31.99 ± 2.78 ^{a}	57.20 ± 1.80 ^{b}	5.40 ± 0.26 ^{b}	815.25 ± 53.59 ^{a}	74.48 ± 3.94 ^{b}	C
	133.15 ± 10.39 ^{c}	16.88 ± 1.31 ^{c}	3.10 ± 0.13 ^{c}	12.04 ± 0.83 ^{b}	0.00 ± 0.00 ^{b}	0.00 ± 0.00 ^{b}	2.39 ± 0.02 ^{c}	65.51 ± 3.73 ^{a}	1.70 ± 0.04 ^{c}	434.63 ± 38.38 ^{b}	18.67 ± 1.33 ^{c}	SUM
	259.56 ± 24.37 ^{b}	21.25 ± 0.75 ^{b}	32.05 ± 2.53 ^{a}	55.52 ± 4.94 ^{a}	16.32 ± 0.29 ^{a}	14.56 ± 0.92 ^{a}	30.87 ± 0.79 ^{a}	44.31 ± 2.82 ^{c}	35.04 ± 1.48 ^{a}	457.67 ± 40.57 ^{b}	93.58 ± 6.11 ^{a}	CUM
	181.69 ± 7.46 ^{c}	13.18 ± 1.03 ^{d}	8.08 ± 0.14 ^{b}	4.67 ± 0.23 ^{c}	0.30 ± 0.01 ^{b}	0.00 ± 0.00 ^{b}	11.13 ± 0.46 ^{b}	6.27 ± 0.60 ^{d}	5.06 ± 0.21 ^{b}	97.64 ± 4.07 ^{c}	25.20 ± 1.11 ^{c}	BP
	267.97 ± 18.00 ^{b}	24.88 ± 0.96 ^{a}	9.14 ± 0.01 ^{b}	16.63 ± 0.41 ^{b}	0.39 ± 0.55 ^{b}	0.66 ± 0.08 ^{b}	31.02 ± 0.18 ^{a}	6.48 ± 0.10 ^{d}	0.91 ± 0.02 ^{c}	160.19 ± 5.44 ^{c}	22.55 ± 1.58 ^{c}	RP
Pseudomonas aeruginosa (ATCC27853)	844.24 ± 46.07 ^{a}	2.20 ± 0.18 ^{c}	3.42 ± 0.33 ^{c}	46.77 ± 2.93 ^{b}	0.00 ± 0.00 ^{c}	0.28 ± 0.03 ^{b}	21.51 ± 0.47 ^{c}	16.91 ± 0.69 ^{c}	20.83 ± 1.54 ^{b}	668.71 ± 55.08 ^{a}	64.32 ± 4.17 ^{b}	C
	240.06 ± 8.52 ^{c}	1.07 ± 0.02 ^{c}	3.26 ± 0.26 ^{c}	2.88 ± 0.17 ^{d}	0.00 ± 0.00 ^{c}	0.00 ± 0.00 ^{b}	11.87 ± 0.39 ^{cd}	51.04 ± 1.59 ^{a}	4.34 ± 0.02 ^{c}	453.12 ± 41.40 ^{b}	24.99 ± 0.05 ^{d}	SUM
	328.78 ± 9.13 ^{b}	60.45 ± 7.54 ^{a}	59.49 ± 0.92 ^{a}	104.70 ± 8.03 ^{a}	21.56 ± 1.46 ^{a}	25.06 ± 1.62 ^{a}	67.97 ± 3.48 ^{b}	46.76 ± 2.20 ^{b}	48.89 ± 0.29 ^{a}	357.57 ± 12.44 ^{c}	92.37 ± 4.40 ^{a}	CUM
	166.48 ± 6.82 ^{d}	16.40 ± 1.12 ^{b}	3.44 ± 0.05 ^{c}	0.77 ± 0.10 ^{d}	0.00 ± 0.00 ^{c}	0.00 ± 0.00 ^{b}	7.69 ± 0.15 ^{d}	4.52 ± 0.19 ^{d}	1.34 ± 0.08 ^{d}	118.81 ± 6.11 ^{d}	33.98 ± 2.08 ^{c}	BP
	193.23 ± 18.26 ^{d}	3.30 ± 0.33 ^{c}	11.09 ± 0.52 ^{b}	35.86 ± 1.68 ^{c}	4.05 ± 0.01 ^{b}	0.36 ± 0.05 ^{b}	109.52 ± 10.36 ^{a}	14.83 ± 0.39 ^{c}	2.01 ± 0.07 ^{d}	175.72 ± 4.57 ^{d}	36.93 ± 0.56 ^{c}	RP
Enterococcus faecalis (ATCC29212)	689.23 ± 56.42 ^{a}	4.88 ± 0.45 ^{c}	3.50 ± 0.16 ^{c}	90.22 ± 4.82 ^{a}	0.95 ± 0.07 ^{b}	0.35 ± 0.01 ^{b}	7.50 ± 0.09 ^{c}	20.98 ± 0.94 ^{c}	36.16 ± 2.58 ^{b}	998.43 ± 15.67 ^{a}	55.08 ± 4.19 ^{b}	C
	87.75 ± 1.91 ^{d}	1.46 ± 0.05 ^{c}	2.24 ± 0.09 ^{c}	4.43 ± 0.04 ^{cd}	0.88 ± 0.04 ^{b}	0.00 ± 0.00 ^{b}	2.97 ± 0.04 ^{d}	75.87 ± 2.93 ^{a}	5.56 ± 0.28 ^{d}	517.62 ± 38.93 ^{b}	22.97 ± 1.63 ^{d}	SUM
	290.61 ± 7.15 ^{bc}	23.78 ± 1.44 ^{b}	50.75 ± 3.56 ^{a}	0.00 ± 0.00 ^{d}	20.35 ± 2.08 ^{a}	15.75 ± 0.52 ^{a}	57.36 ± 4.62 ^{a}	25.57 ± 0.80 ^{b}	49.73 ± 0.54 ^{a}	532.55 ± 25.63 ^{b}	112.61 ± 4.92 ^{a}	CUM
	329.76 ± 4.75 ^{b}	26.72 ± 2.27 ^{b}	9.12 ± 0.11 ^{b}	8.14 ± 0.61 ^{c}	0.24 ± 0.00 ^{b}	0.51 ± 0.01 ^{b}	7.90 ± 0.50 ^{c}	13.65 ± 1.38 ^{d}	3.86 ± 0.28 ^{d}	94.04 ± 4.44 ^{c}	22.39 ± 1.63 ^{d}	BP
	234.01 ± 14.94 ^{c}	33.09 ± 1.69 ^{a}	3.74 ± 0.42 ^{c}	27.99 ± 0.94 ^{b}	0.37 ± 0.05 ^{b}	0.00 ± 0.00 ^{b}	41.42 ± 0.93 ^{b}	9.92 ± 0.71 ^{d}	10.79 ± 0.09 ^{c}	516.60 ± 29.40 ^{b}	43.36 ± 1.28 ^{c}	RP
Yersinia enterocolitica (NCTC 11175)	570.71 ± 10.84 ^{a}	27.70 ± 1.05 ^{a}	4.95 ± 0.08 ^{bc}	48.16 ± 0.45 ^{a}	0.00 ± 0.00 ^{c}	0.00 ± 0.00 ^{b}	24.81 ± 0.65 ^{c}	135.38 ± 10.36 ^{a}	8.23 ± 0.53 ^{b}	1159.63 ± 114.29 ^{a}	65.27 ± 4.36 ^{b}	C
	55.10 ± 1.52 ^{d}	2.81 ± 0.27 ^{c}	2.27 ± 0.00 ^{d}	0.00 ± 0.00 ^{e}	0.00 ± 0.00 ^{c}	0.00 ± 0.00 ^{b}	2.59 ± 0.11 ^{e}	14.06 ± 0.29 ^{c}	1.28 ± 0.08 ^{d}	440.95 ± 28.32 ^{c}	25.79 ± 0.08 ^{d}	SUM
	278.62 ± 5.22 ^{b}	26.75 ± 0.58 ^{ab}	26.27 ± 1.62 ^{a}	38.72 ± 1.66 ^{b}	17.19 ± 0.27 ^{a}	14.43 ± 0.68 ^{a}	66.23 ± 1.43 ^{a}	97.55 ± 1.46 ^{b}	16.08 ± 1.42 ^{a}	599.80 ± 23.20 ^{b}	97.11 ± 1.07 ^{a}	CUM
	135.34 ± 6.13 ^{c}	2.67 ± 0.12 ^{c}	1.63 ± 0.04 ^{d}	4.41 ± 0.15 ^{d}	0.38 ± 0.03 ^{b}	0.00 ± 0.00 ^{b}	8.70 ± 0.47 ^{d}	6.28 ± 0.44 ^{c}	1.79 ± 0.10 ^{d}	102.60 ± 3.14 ^{d}	10.70 ± 0.10 ^{e}	BP
	143.63 ± 14.69 ^{c}	25.85 ± 0.65 ^{b}	3.67 ± 0.26 ^{c}	21.71 ± 2.23 ^{c}	0.63 ± 0.01 ^{b}	0.00 ± 0.00 ^{b}	51.04 ± 0.82 ^{b}	15.28 ± 3.22 ^{c}	5.67 ± 3.73 ^{c}	214.46 ± 120.13 ^{d}	31.45 ± 7.25 ^{c}	RP
Campylobacter jejuni (ATCC 33560)	691.20 ± 66.03 ^{a}	35.49 ± 3.54 ^{b}	3.23 ± 0.00 ^{c}	37.71 ± 2.94 ^{a}	1.07 ± 0.09 ^{b}	0.00 ± 0.00 ^{b}	46.78 ± 2.95 ^{b}	15.33 ± 1.02 ^{b}	14.35 ± 0.12 ^{b}	636.14 ± 24.06 ^{a}	81.69 ± 7.40 ^{b}	C
	309.57 ± 8.03 ^{b}	7.84 ± 0.24 ^{c}	3.27 ± 0.01 ^{c}	4.22 ± 0.39 ^{c}	0.00 ± 0.00 ^{c}	0.00 ± 0.00 ^{b}	6.34 ± 0.21 ^{c}	11.06 ± 1.31 ^{bc}	6.39 ± 0.28 ^{c}	279.47 ± 7.78 ^{c}	19.04 ± 1.00 ^{cd}	SUM
	317.87 ± 9.43 ^{b}	71.68 ± 2.93 ^{a}	24.72 ± 0.56 ^{a}	0.00 ± 0.00 ^{d}	20.85 ± 0.29 ^{a}	26.80 ± 1.32 ^{a}	70.20 ± 2.55 ^{a}	125.78 ± 4.12 ^{a}	22.05 ± 0.29 ^{a}	628.98 ± 61.35 ^{a}	95.69 ± 0.04 ^{a}	CUM
	168.40 ± 9.99 ^{c}	13.43 ± 1.16 ^{c}	13.98 ± 0.24 ^{b}	3.96 ± 0.12 ^{c}	0.89 ± 0.01 ^{b}	0.00 ± 0.00 ^{b}	7.65 ± 0.61 ^{c}	10.91 ± 0.53 ^{bc}	6.23 ± 0.32 ^{c}	16.78 ± 1.17 ^{d}	11.24 ± 0.01 ^{d}	BP
	211.38 ± 17.40 ^{c}	9.30 ± 0.37 ^{c}	2.58 ± 0.08 ^{c}	20.91 ± 0.99 ^{b}	0.21 ± 0.01 ^{c}	0.00 ± 0.00 ^{b}	4.89 ± 0.27 ^{c}	7.84 ± 0.29 ^{c}	2.15 ± 0.06 ^{d}	494.09 ± 3.03 ^{b}	26.79 ± 0.87 ^{c}	RP
Aeromonas hydrophila (NCIMB1135)	543.77 ± 52.09 ^{a}	3.29 ± 0.10 ^{b}	4.61 ± 0.77 ^{c}	78.29 ± 3.54 ^{a}	0.66 ± 0.08 ^{b}	0.97 ± 0.11 ^{b}	37.21 ± 1.51 ^{b}	36.74 ± 3.60 ^{b}	53.66 ± 0.98 ^{a}	746.78 ± 5.75 ^{a}	254.91 ± 17.94 ^{a}	C
	99.15 ± 2.96 ^{c}	5.71 ± 6.63 ^{b}	5.33 ± 0.28 ^{c}	6.36 ± 0.44 ^{d}	0.00 ± 0.00 ^{c}	0.00 ± 0.00 ^{c}	6.73 ± 0.22 ^{d}	7.36 ± 0.17 ^{d}	2.76 ± 0.23 ^{d}	578.68 ± 34.14 ^{b}	51.44 ± 1.79 ^{c}	SUM
	58.50 ± 3.53 ^{c}	25.64 ± 1.16 ^{a}	9.06 ± 1.11 ^{b}	32.21 ± 1.76 ^{b}	5.98 ± 0.28 ^{a}	4.04 ± 0.04 ^{a}	31.71 ± 1.03 ^{c}	66.56 ± 2.16 ^{a}	8.27 ± 0.64 ^{b}	707.41 ± 69.65 ^{a}	164.72 ± 4.99 ^{b}	CUM
	83.68 ± 5.21 ^{c}	1.74 ± 0.06 ^{b}	4.57 ± 0.14 ^{c}	1.29 ± 0.13 ^{e}	0.28 ± 0.02 ^{c}	0.00 ± 0.00 ^{c}	8.91 ± 0.19 ^{d}	5.07 ± 0.07 ^{d}	4.89 ± 0.42 ^{c}	57.53 ± 4.14 ^{c}	13.96 ± 0.95 ^{d}	BP
	256.36 ± 11.04 ^{b}	3.76 ± 0.21 ^{b}	11.18 ± 0.23 ^{a}	20.45 ± 0.68 ^{c}	0.70 ± 0.07 ^{b}	0.00 ± 0.00 ^{c}	93.57 ± 2.23 ^{a}	24.02 ± 0.25 ^{c}	1.42 ± 0.11 ^{d}	144.53 ± 6.51 ^{c}	29.24 ± 1.79 ^{d}	RP

Different superscript lowercase letters (^{a}–^{e}) in a column indicate significant differences ($p < 0.05$) between the control (C) and bacteria treated with 1% extracts (w/v). Abbreviations of extracts, ammonia, and biogenic amines (BAs): C—control; SUM—sumac; CUM—cumin; BP—black pepper; RP—red pepper. AMN—ammonia; PUT—putrescine—CAD—cadaverine; SPD—spermidine; TRP—tryptamine; PHEN—phenylethylamine; SPM—spermine; SER—serotonin; TMA—trimethylamine; DOP—dopamine; AGM—agmatine. Spice extracts were added to the in tyrosine decarboxylase broth at a concentration of 1% (w/v).

3.2.1. Histamine Production by Food-Borne Pathogen Bacteria in Tyrosine Decarboxylase Broth

Histamine is regarded as BAs that is dangerous when taken into the human body at high proportions—(100 mg/kg) [32]. Histamine is usually produced from histidine converted by microorganisms or their enzymes [43]. Histamine is generally used for the estimation of quality and freshness indexes for meat-based foods [3]. Among all of the microorganisms evaluated, sumac extract had the strongest inhibitory effect (<10 mg/L) on histamine production (Figure 2). On the other hand, *Y. enterocolitica* was completely inhibited. This was followed by red pepper, which exhibited significant inhibition (<10, mg/L), for *S.* Paratyphi A, *K. pneumoniae*, *Y. enterocolitica*, *P. aeruginosa*, and *A. hydrophila*. As for the rest of the samples, production was above 10 mg/L but below the maximum allowable level. On the other hand, black and red pepper extracts were observed to promote the production of histamine by *S. aureus*, *S.* Paratyphi A, and *Y. enterocolitica*. Cumin extracts increased production by all FBP except for *E. faecalis* and *A. hydrophila*. Therefore, we can speculate that spice extracts can be used against the production of histidine due to their ability to inhibit bacterial growth, arrest biogenic amine synthesis, and inhibit amino acid decarboxylation, specifically enzyme (histidine decarboxylase) activity [44,45].

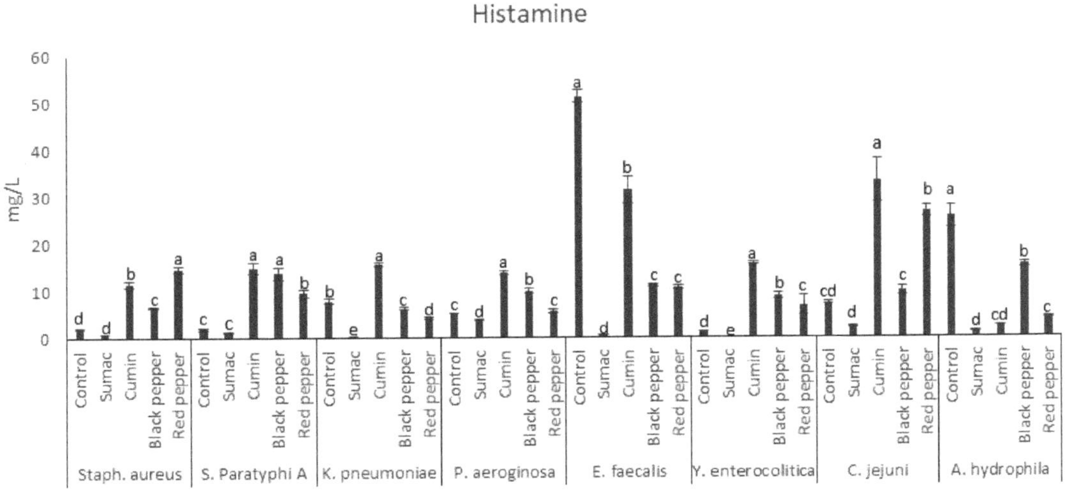

Figure 2. Histamine production by food-borne pathogen bacteria in the presence of spice extracts in tyrosine decarboxylase broth. a–e indicate significant differences ($p < 0.05$) among groups. Spice extracts were added to the in tyrosine decarboxylase broth at a concentration of 1% (*w*/*v*).

Similar results were demonstrated by Shakila, Vasundhara [46] who detected the efficacy of spice extracts (cinnamon, clove, turmeric, and cardamom) on in vitro histamine production by *Morganella morganii*. Based on the proposed inhibition of histamine decarboxylation activity exhibited by spices, the aforementioned results were obtained. Some extracts, however, were also found to promote amine production due to their lower activity in inhibiting histamine decarboxylation, corresponding to a delay in amine production [42].

3.2.2. Tyramine Production by Food-Borne Pathogen Bacteria in Tyrosine Decarboxylase Broth

The tyramine production profile of microorganisms in TBD is shown in Figure 3. The sumac extract significantly increased the formation of tyramine by *S. aureus*, *S.* Paratyphi A, and *P. aeruginosa*. The black and red peppers were found to significantly inhibit tyramine production by *S. aureus*, *S.* Paratyphi A, *E. faecalis*, *Y. enterocolitica*, and *A. hydrophila*. Also, cumin extracts were observed to promote the production of tyramine at levels higher than those produced in the control sample, with the exception of the strain *C. jejuni*, *A. hydrophila*, and *S. aureus*. Tyrosine was observed to be produced by the tyrosine amino acid by the action of a microbial enzyme [3]. Tyramine is associated with several disorders in humans, and in some cases its toxicity was reported to be higher as compared to histamine [47]. There have been similar results regarding the impacts of spice extracts on BAs (tyramine) production in minced meat [48]. It was concluded that spice extract inhibited microorganism growth and amino acid decarboxylase activity, lowering the production of amine. A recent molecular docking study has shown the ability of spices to bind with the amino acid decarboxylase active site and inhibit the enzyme, resulting in less BAs production and accumulation [49].

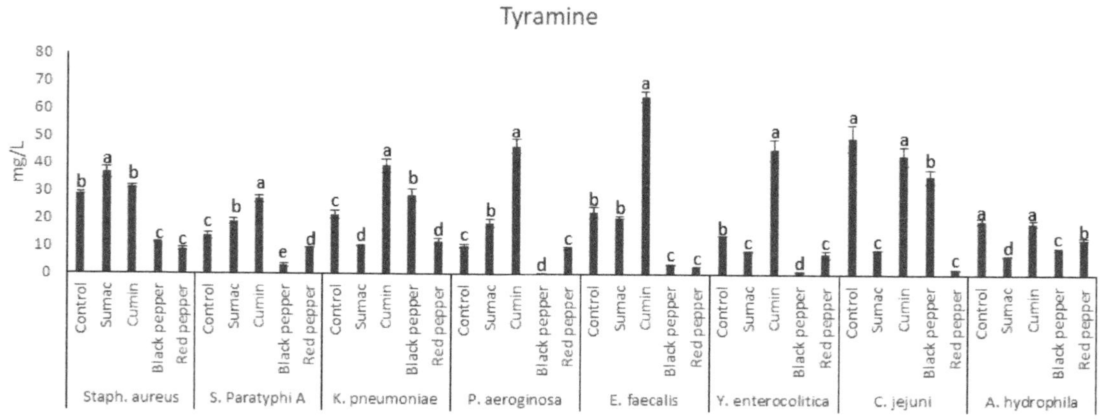

Figure 3. Tyramine production by food-borne pathogen bacteria in the presence of spice extracts in tyrosine decarboxylase broth. a–e indicate significant differences ($p < 0.05$) among groups. Spice extracts were added to the in tyrosine decarboxylase broth at a concentration of 1% (*w*/*v*).

4. Conclusions

Sumac extract exhibited significant inhibition of Gram-positive and Gram-negative bacteria, while cumin, black pepper, and red pepper spice extracts had lower bactericidal activities. This study also demonstrated the inhibitory effects of four spice extracts (sumac, black pepper, red pepper, and cumin) on the production of ammonia and BAs (histamine, tyramine, putrescine, cadaverine, spermidine, tryptamine, phenylethylamine, spermine, trimethylamine, serotonin, dopamine, and agmatine) by eight food-borne pathogen bacteria using tyrosine decarboxylase broth. Results showed that sumac was the most effective inhibitor, followed by black pepper, red pepper, and cumin extracts. All evaluated microorganisms produced less histamine when sumac extract was used, while cumin extract induced histamine production. In order to ensure food safety, sumac extract is recommended as a food preservative for controlling biogenic amine production. Further research should be conducted on the various methods of extracting these materials, particularly sumac extract. A variety of foods, their antioxidant and antibacterial properties, and their safety aspects should be discussed in addition to their integrated or combined use with other technologies. Integrating sumac extract with other technologies has the potential to enhance its effectiveness and expand its applications. By combining it with innovative delivery systems or processing techniques, we can unlock new possibilities for preserving food, improving health, and combating bacterial infections. This integration could lead to synergistic effects and create unique solutions in various fields such as food science, medicine, and environmental sustainability.

Author Contributions: F.K., investigation, methodology, software- SPSS version 19 (microbiological and statistical analyses); E.K., software (microbiological and chemical analysis), review and editing; N.B.R., original draft preparation, writing; M.T.Y., original draft preparation, writing, manuscript submission and editing; F.O., Project administration, conceptualization, review and editing. All authors have read and agreed to the published version of the manuscript.

Funding: The project was funded by the Scientific Research Projects Unit of Cukurova University (Project No: FYL-2015-4813). The authors are grateful for their financial support.

Institutional Review Board Statement: Not applicable.

Informed Consent Statement: Not applicable.

Data Availability Statement: The original contributions presented in the study are included in the article, further inquiries can be directed to the corresponding author.

Conflicts of Interest: The authors declare no conflicts of interest.

References

1. Ramachandran, G. Gram-positive and gram-negative bacterial toxins in sepsis: A brief review. *Virulence* **2014**, *5*, 213–218. [CrossRef] [PubMed]
2. Gourama, H. Foodborne pathogens. In *Food Safety Engineering*; Demirci, A., Feng, H., Krishnamurthy, K., Eds.; Food Engineering Series; Springer: Cham, Switzerland, 2020; pp. 25–49.
3. Özogul, Y.; Özogul, F. Biogenic amines formation, toxicity, regulations in food. In *Biogenic Amines in Food Analysis, Occurrence and Toxicity*; The Royal Society of Chemistry: London, UK, 2020; pp. 1–17.
4. Gardini, F.; Özogul, Y.; Suzzi, G.; Tabanelli, G.; Özogul, F. Technological factors affecting biogenic amine content in foods: A review. *Front. Microbiol.* **2016**, *7*, 1218. [CrossRef] [PubMed]
5. Tsafack, P.B.; Tsopmo, A. Effects of bioactive molecules on the concentration of biogenic amines in foods and biological systems. *Heliyon* **2022**, *8*, e10456. [CrossRef] [PubMed]
6. Shalaby, A.R.; Anwar, M.M.; Sallam, E.M.; Emam, W.H. Quality and safety of irradiated food regarding biogenic amines: Ras cheese. *Int. J. Food Sci. Technol.* **2016**, *51*, 1048–1054. [CrossRef]
7. Jaguey-Hernandez, Y.; Aguilar-Arteaga, K.; Ojeda-Ramirez, D.; Anorve-Morga, J.; González-Olivares, L.G.; Castaneda-Ovando, A. Biogenic amines levels in food processing: Efforts for their control in foodstuffs. *Food Res. Int.* **2021**, *144*, 110341. [CrossRef]
8. Rathod, N.B.; Ranveer, R.C.; Benjakul, S.; Kim, S.K.; Pagarkar, A.U.; Patange, S.; Ozogul, F. Recent developments of natural antimicrobials and antioxidants on fish and fishery food products. *Compr. Rev. Food Sci. Food Saf.* **2021**, *20*, 4182–4210. [CrossRef]
9. El-Sayed, S.M.; Youssef, A.M. Potential application of herbs and spices and their effects in functional dairy products. *Heliyon* **2019**, *5*, e01989. [CrossRef]
10. Batiha, G.E.S.; Ogunyemi, O.M.; Shaheen, H.M.; Kutu, F.R.; Olaiya, C.O.; Sabatier, J.M.; De Waard, M. *Rhus coriaria* L.(Sumac), a versatile and resourceful food spice with cornucopia of polyphenols. *Molecules* **2022**, *27*, 5179. [CrossRef]
11. Pervez, M.K.; Ahmed, F.; Dewani, R.; Ayaz, T.; Mehboob, S.J.; Soomro, S.A. Qualitative investigation of prohibited food colors in red hot chilli & curry collected from Karachi City. *Pak. J. Pharmacol.* **2017**, *34*, 17–22.
12. Modupalli, N.; Naik, M.; Sunil, C.K.; Natarajan, V. Emerging non-destructive methods for quality and safety monitoring of spices. *Trends Food Sci. Technol.* **2021**, *108*, 133–147. [CrossRef]
13. Ullah, A.; Chan, M.W.H.; Aslam, S.; Khan, A.; Abbas, Q.; Ali, S.; Hussain, A.; Mirania, Z.A.; Sibt-e-Hassan, S.; Kazmik, M.R.; et al. Banned Sudan dyes in spices available at markets in Karachi, Pakistan. *Food Addit. Contam. Part B* **2023**, *16*, 69–76. [CrossRef] [PubMed]
14. Daud, N.M.; Putra, N.R.; Jamaludin, R.; Norodin, N.S.M.; Sarkawi, N.S.; Hamzah, M.H.S.; Nasir, H.M.; Zaidel, D.N.A.; Yunus, M.A.C.; Salleh, L.M. Valorisation of plant seed as natural bioactive compounds by various extraction methods: A review. *Trends Food Sci. Technol.* **2022**, *119*, 201–214. [CrossRef]
15. Kuley, E.; Özogul, F. Synergistic and antagonistic effect of lactic acid bacteria on tyramine production by food-borne pathogenic bacteria in tyrosine decarboxylase broth. *Food Chem.* **2011**, *127*, 1163–1168. [CrossRef] [PubMed]
16. Yazgan, H.; Kuley, E.; Güven Gökmen, T.; Regenstein, J.M.; Özogul, F. The antimicrobial properties and biogenic amine production of lactic acid bacteria isolated from various fermented food products. *J. Food Process. Preserv.* **2021**, *45*, e15085. [CrossRef]
17. Klausen, N.K.; Huss, H.H. A rapid method for detection of histamine-producing bacteria. *Int. J. Food Microbiol.* **1987**, *5*, 137–146. [CrossRef]
18. Özogul, F. Effects of specific lactic acid bacteria species on biogenic amine production by foodborne pathogen. *Int. J. Food Sci. Technol.* **2011**, *46*, 478–484. [CrossRef]
19. Özoğul, F. Production of biogenic amines by *Morganella morganii*, *Klebsiella pneumoniae* and *Hafnia alvei* using a rapid HPLC method. *Eur. Food Res. Technol.* **2004**, *219*, 465–469. [CrossRef]
20. Fazeli, M.R.; Amin, G.; Attari MM, A.; Ashtiani, H.; Jamalifar, H.; Samadi, N. Antimicrobial activities of Iranian sumac and avishan-e shirazi (*Zataria multiflora*) against some food-borne bacteria. *Food Control* **2007**, *18*, 646–649. [CrossRef]

21. Nasar-Abbas, S.; Halkman, A.K. Antimicrobial effect of water extract of sumac (*Rhus coriaria* L.) on the growth of some food borne bacteria including pathogens. *Int. J. Food Microbiol.* **2004**, *97*, 63–69. [CrossRef]
22. Rayne, S.; Mazza, G. Biological activities of extracts from sumac (*Rhus* spp.): A review. *Nat. Preced.* **2007**, *18*, 646–649.
23. Mostafa, A.A.; Al-Askar, A.A.; Almaary, K.S.; Dawoud, T.M.; Sholkamy, E.N.; Bakri, M.M. Antimicrobial activity of some plant extracts against bacterial strains causing food poisoning diseases. *Saudi J. Biol. Sci.* **2018**, *25*, 361–366. [CrossRef] [PubMed]
24. Mohammad Salamatullah, A.; Hayat, K.; Mabood Husain, F.; Asif Ahmed, M.; Arzoo, S.; Musaad Althbiti, M.; Alzahrani, A.; Al-Zaied, B.A.M.; Alyahya, H.K.; Albader, N.; et al. Effects of different solvents extractions on total polyphenol content, HPLC analysis, antioxidant capacity, and antimicrobial properties of peppers (red, yellow, and green (*Capsicum annum* L.). *Evid.-Based Complement. Altern. Med.* **2022**, *2022*, 7372101. [CrossRef] [PubMed]
25. Pundir, R.K.; Jain, P. Comparative studies on the antimicrobial activity of black pepper (*Piper nigrum*) and turmeric (*Curcuma longa*) extracts. *Int. J. Appl. Biol. Pharm. Technol.* **2010**, *1*, 492–500.
26. Tırıs, G.; Yanıkoğlu, R.S.; Ceylan, B.; Egeli, D.; Tekkeli, E.K.; Önal, A. A review of the currently developed analytical methods for the determination of biogenic amines in food products. *Food Chem.* **2023**, *398*, 133919. [CrossRef]
27. Mallik, S.; Sharangi, A.; Sarkar, T. Phytochemicals of coriander, cumin, fenugreek, fennel and black cumin: A preliminary study. *Natl. Acad. Sci. Lett.* **2020**, *43*, 477–480. [CrossRef]
28. Merah, O.; Sayed-Ahmad, B.; Talou, T.; Saad, Z.; Cerny, M.; Grivot, S.; Evon, P.; Hijazi, A. Biochemical Composition of Cumin Seeds, and Biorefining Study. *Biomolecules* **2020**, *10*, 1054. [CrossRef]
29. Sharma, A.; Verma, P.; Mathur, A.; Mathur, A.K. Genetic engineering approach using early Vinca alkaloid biosynthesis genes led to increased tryptamine and terpenoid indole alkaloids biosynthesis in differentiating cultures of *Catharanthus roseus*. *Protoplasma* **2018**, *255*, 425–435. [CrossRef]
30. Abu-Reidah, I.M.; Ali-Shtayeh, M.S.; Jamous, R.M.; Arráez-Román, D.; Segura-Carretero, A. HPLC–DAD–ESI-MS/MS screening of bioactive components from *Rhus coriaria* L.(Sumac) fruits. *Food Chem.* **2015**, *166*, 179–191. [CrossRef]
31. Burgut, A.; Kuley, E.; Ucar, Y.; Özogul, F. Suppression effects of aqueous and ethanolic extracts of propolis on biogenic amine production by *Morganella psychrotolerans*. *LWT* **2020**, *131*, 109771. [CrossRef]
32. Houicher, A.; Bensid, A.; Regenstein, J.M.; Özogul, F. Control of biogenic amine production and bacterial growth in fish and seafood products using phytochemicals as biopreservatives: A review. *Food Biosci.* **2021**, *39*, 100807. [CrossRef]
33. Rathod, N.B.; Elabed, N.; Punia, S.; Ozogul, F.; Kim, S.K.; Rocha, J.M. Recent developments in polyphenol applications on human health: A review with current knowledge. *Plants* **2023**, *12*, 1217. [CrossRef] [PubMed]
34. Wang, S.; Zhu, F. Chemical composition and biological activity of staghorn sumac (*Rhus typhina*). *Food Chem.* **2017**, *237*, 431–443. [CrossRef] [PubMed]
35. Ghorbani, P.; Soltani, M.; Homayouni-Tabrizi, M.; Namvar, F.; Azizi, S.; Mohammad, R.; Boroumand Moghaddam, A. *Sumac* silver novel biodegradable nano composite for bio-medical application: Antibacterial activity. *Molecules* **2015**, *20*, 12946–12958. [CrossRef] [PubMed]
36. Harich, M.; Maherani, B.; Salmieri, S.; Lacroix, M. Evaluation of antibacterial activity of two natural bio-preservatives formulations on freshness and sensory quality of ready to eat (RTE) foods. *Food Control* **2018**, *85*, 29–41. [CrossRef]
37. Karsha, P.V.; Lakshmi, O.B. Antibacterial activity of black pepper (*Piper nigrum* Linn.) with special reference to its mode of action on bacteria. *Indian J. Nat. Prod. Resour.* **2010**, *1*, 213–215.
38. Milenković, A.N.; Stanojević, L.P. Black pepper: Chemical composition and biological activities. *Adv. Technol.* **2021**, *10*, 40–50. [CrossRef]
39. Zou, L.; Hu, Y.-Y.; Chen, W.-X. Antibacterial mechanism and activities of black pepper chloroform extract. *J. Food Sci. Technol.* **2015**, *52*, 8196–8203. [CrossRef]
40. Jia, W.; Zhang, R.; Shi, L.; Zhang, F.; Chang, J.; Chu, X. Effects of spices on the formation of biogenic amines during the fermentation of dry fermented mutton sausage. *Food Chem.* **2020**, *321*, 126723. [CrossRef]
41. Trifan, A.; Zengin, G.; Brebu, M.; Skalicka-Woźniak, K.; Luca, S.V. Phytochemical characterization and evaluation of the antioxidant and anti-enzymatic activity of five common spices: Focus on their essential oils and spent material extractives. *Plants* **2021**, *10*, 2692. [CrossRef]
42. Wendakoon, C.N.; Sakaguchi, M. Inhibition of amino acid decarboxylase activity of *Enterobacter aerogenes* by active components in spices. *J. Food Prot.* **1995**, *58*, 280–283. [CrossRef]
43. Oktariani, A.F.; Ramona, Y.; Sudaryatma, P.E.; Dewi, I.A.M.M.; Shetty, K. Role of marine bacterial contaminants in histamine formation in seafood products: A review. *Microorganisms* **2022**, *10*, 1197. [CrossRef] [PubMed]
44. Jeon, A.R.; Lee, J.H.; Mah, J.-H. Biogenic amine formation and bacterial contribution in Cheonggukjang, a Korean traditional fermented soybean food. *LWT* **2018**, *92*, 282–289. [CrossRef]
45. Kim, D.H.; Kim KB, W.R.; Cho, J.Y.; Ahn, D.H. Inhibitory effects of brown algae extracts on histamine production in mackerel muscle via inhibition of growth and histidine decarboxylase activity of *Morganella morganii*. *J. Microbiol. Biotechnol.* **2014**, *24*, 465–474. [CrossRef] [PubMed]
46. Shakila, R.J.; Vasundhara, T.; Rao, D.V. Inhibitory effect of spices on in vitro histamine production and histidine decarboxylase activity of *Morganella morganii* and on the biogenic amine formation in mackerel stored at 30 C. *Z. Für Lebensm.-Unters. Und Forsch.* **1996**, *203*, 71–76. [CrossRef]

47. Linares, D.M.; del Rio, B.; Redruello, B.; Ladero, V.; Martin, M.C.; Fernandez, M.; Ruas-Madiedo, P.; Alvarez, M.A. Comparative analysis of the in vitro cytotoxicity of the dietary biogenic amines tyramine and histamine. *Food Chem.* **2016**, *197*, 658–663. [CrossRef]
48. Mahmoud AF, A.; Elshopary, N.; El-Naby, G.H.; El Bayomi, R. Reduction of biogenic amines production in chilled minced meat using antimicrobial seasonings. *J. Microbiol. Biotechnol. Food Sci.* **2021**, *10*, e3663. [CrossRef]
49. Lin, Z.; Wu, Z.-Y.; Zhang, W.-X. Bioinformatics analysis of amino acid decarboxylases related to four major biogenic amines in pickles. *Food Chem.* **2022**, *393*, 133339. [CrossRef]

foods

Article

Characterization of the Chemical Composition and Biological Activities of Bog Bilberry (*Vaccinium uliginosum* L.) Leaf Extracts Obtained via Various Extraction Techniques

Bianca Eugenia Ştefănescu [1], Sonia Ancuţa Socaci [2], Anca Corina Fărcaş [2], Silvia Amalia Nemeş [1,2], Bernadette Emőke Teleky [1,2], Gheorghe Adrian Martău [1,3], Lavinia Florina Călinoiu [1,*], Laura Mitrea [2,*], Floricuţa Ranga [1,2], Dan Grigoroaea [4], Dan Cristian Vodnar [1,2] and Carmen Socaciu [2]

[1] Life Science Institute, University of Agricultural Sciences and Veterinary Medicine, 400372 Cluj-Napoca, Romania; bianca.vodnar@usamvcluj.ro (B.E.Ş.); amalia.nemes@usamvcluj.ro (S.A.N.); bernadette.teleky@usamvcluj.ro (B.E.T.); adrian.martau@usamvcluj.ro (G.A.M.); floricutza_ro@yahoo.com (F.R.); dan.vodnar@usamvcluj.ro (D.C.V.)

[2] Department of Food Science, Faculty of Food Science and Technology, University of Agricultural Sciences and Veterinary Medicine, 400372 Cluj-Napoca, Romania; sonia.socaci@usamvcluj.ro (S.A.S.); anca.farcas@usamvcluj.ro (A.C.F.); carmen.socaciu@usamvcluj.ro (C.S.)

[3] Department of Food Engineering, Faculty of Food Science and Technology, University of Agricultural Sciences and Veterinary Medicine, 400372 Cluj-Napoca, Romania

[4] Călimani National Park Administration, Şaru Dornei, 727515 Suceava, Romania; danranger1966@yahoo.com

* Correspondence: lavinia.calinoiu@usamvcluj.ro (L.F.C.); laura.mitrea@usamvcluj.ro (L.M.); Tel.: +40-740-157-575 (L.F.C.); +40-745-897-829 (L.M.)

Abstract: This investigation aimed to assess the chemical composition and biological activities of bog bilberry (*Vaccinium uliginosum* L.) leaves. Hydroethanolic extracts were obtained using four extraction techniques: one conventional (CE) and three alternative methods; ultrasound (UAE), microwave (MAE) and high-pressure (HPE) extractions. Spectrophotometric analysis was conducted to determine their chemical content, including the total phenolic content (TPC) and total flavonoid content (TFC). Furthermore, their antioxidative and antimicrobial properties were evaluated. HPLC (high performance liquid chromatography) analysis identified and quantified 17 phenolic compounds, with chlorogenic acid being the predominant compound, with the lowest level (37.36 ± 0.06 mg/g) for the bog bilberry leaf extract obtained by CE and the highest levels (e.g., HPE = 44.47 ± 0.08 mg/g) for the bog bilberry leaf extracts obtained by the alternative methods. Extracts obtained by HPE, UAE and MAE presented TPC values (135.75 ± 2.86 mg GAE/g; 130.52 ± 1.99 mg GAE/g; 119.23 ± 1.79 mg GAE/g) higher than those obtained by the CE method (113.07 ± 0.98 mg GAE/g). Regarding the TFC values, similar to TPC, the highest levels were registered in the extracts obtained by alternative methods (HPE = 43.16 ± 0.12 mg QE/g; MAE = 39.79 ± 0.41 mg QE/g and UAE = 33.89 ± 0.35 mg QE/g), while the CE extract registered the lowest level, 31.47 ± 0.28 mg QE/g. In the case of DPPH (1,1-diphenyl-2-picrylhydrazyl) antioxidant activity, the extracts from HPE, UAE and MAE exhibited the strongest radical scavenging capacities of 71.14%, 63.13% and 60.84%, respectively, whereas the CE extract registered only 55.37%. According to Microbiology Reader LogPhase 600 (BioTek), a common MIC value of 8.88 mg/mL was registered for all types of extracts against *Staphylococcus aureus* (Gram-positive bacteria) and *Salmonella enterica* (Gram-negative bacteria). Moreover, the alternative extraction methods (UAE, HPE) effectively inhibited the growth of *Candida parapsilosis*, in comparison to the lack of inhibition from the CE method. This study provides valuable insights into bog bilberry leaf extracts, reporting a comprehensive evaluation of their chemical composition and associated biological activities, with alternative extraction methods presenting greater potential for the recovery of phenolic compounds with increased biological activities than the conventional method.

Keywords: bog bilberry leaves; biological activities; conventional extraction; high-pressure extraction; microwave-assisted extraction; polyphenolic compounds; ultrasound-assisted extraction

1. Introduction

Significant interest has grown in using bioactive compounds obtained from plants for the treatment and/or prevention of various non-communicable diseases. Secondary metabolites, including phenolic compounds, carotenoids, and other naturally occurring plant-derived molecules, have been the subject of numerous studies, revealing their beneficial impact on health [1]. Additionally, there has been a considerable focus on researching and developing novel plant-derived functional products and dietary supplements, characterized as nutrient-rich foods with a heightened concentration of antioxidants, a subject that has been extensively investigated in recent years [2]. Many species of the *Vaccinium* genus are renowned for containing a large amount and variety of phenolic compounds [3–6].

Vaccinium uliginosum L. (bog bilberry) is a wild bush indigenous to many parts of the Northern Hemisphere, particularly at higher altitudes in Asia, and in North America and Europe. This small shrub is circumpolar in the Arctic and boreal regions, and it grows on moist and acidic ground, and many different types of wildlife animals consume both the leaves and the fruits [7].

The fruits of *V. uliginosum* are characterized by the presence of anthocyanins and flavonols. Bog bilberries have a distinct flavonol and anthocyanidin profile compared to other *Vaccinium* berries. As a result, it appears that their phenolic profile could be utilized to distinguish them from other berries [6,8].

The leaves of *V. uliginosum* are characterized by several groups of phenolic compounds. In a study by Stanoeva et al. [9], their results revealed the presence of 20 phenolics in the leaf extract from five groups of phenolic compounds: phenolic acids, flavonols, flavanols, iridoids and cinchonain. The extract obtained from *V. uliginosum* leaves is abundant in chlorogenic acid, comprising 64% of the total amount of phenolic acids. The leaves also contain several other phenolic acids, like caffeoylquinic acid, p-coumaroylquinic acid derivatives, feruloylquinic acid, gallic acid derivatives, and various flavonols, derivatives of quercetin, kaempferol and isorhamnetin [9].

The plant material contains a wide range of polyphenolic structures, including simple, complex or polymerized phenolic compounds. These compounds often interact with other molecules naturally present in plants, such as polysaccharides and lipids, making the recovery of these polyphenols a challenging process. For this reason, finding an optimum extraction technique that results in the outstanding recovery of phenolic compounds from plant material is an essential stage in the research's success, and various methods may be employed to achieve the purpose [10].

Conventional extraction (CE) methods are standard and easy to perform, although they require a significant volume of solvents and are often demanding in terms of time and energy. These methods could also lead to the deterioration of thermally sensitive polyphenolics and are frequently challenging, rendering them unsuitable for large-scale applications [10,11]. Due to these limitations, there is an increasing concern about using alternative and environmentally friendly methods for the recovery of polyphenolic compounds from plant materials. The main objective of studying alternative extraction techniques is to minimize the extraction time, decrease the amount of energy consumed and the volume of solvent, increase the extraction yield, and reduce negative environmental impact [11].

Ultrasound-assisted extraction (UAE), microwave-assisted extraction (MAE) and high-pressure extraction (HPE) have gained attention because of their multiple benefits, including increased yields of extracted compounds and reduced extraction duration and solvent utilization [12].

UAE is generally considered a highly productive and economical method for extracting phenolic compounds from plant-based materials. This is due to the widespread use and effectiveness of ultrasonic equipment. In addition, UAE allows the use of lower temperatures and the preservation of thermally sensitive compounds [13,14]. UAE utilizes high-frequency mechanical waves to generate the cavitation phenomenon. Cavitation is the occurrence of the development and subsequent destruction of cavities in a liquid due to the passage of ultrasonic waves, subject to certain conditions. This effect results in enhanced interaction between the solvent and the cell content, leading to the improved extraction of phenolic compounds [15,16].

The defining aspect of MAE is the interdependent interaction of the processes related to the transfer of heat and mass, in which the two gradients are acting in a single direction, and associated with the volumetric dispersal of heat within the radiated environment. In addition, it has been observed that the heating process results in interior pressure, leading cell walls to break down, facilitating the solvent's access inside and promoting the extraction of bioactive compounds [17].

HPE is one of the developing technologies that has been effectively used to extract biologically active compounds from plant-based materials [18]. HPE induces a significant pressure gradient between the cell's interior and exterior, inducing the structural deformation of the cell walls and membranes. This deformation increases the cells' permeability and, consequently, increases bioactive compound extraction into the solvent [18,19].

To our knowledge, investigations on the chemical composition of Romanian bog bilberry leaves are poor and have yet to be performed. Hence, this research aimed to extract phenolic compounds from bog bilberry leaves using various extractions techniques: CE, UAE, MAE and HPE. Additionally, the study evaluated their biological activities (antioxidant and antimicrobial) along with assessing their phenolic and flavonoid contents.

2. Materials and Methods

2.1. Chemicals and Reagents

The reference substances (catechin, chlorogenic acid, quercetin and gallic acid) were bought from Sigma-Aldrich (Steinheim, Germany). The same source was used to obtain the chemical reagents needed for the relevant analytical techniques (extraction, chemical and biological characterization). The culture media for the antimicrobial activity were purchased from BioMerieux (Craponne, France) and Sigma-Aldrich (Steinheim, Germany).

2.2. Plant Material

The leaves of *V. uliginosum* L. were harvested in the autumn of 2021 from the spontaneous wild flora of 12 Apostoli, Suceava county, Romania, and, afterwards, the leaves were dried (7–10 days, room temperature, darkroom) as previously described in our other studies [5,20,21]. The dried leaves were ground to achieve a fine powder and stored in a dark, cool, dry place until the analyses were completed. The extraction solvent, namely ethanol/water (40% *v*/*v*) and the solid/liquid ratio adopted for all the extraction methods (1:14) were based on our previous studies [5,20], whereas in the present study we have doubled the quantity of plant material for concentration purposes, which should be reflected in the increased biological activities.

2.3. Extraction Procedures

2.3.1. Conventional Extraction Protocol

For CE, the previously validated method by Dahmoune et al. [22] was followed, with slight modifications: over 1.5 g of plant powder was added 21 mL of ethanol/water (40%, *v*/*v*) in a glassware-type container (Erlenmeyer) that was closed during the extraction. After stirring for 2 h (750 rpm, room temperature) using magnetic stirrer equipment (Heidolph MR-Hei-Standard, Schwabach, Germany), the mixture was centrifuged at 10,000 rpm for 10 min at 24 °C, and the supernatant was filtered and stored at −18 °C until further analyses.

2.3.2. Ultrasound-Assisted Extraction Protocol

For the UAE, our previous method [5] was used, with slight modifications: the leaf powder (1.5 g) was extracted with 21 mL 40% *v*/*v* ethanol/water for 30 min in a closed glassware-type container (Erlenmeyer) using an ultrasonic bath (Elmasonic E15H, Elma, Singen, Germany) at room temperature. After centrifugation at 10,000 rpm for 10 min at 24 °C, the supernatant was filtered and stored (−18 °C) until further analyses.

2.3.3. Microwave-Assisted Extraction Protocol

MAE of the phenolic compounds from bog bilberry leaves was conducted following the previous method of Nisca et al. [23], with slight modifications. Briefly, a Milestone ETHOS-X microwave oven (Milestone srl, Bergamo, Italy) system with 40% *v*/*v* ethanol/water solvent at 280 W for 5 min and a closed glassware-type container (Erlenmeyer) were used. The amount of powdered sample used for extraction was 1.5 g, along with 21 mL of extraction solvent. After the MAE, the extract was cooled at room temperature (from 85 °C), centrifuged for 10 min at 10,000 rpm, 24 °C, and the supernatant was recovered and stored at −18 °C until further analyses.

2.3.4. High-Pressure Extraction Protocol

The HPE procedure was conducted following the study of Ben Hamissa et al. [24], with slight modifications as follows: a Parr 4790 reactor (PARR Instrument Company, Moline, IL, USA) was used, outfitted with a Controller 4838 and modified with dual valves and pressure regulators to enable the controlled introduction and evacuation of CO_2 and N_2 gases within the reaction chamber. In two steps, 1.5 g of bog bilberry leaf powder was extracted in the static mode for 60 min using 21 mL of 40% *v*/*v* ethanol/water as a solvent. In the first stage, the experiment was carried out by replacing air by flushing carbon dioxide through the hermetically closed stainless steel chamber, before increasing and maintaining the pressure at 1000 kPa for 10 min. The second step involved introducing nitrogen until the gas mixture reached a stable pressure of 4000 kPa at 50 °C. After 50 min, the gas mixture was carefully ejected from the reaction chamber, and the extract was separated by centrifugation at 10,000 rpm for 10 min at 24 °C. The resulting supernatant was stored at a temperature of −18 °C, pending further analyses.

2.4. Analysis of Phenolic Compounds

2.4.1. HPLC-DAD-ESI-MS Analysis

Phenolic content identification and quantification were conducted via High Performance Liquid-Chromatography, HPLC-DAD-ESI-MS analysis, using an Agilent 1200 HPLC system equipped with a DAD detector linked to an MS-detector single-quadrupole Agilent 6110. The separation of phenolic compounds employed an Eclipse XDB C18 column (4.6 × 150 mm, particle size 5 μm) from Agilent Technologies, Santa Clara, CA, USA. Two gradients were utilized: the first comprised 0.1% acetic acid/acetonitrile (99:1) in distilled water (*v*/*v*) (solvent A), and the second contained 0.1% acetic acid in acetonitrile (*v*/*v*) (solvent B). The elution followed the procedure defined by Dulf et al. [25] at a flow rate of 0.5 mL/min. For MS fragmentation, the ESI (+) mode scanned a range of 100–1200 *m*/*z*, with the capillary voltage set at 3000 V, temperature at 350 °C and nitrogen flow at 8 L/min. DAD was utilized to measure the eluent, recording absorbance spectra (200–600 nm) throughout each run. Data examination was performed using Agilent ChemStation Software (Rev B.02.01–SR2 [260], Palo Alto, CA, USA). The identification and quantification of phenolic compounds involved comparing retention times, UV–Vis absorbance spectra, and mass spectra of peaks with three reference standards. The flavanol subclass compounds were measured using a calibration curve generated with a catechin standard within the concentration range of 10–200 μg/mL and presented as catechin equivalents (mg catechin/g plant material) ($y = 15.224x - 130.24$, $r^2 = 0.9985$). For the hydroxycinnamic acid subclass, quantification relied on a calibration curve established using chlorogenic acid in the range of 10–50 μg/mL, denoted as chlorogenic equivalents (mg chlorogenic acid/g plant material) ($y = 22.585x - 36.728$, $r^2 = 0.9937$). Quantification of flavonols was achieved using a calibration curve constructed using quercetin, in the concentration range of 10–200 μg/mL, expressed as quercetin equivalents (mg quercetin/g plant material) (y = 87.392x + 78.795, r^2 = 0.9951).

2.4.2. Total Phenolic Content

Total phenolic content (TPC) was assessed using the Folin–Ciocalteu method [26]. Aliquots of 25 μL of sample were mixed with 1.8 mL distilled water in a 24-well microplate. The extracts were mixed with 125 μL Folin–Ciocalteu reagent (0.2 N) and maintained at room temperature for 5 min. Afterward, the mixture was supplemented with 340 μL of a 7.5% (m/z) Na_2CO_3 solution to establish the initial conditions (pH ~ 10) facilitating the redox interaction between phenolic compounds and the Folin–Ciocalteu reagent. Subsequently, the solution was incubated in darkness at 25 °C for 2 h. A blank was prepared using ethanol, and the absorbance was read at 760 nm using a microplate reader (BioTek Instruments, Winooski, VT, USA). Gallic acid (0.01–1.00 mg/mL) was used to create the standard curve, and the TPC in the samples was recorded as gallic acid equivalent (GAE) (mg GAE/g of plant material).

2.4.3. Total Flavonoid Content

The extracts' total flavonoid content (TFC) was assessed using the aluminum chloride colorimetric method, following the protocol described in a previously published study [27], with slight modifications. The extracts were diluted with 720 μL of distilled water, and 90 μL of 5% $NaNO_2$ was added. After a 5 min incubation, the mixture was treated with 90 μL of 10% $AlCl_3$, followed by an additional 5 min incubation. Next, the mixture received an addition of 600 μL of 1 N NaOH. Subsequently, the absorbance was measured at 510 nm, employing quercetin as the reference standard. Each determination was performed in triplicate. The total flavonoid content was recorded as quercetin equivalent (QE) (mg QE/g plant material).

2.5. *DPPH Antioxidant Capacity*

The antioxidant capacity of the extracts was assessed through the utilization of the DPPH (1,1-diphenyl-2-picrylhydrazyl) method for assessing free radical scavenging capacity, following the protocol described previously [28]. To assess the antioxidant activity of the samples, we created triplicate preparations by combining 35 μL of previously hydroethanol-extracted samples with 250 μL of ethanol-based DPPH solution. After incubating the solution for 30 min at room temperature, in darkness, we measured the absorbance at 515 nm using a multi-mode plate reader (BioTek, Winuschi, VT, USA). The DPPH inhibition percentage (I%) was calculated as follows: $I\% = [(A_0 - A_E)/A_0] \times 100$, where A_0 = absorbance of blank and A_E = absorbance of the extract.

2.6. *Antimicrobial Activity*

2.6.1. Microbial Strains

The following standard microbial strains (obtained from the Food Biotechnology Laboratory, UASVM, Cluj-Napoca, Romania) were tested: *Staphylococcus aureus* subsp. *aureus* ATCC 29213, *Enterococcus faecalis* ATCC 29212, *Staphylococcus epidermidis* ATCC 12228, *Candida parapsilosis* ATCC 22019, *Candida zeylanoides* ATCC 20367, *Escherichia coli* ATCC 25922, *Pseudomonas aeruginosa* ATCC 27853 and *Salmonella enterica* (*S. typhimurium*) ATCC 14028. The strains were grown in test tubes containing 9 mL sterile TSB (tryptic soy broth), MH (Mueller–Hinton), BHI (brain heart infusion), NB (nutrient broth) and YPD (yeast extract, peptone, dextrose). The tubes with TSB were incubated for 24 h at 37 °C for *E. coli* and at 30 °C for *C. parapsilosis*. The tubes with MH were incubated for 24 h at 37 °C for *S. aureus* and *P. aeruginosa*. The tubes with BHI were incubated for 24 h at 37 °C for *E. faecalis*. The tubes with NB were incubated for 24 h at 37 °C for *S. epidermidis* and *S. enterica* (*S. typhimurium*). The tubes with YPD were incubated for 24 h at 30 °C for *C. zeylanoides*. A loopful of inoculum was transferred to a growth agar medium. Plates were incubated for 24 h at 37 °C or 30 °C, respectively. Bacterial morphology was confirmed by optical microscopy (Nikon ECLIPSE Ci-L, Tokyo, Japan) for an accurate interpretation of results and extra justifications of the bacteria response. Multiple colonies of each strain grown on the mentioned media were moved into 9 mL of sterile saline solution (8.5 g/L NaCl) and adjusted to match the turbidity of McFarland 0.5 standard (10^8 CFU/mL). Subsequently, microbial suspensions of 10^5 CFU/mL for bacteria and 10^6 CFU/mL for *Candida* spp., after suitable dilution, were prepared to be added to individual wells of the microplate.

2.6.2. Determination of the Minimum Inhibitory Concentration (MIC)

The MIC was determined through the resazurin microtiter plate-based antibacterial assay [29]. Initially, 100 µL of a specific sterile medium for each strain was dispensed into the wells of a 96-well microplate. Subsequently, 100 µL of each extract (71.43 mg/mL concentration) was introduced into the first well, with consecutive 2-fold dilutions prepared across each row by transferring 100 µL from well to well. The excess 100 µL in the final well of the row was removed. Following this, 10 µL of inoculum ($10^5/10^6$ CFU/mL) was added to all wells. Positive controls (C+) consisted of Gentamicin (0.4 mg/mL in saline solution) or Ketoconazole (1 mg/mL in DMSO), while the negative control (C−) involved the extraction solvent (ethanol 40%). The microplates were then incubated for 20–22 h at 37 °C or 30 °C, after which 20 µL of sterile 0.2 mg/mL resazurin aqueous solution was added to all wells. Subsequent incubation for 2 h at 37 °C or 30 °C ensued. At the end of this period, the viable bacterial cells caused the resazurin (initially blue and non-fluorescent) to oxidize into resorufin (pink and fluorescent) within the wells. Thus, the concentration in the last well of each row that retained a blue color signified the complete inhibition of bacterial growth, indicating the MIC. Each experiment was performed in triplicate.

2.6.3. Determination of the Minimum Inhibitory Concentration (MIC) Using a Microbiology Reader LogPhase 600

The MIC was determined for *Staphylococcus aureus* subsp. *aureus* ATCC 29213 and *Salmonella enterica* (*S. typhimurium*) ATCC 14028 using a Microbiology Reader LogPhase 600 (Agilent BioTek, Santa Clara, CA, USA). A volume of 100 µL of sterile medium specific to each strain was added to the wells of a 96-well microplate. Then, 100 µL of each extract (71.43 mg/mL concentration) was added in the first well, and serial 2-fold dilutions were made in the subsequent wells of each row by transferring 100 µL from well to well. The surplus of 100 µL in the last well of the row was discarded. Then, 10 µL of inoculum (10^5 CFU/mL) was added to all the wells. Gentamicin (0.4 mg/mL in saline solution) was used as the positive control (C+), and the extraction solvent (ethanol 40%) was the negative control (C−). The microplates were incubated in the Microbiology Reader LogPhase 600 for 24 h at 37 °C and 600 rpm, and plates were read at 600 nm absorbance to determine their optical density (OD). The increase in OD versus the initial load of each microorganism added to the extracts (since these loads were not visually detected during observation) was considered a consequence of bacterial growth, indicating no antimicrobial effect. Therefore, the MIC was defined as the concentration at which no OD increase was observed in comparison with the initial loads/values [30].

2.7. Statistical Analysis

The outcomes of each study (each with its three or four replicates) were presented as the mean value ± SD. Statistical analysis was conducted using Graph Prism Version 8.0.1 (GraphPad Software Inc., San Diego, CA, USA) via a one-way ANOVA, followed by Tukey's multiple comparison tests. Significant distinctions between means were considered statistically significant at a 5% significance level.

3. Results and Discussions

3.1. Phenolic Profile of Bog Bilberry Leaves under Different Extraction Methods

Seventeen phenolic compounds were detected in the leaves of the Romanian bog bilberry. They belong to three phenolic groups: hydroxycinnamic acids, flavanols and flavonols. It is essential to mention that all 17 phenolic compounds were identified in the samples obtained by all four extraction methods (Table 1). The identified phenolic acids include chlorogenic acid (5-caffeoylquinic acid), neochlorogenic acid (3-caffeoylquinic acid) and caffeic acid. Within the flavanols class, four compounds were identified: gallocatechin, epicatechin, procyanidin dimer and procyanidin trimer. Additionally, in the flavonols class, ten compounds were identified, including quercetin, kaempferol and derivatives of quercetin, kaempferol and isorhamnetin. These findings are aligned with the research of Stanoeva et al. [9], in which the bog bilberry fruits and leaves harvested from northwestern Macedonia were investigated.

Table 1. The phenolic compounds detected in the leaf extracts of the bog bilberry using HPLC.

Peak No.	Retention Time R_t(min)	UV λ_{max}(nm)	$[M + H]^+$ (m/z)	Compound	Subclass
1	3.16	279	307, 290	Gallocatechin	Flavanol
2	11.52	330	355, 163	3-Caffeoylquinic acid (Neochlorogenic acid)	Hydroxycinnamic acid
3	12.27	330	355, 163	5-Caffeoylquinic acid (Chlorogenic acid)	Hydroxycinnamic acid
4	13.56	280	579, 291	Procyanidin dimer	Flavanol
5	13.80	282, 329	181, 163	Caffeic acid	Hydroxycinnamic acid
6	14.01	280	291	Epicatechin	Flavanol
7	14.43	280	867, 291	Procyanidin trimer	Flavanol
8	15.78	263, 355	611, 303	Quercetin-rutinoside (Rutin)	Flavonol
9	16.20	263, 355	465, 303	Quercetin-glucoside	Flavonol
10	16.44	263, 355	479, 303	Quercetin-glucuronide	Flavonol
11	17.24	263, 355	435, 303	Quercetin-arabinoside	Flavonol
12	17.43	260, 340	463, 287	Kaempferol-glucuronide	Flavonol
13	17.77	260, 360	493, 317	Isorhamnetin-glucuronide	Flavonol
14	18.44	260, 360	463, 317	Isorhamnetin-rhamnoside	Flavonol
15	18.79	260, 340	419, 287	Kaempferol-arabinoside	Flavonol
16	21.79	261, 355	303	Quercetin	Flavonol
17	23.39	260, 340	287	Kaempferol	Flavonol

Of the phenolic compounds, flavonols exhibited the highest presence in terms of number, but ranked second in terms of the highest amounts for all four extraction methods (Table 2). Moreover, hydroxycinnamic acids were the most prevalent subclass in terms of the highest levels, ranging from 68.99 $\pm$ 0.25 mg/g (in the CE extract) to 85.41 $\pm$ 0.22 mg/g (in the HPE extract). It is significant to mention that our results are in agreement with the literature [10,31]. In the studies of Dobroslavić et al. [10,31], according to their UPLC-MS/MS results, the content of phenolic acids of *Laurus nobilis* L. leaf extract obtained by pressurized liquid extraction was higher than that obtained by other extraction methods (conventional heat reflux, UAE and MAE extractions). These results may be due to the demonstrated thermal stability of phenolic acids, particularly hydroxycinnamic acids [32]. Similar results were presented in our previous study, where the extracts from blueberry leaves obtained by UAE presented the highest amounts of hydroxycinnamic acids compared to the other phenolic groups [20].

Among the identified phenolic compounds, chlorogenic acid, belonging to the hydroxycinnamic group, showed the highest concentration, with its lowest level (37.36 $\pm$ 0.06 mg/g) for the bog bilberry leaf extract obtained by CE and its highest levels (e.g., HPE = 44.47 $\pm$ 0.08 mg/g) for the bog bilberry leaf extract obtained by alternative methods. Neochlorogenic acid, also a hydroxycinnamic acid, was the second most abundant phenolic compound identified in the leaves of bog bilberries, with a concentration ranging from 24.16 ± 0.06 mg/g (for CE method) to 32.50 ± 0.04 mg/g (for UAE method). An increase in photosynthetic active radiation significantly improved the total amount of hydroxycinnamic acids, as observed by Bidel et al. [33]. To protect vital cells against damaging UV radiation, hydroxycinnamic acids will probably accumulate more when exposed to intense light [34]. In addition, the accumulation of plants' secondary metabolites, specifically hydroxycinnamic acids, is sustained at higher altitudes and cooler temperatures [35]. Regarding caffeic acid, the extract obtained by HPE had a significant concentration (9.87 ± 0.07 mg/g).

Table 2. The concentration of individual phenolic compounds in the extracts of the bog bilberry leaves using different extraction methods, expressed as mg/g.

Phenolic Compounds		Extraction Methods			
		CE	UAE	MAE	HPE
Hydroxycinnamic acids	Neochlorogenic acid	24.16 ± 0.06	32.50 ± 0.04 ***	27.12 ± 0.08 ***	31.07 ± 0.07 ***
	Chlorogenic acid	37.36 ± 0.06	43.22 ± 0.08 ***	39.65 ± 0.01 ***	44.47 ± 0.08 ***
	Caffeic acid	7.47 ± 0.13	8.36 ± 0.08 ***	7.32 ± 0.08 *	9.87 ± 0.07 ***
Flavanols	Gallocatechin	9.92 ± 0.03	7.67 ± 0.02 ***	9.75 ± 0.07 ***	11.23 ± 0.09 ***
	Epicatechin	5.80 ± 0.05	5.93 ± 0.04 N.S.	7.69 ± 0.05 ***	5.15 ± 0.05 ***
	Procyanidin dimer	4.59 ± 0.03	4.58 ± 0.02 N.S.	4.86 ± 0.03 ***	3.22 ± 0.09 ***
	Procyanidin trimer	5.54 ± 0.03	4.73 ± 0.03 ***	6.16 ± 0.04 ***	4.44 ± 0.04 ***
Flavonols	Quercetin-rutinoside (Rutin)	0.37 ± 0.02	0.26 ± 0.02 ***	0.45 ± 0.02 ***	0.69 ± 0.03 ***
	Quercetin-glucoside	9.37 ± 0.03	9.07 ± 0.05 ***	9.36 ± 0.07 N.S.	9.78 ± 0.04 ***
	Quercetin-glucuronide	13.54 ± 0.04	13.49 ± 0.05 N.S.	13.65 ± 0.04 **	16.09 ± 0.04 ***
	Quercetin-arabinoside	1.84 ± 0.05	1.92 ± 0.02 ***	1.75 ± 0.01 ***	2.38 ± 0.02 ***
	Kaempferol-glucuronide	8.71 ± 0.07	9.22 ± 0.11 ***	8.70 ± 0.11 N.S.	10.35 ± 0.15 ***
	Isorhamnetin-glucuronide	3.27 ± 0.14	3.81 ± 0.11 ***	3.25 ± 0.08 N.S.	4.533 ± 0.076 ***
	Isorhamnetin-rhamnoside	0.27 ± 0.01	0.28 ± 0.0.1 **	0.34 ± 0.02 ***	0.93 ± 0.02 ***
	Kaempferol-arabinoside	0.24 ± 0.01	0.25 ± 0.02 N.S.	0.29 ± 0.02 ***	0.58 ± 0.02 ***
	Quercetin	0.62 ± 0.02	0.51 ± 0.02 ***	0.39 ± 0.01 ***	0.70 ± 0.02 ***
	Kaempferol	0.41 ± 0.02	0.47 ± 0.03 ***	0.14 ± 0.01 ***	0.49 ± 0.01 ***

Phenolic compounds, including flavonoids such as flavanols and flavonols, as well as hydroxycinnamic acids, were quantified as a concentration of miligrams per gram (mg/g). The experiments were replicated four times, and the reported values represent the average and standard deviation (SD) of these replicates. Data normality was assessed using the Shapiro–Wilk test, where a p-value greater than 0.05 indicated normally distributed data. The mean ± SD is presented in the descriptive statistics table. To investigate significant differences between the four extraction methods for each compound, a two-way ANOVA was conducted, followed by Tukey's multiple comparisons test. In this analysis, the second column compares CE and UAE, the third column compares CE and MAE, and the fourth column assesses the distinctions between CE and HPE. Significance levels are denoted using the following symbols: *** $p < 0.001$, ** $p < 0.01$, * $p < 0.05$, and N.S. (not significant).

Interestingly, in the research of Stanoeva et al. [9], similar to our results, chlorogenic acid was the predominant compound found in the leaves of bog bilberry, representing 64% of the total phenolic acid derivatives. However, they did not report the presence of neochlorogenic and caffeic acids in their extract from bog bilberry leaves. To our knowledge, there have been no reports of these compounds' occurrence in bog bilberry leaves, only in the fruits and leaves of other *Vaccinium* species [3,20,21,36].

Considering the flavonols class, quercetin-glucuronide was the predominant phenolic compound measured, for which the extracts obtained by alternative methods (e.g., HPE, MAE) registered the highest concentrations (16.09 ± 0.04 mg/g; 13.65 ± 0.04 mg/g) in comparison to CE (13.54 ± 0.04 mg/g). The second most abundant flavonol identified in the bog bilberry extracts was kaempferol-glucuronide, where again the modern techniques registered the most significant quantities: HPE = 10.35 ± 0.15 mg/g and UAE = 9.22 ± 0.11 mg/g, in comparison with CE (8.71 ± 0.07 mg/g). Traces of quercetin-rutinoside (Rutin) were found in all the extracts. In the study of Stanoeva et al. [9], quercetin-rutinoside (Rutin) was not detected. However, in our previous studies [5,20] on other

Vaccinium spp. leaves, quercetin-rutinoside (Rutin) quantities were 50 to 70-fold higher than in the bog bilberry leaves analyzed in the present study.

Moreover, quercetin and kaempferol aglycones were also quantified in small amounts. These flavonol aglycones were not detected in the research of Stanoeva et al. [9], where bog bilberry leaves were analyzed, but they have been quantified in other *Vaccinium* spp. leaves [3–5,20]. Additionally, two glycosides of isorhamnetin were identified in bog bilberry leaves: isorhamnetin-glucuronide and isorhamnetin-rhamnoside. At the same time, the study of Stanoeva et al. [9] reported only one glycoside of isorhamnetin. As previously reported [3,4,9], and similar to our results, kaempferol glycosides were considerably less prevalent than quercetin glycosides in the *Vaccinium* spp.

In the flavanols group, four compounds were detected, with gallocatechin as the dominant phenolic compound, with its highest concentration found in the extract obtained by the modern method HPE, namely 11.23 ± 0.09 mg/g. Epicatechin was identified in all four extracts, ranging from 5.15 ± 0.05 mg/g to 7.69 ± 0.05 mg/g. The two procyanidins were detected in all the extracts, with levels comparable to the amount of epicatechin. Stanoeva et al. [9] reported that in the leaves of bog bilberry from Macedonia, there was only one compound from the flavanol group, namely procyanidin dimer.

Significant selectivity was not induced by our extraction techniques, as indicated by the absence of noticeable variations in the characteristics of the extracts in their HPLC profiles. The extracts acquired using all four extraction methods displayed equivalent qualitative contents, respectively; the same 17 phenolic compounds were identified in all the extracts.

According to previous research [37], utilizing various types of energy, such as ultrasound, microwaves and high pressure, may provide a potentially beneficial alternative for enhancing the levels of phenolic extraction. For example, the results reported by Caldas et al. [37] in extracting phenolic compounds from grape skin showed that the UAE and MAE methods provided greater phenolic recovery within a reduced time period.

Additionally, in the research study of Mróz et al. [38] all of the alternative extraction techniques investigated, MAE, UAE and HPE, enhanced the total recovery of phytochemicals from the flowering aerial parts of *Sideritis scardica* and *Sideritis raeseri*.

Quantitatively, in the present study, it was observed that the highest concentrations were obtained in the bog bilberry leaf extracts obtained by modern techniques (HPE, UAE, MAE) for almost all phenolic compounds, in comparison to the extract obtained by CE. Given that distinct extraction mechanisms are employed in the HPE, UAE and MAE methods, it is appropriate to expect diverse secondary metabolites to be collected from the bog bilberry leaves when utilizing these extractions techniques. Furthermore, these extraction methods will likely uniquely impact the amounts of the individual phenolic compounds extracted [12].

3.2. *Total Phenolic and Total Flavonoid Content*

The TPC and TFC values of the bog bilberry leaf extracts obtained by various extraction methods are presented in Table 3. Our initial remark concerns the variation in the levels of phenolic and flavonoid constituents. In all the extracts obtained, the levels of TPC were consistently higher than those of TFC, independent of the extraction technique utilized.

Table 3. Total phenolic content, total flavonoid content and DPPH activity of the extracts.

Extraction Methods	TPC (mg GAE/g Plant Material)	TFC (mg QE/g Plant Material)	DPPH (I%)
CE	113.07 ± 0.98	31.47 ± 0.28	55.37%
UAE	130.52 ± 1.99 ***	33.89 ± 0.35 *	63.13%
MAE	119.23 ± 1.79 ***	39.79 ± 0.41 ***	60.84%
HPE	135.75 ± 2.86 **	43.16 ± 0.12 ***	71.14%

This study reported significance levels as *** $p < 0.001$, ** $p < 0.01$, * $p < 0.05$. Results, presented as mean ± standard deviation, were derived from three replicates. Two-way ANOVA followed by Turkey's multiple comparisons test explored differences between the four extraction methods.

Previous research studies have noted a wide range of variations in TPC for extracts derived from various *Vaccinium* species' leaves. In our previous study [5], we observed variations in the TPC of the extracts from bilberry (*Vaccinium myrtilus* L.) leaves collected from three different habitats; the values varied from 132.47 to 135.8 mg GAE/g plant material. In the same study [5], we reported comparable TPC values in the extracts from lingonberries (*Vaccinium vitis-idaea* L.) leaves collected from three distinct habitats. Moreover, in another study on bilberry leaves [39], the reported values of the TPC were higher than our results, with values varying between 196.48 and 280.69 mg GAE/g extract. In Bujor et al.'s study [4], the TPC values of lingonberry leaf extracts were from 135 to 158 mg GAE/g dry extract, depending on the harvest period. Furthermore, in blueberry leaf (*Vaccinium corymbosum* L.) extracts the values of the TPC ranged from 98.00 to 135.55 mg GAE/g plant material, depending on the cultivar [20].

In the research study of Páscoa et al. [40], an extract from winter leaves of the blueberry (*V. corymbosum* L.) cultivar Aurora contained the highest TPC (227.4 mg GAE/g dry leaf) of all harvest seasons (spring, autumn and winter). Conversely, their extract from the blueberry cultivar Huron (autumn leaves) exhibited the lowest TPC (39.6 mg GAE/g dry leaf). Within the research study of Gao et al. [41], the extract obtained using 80% ethanol from the leaves of *Vaccinium dunalianum* presented a TPC value of 257.11 mg GAE/g dry extract. Our present findings exhibited a strong similarity to those reported by the abovementioned studies, despite the utilization of different extraction methods and solvents, *Vacciunium* species, harvest seasons and geographic regions.

The extracts obtained by HPE, UAE and MAE presented TPC values (135.75 ± 2.86 mg GAE/g; 130.52 ± 1.99 mg GAE/g; 119.23 ± 1.79 mg GAE/g) higher than those obtained by the CE method (113.07 ± 0.98 mg GAE/g).

Additionally, a high efficacy of non-conventional extraction methods has been observed for a variety of plant material. For instance, in the research study of Cheng et al. [42], the TPC values of the water and 60% methanol extracts obtained from jackfruit pulp using a UAE method were high. In the same study, they reported a higher TPC of the 60% ethanol extract obtained by MAE. Similar results were reported by Routray et al. [43], where their blueberry leaf extracts obtained by UAE and MAE presented high TPC values. Analogous results were reported by Alexandre et al. [44], in which the TPC value of the extract obtained by HPE from prickly pear peel was quite significant.

Regarding the TFC values, similar to TPC, the highest levels were registered in the extracts obtained by alternative methods (HPE = 43.16 ± 0.12 mg QE/g; MAE = 39.79 ± 0.41 mg QE/g; and UAE = 33.89 ± 0.35 mg QE/g), while the CE extract registered the lowest level, 31.47 ± 0.28 mg QE/g. These values are consistent with our previous results (31.36–67.88 mg QE/g plant material) reported for the 40% ethanol extracts of blueberry leaves obtained by UAE [20]. Moreover, our results are higher than Brezoiu et al.'s previous results (2.20–10.36 mg QE/g plant) [39] for bilberry leaves obtained by CE and UAE using ethanol or 50% ethanol. Additionally, using 80% ethanol and UAE, Gao et al. [41] reported higher values of the TFC in the leaves of *V. dunalianum*. These differences between TFC values were likely related to the composition of the solvent, the different parameters of the extraction techniques and the species of plants utilized. It is relevant to highlight that while earlier research has produced berry leaf extracts with increased or decreased TPCs and TFCs, the plant's location and surrounding factors affect the polyphenolic composition of the leaves. It has been indicated that the temperature's limiting effect on photosynthesis results in nearly two times higher TPC concentrations in the leaves of bilberry bushes growing in high-light locations, at higher latitudes, and/or at higher altitudes than in those growing at lower latitudes or altitudes. Furthermore, depending on biotic and abiotic pressures, there are seasonal fluctuations in the quantity and variety of phenolic compounds and the berry leaves' antioxidant activity [4,34,36,45,46].

3.3. DPPH Antioxidant Capacity

The DPPH radical scavenging capacity of extracts obtained from the leaves of bog bilberry by various extraction techniques was assessed in order to determine their antioxidant activity. The results (Table 3) indicated statistically significant differences in the antioxidant activity of the extracts derived from non-conventional extraction in comparison with those from the CE. According to the DPPH assay, the extracts from HPE, UAE and MAE exhibited a strongest radical scavenging capacity of 71.14%, 63.13% and 60.84%, respectively, whereas the CE extract registered only 55.37%.

Previous research studies have shown a notable direct relationship between overall phenolic and flavonoid contents and antioxidant activity [41,42,47,48]. These findings indicate that the antioxidant activity observed in bog bilberry leaf extracts could be attributed to their higher levels of TPC and TFC. As a consequence, in our previous study [20], the extracts obtained from the Toro, Elliot and Nelson varieties' leaves (*V. corymbosum* L.), which revealed a higher polyphenolic content, displayed the greatest antioxidant activity, expressed as a percentage of inhibition (70.41%, 68.42% and 58.69%, respectively). Brezoiu et al. [39] reported comparable results in their extracts obtained from bilberry leaves; the antioxidant activity increased with the increase in the TPC values.

The DPPH test revealed different results for bog bilberry leaf extracts obtained by various extraction techniques, and this may be explained by the different amounts of polyphenols with dihydroxyphenyl moieties in each extract, considering the different extraction parameters. The results of the HPLC revealed that the extracts from bog bilberry leaves contain phenolic acids and derivatives, including chlorogenic acid, which is known for its antioxidant properties. The antioxidant properties of these compounds emanate from o-diphenolic functionality and the presence of hydroxyl groups within their molecular structure. According to these characteristics, the molecule can donate electrons and hydrogen atoms [49]. Additionally, quercetin and quercetin derivatives, as well as other flavonols, exhibit the capacity to counteract free radicals due to the hydroxyl groups that constitute the molecule [50]. Moreover, among the polyphenolic compounds, the flavanol group as a whole, and the proanthocyanidin (procyanidin dimer and trimer) subgroup, equally possess the highest antioxidant activity [51] because of the catechol structures present in those molecules, linked by C3-OH and C4-C8 bonds, which greatly decrease the production of free radicals [52].

3.4. Antimicrobial Activity of Bog Bilberry Leaf Extracts

All the bog bilberry leaf extracts have been tested for their antimicrobial activity against three Gram-positive and three Gram-negative bacterial strains and against two fungi. The results of the minimum inhibitory concentration are presented in Table 4.

Table 4. The results of the determination of the minimum inhibitory concentration (MIC) (mg/mL) of the extracts against *Staphylococcus aureus* subsp. *aureus* ATCC 29213, *Enterococcus faecalis* ATCC 29212, *Staphylococcus epidermidis* ATCC 12228, *Candida parapsilosis* ATCC 22019, *Candida zeylanoides* ATCC 20367, *Escherichia coli* ATCC 25922, *Pseudomonas aeruginosa* ATCC 27853 and *Salmonella enterica* (*S. typhimurium*) ATCC 14028.

Extraction Methods	Gram (+) Bacteria			Fungi		Gram (−) Bacteria		
	S. aureus	*E. faecalis*	*S. epidermidis*	*C. parapsilosis*	*C. zeylanoides*	*E. coli*	*P. aeruginosa*	*S. enterica*
CE	17.75	h.c	17.75	h.c	h.c	h.c	h.c	17.75
UAE	17.75	h.c	17.75	8.88	h.c	h.c	h.c	8.88
MAE	17.75	h.c	h.c	h.c	h.c	h.c	h.c	17.75
HPE	17.75	h.c	h.c	8.88	h.c	h.c	h.c	17.75
Gentamicin	0.0001	0.013	0.002	-	-	0.003	0.0001	0.002
Ketoconazole	-	-	-	0.016	0.063	-	-	-

h.c—higher than the highest concentration tested (≤71.43 mg/mL); (-)—not tested.

Although the bog bilberry leaf extracts showed antioxidant activity and presented high TPC and TFC values, the extracts displayed antimicrobial activity only against some of the tested strains.

Regarding Gram-positive bacteria, all the extracts exhibited the same MIC towards *S. aureus*, respectively, 17.75 mg/mL. This result is in agreement with a previous research paper [53] evaluating the antibacterial activity of *Annona cherimola* phytochemicals obtained by UAE and comparing it to CE methods (maceration-MAC and Soxhlet-SE), where all the extracts presented an antimicrobial effect against *S. aureus*, with higher inhibition percentages from the UAE samples. In the present study, the strain *E. faecalis* was the most resistant. The results showed no inhibitory effect against this strain. In our previous study [20], *E. faecalis* was the most resistant strain towards all the blueberry leaf extracts tested. Regarding the *S. epidermidis* strain, only the extracts obtained by UAE and CE registered antimicrobial activity, with a MIC of 17.75 mg/mL. Our results are in line with previous studies. For example, in the study of Saifullah et al. [54], the antibacterial properties of their extracts prepared from modern techniques (MAE, UAE) and CE (SWB–shaking water bath) were not significantly different, a fact that could be due to the similarity of the phenolic compounds and antioxidant properties in the extracts obtained from these extraction techniques. In another study [55] evaluating the efficacy of two methods (agitation as the CE and UAE as the modern technique) at extracting phenolic compounds from 15 native plants, a greater inhibition capacity was obtained through UAE against three of the six bacteria studied: *Listeria monocytogenes*, *Listeria innocua* and *Salmonella choleraesuis*, whereas against the other three bacteria, *S. aureus*, *Bacillus cereus* and *E. coli*, the CE method proved to have a better antimicrobial capacity, therefore underlying the complex relationship between phenolic composition and biological activity. In the research paper of Mašković et al. [56], *Satureja hortensis* L. (summer savory) herb extracts were prepared using CE (MAC and SE) and non-conventional (UAE, MAE and subcritical water extraction-SWE) techniques. The antibacterial activity of their extracts was determined against 15 selected bacterial strains and the results showed MIC values of 7.81 μg/mL, with the SE extract having exhibited the greatest activity towards *S. aureus* and *E. coli*, the MAC extract toward *Enterobacter aerogenes* and the SWE extract toward *Staphylococcus saprophyticus*. On the other hand, the most resistant bacterial strain was *Salmonella enteritidis*. A similar research study [57], dealing with the application of CE methods (MAC and SE) and non-conventional (UAE, MAE and SWE) methods for the isolation of bioactive compounds from *Erica carnea* L. (spring heath), reported that, generally, the best antibacterial result was from the UAE extract, while the MAE and SWE extracts exhibited similar activities. The strongest activity was exerted by the SE extract against *E. coli*, the MAC extract against *E. aerogenes* and *Proteus mirabilis*, the UAE extract against *S. typhimurium* and the SWE extract against *S. saprophyticus*, with a MIC value of 7.81 μg/mL. In the research study of Gutiérrez-Sánchez et al. [58], similar findings to ours were reported regarding the lack of antimicrobial activity against specific strains. They assessed the antimicrobial capacity of their samples from the leaves of *Hamelia patens* against several strains and their results showed no inhibitory effect on the majority of studied Gram-negative and Gram-positive bacterial strains. They reported that 70% dimethyl sulfoxide was used to extract the phenolic compounds in their research study and not 70% ethanol, as was the case in another studies, whereas an explanation of the lack of antimicrobial activity from the extracts could be related to the solvent utilized [58].

When it comes to Gram-negative bacteria, the strain *S. enterica* was the only bacteria sensitive to the extracts, with a MIC of 8.88 mg/mL for the extract obtained by UAE and a MIC of 17.75 mg/mL for the extracts obtained by CE, MAE and HPE. The extracts displayed no antimicrobial activity against *E. coli* and *P. aeruginosa*. These results are comparable to the previous study [44], which compared the antimicrobial activity of prickly pear peel compounds extracted with modern (HPE and OM-ohmic heating) and CE (SE) techniques, whose results showed that for *S. aureus* and *S. enteritidis* the MIC obtained was 125 mg/mL, independent of the extraction method, while for *B. cereus* only the HPE extract exhibited an antimicrobial effect. Moreover, in the paper by Tanase et al. [59], the antibacterial activity of spruce bark (*Picea abies* L.) extracts obtained via CE and UAE methods was tested and the results revealed that both types of extracts had a stronger antimicrobial effect against Gram-positive cocci (*S. aureus*) compared to Gram-negative bacilli (*Klebsiella pneumoniae*, *P. aeruginosa*). However, the UAE extract presented a bactericidal effect on *K. pneumoniae* and *P. aeruginosa* while the CE extract presented a bactericidal effect only on *P. aeruginosa*. Recently, Vilkickyte et al. [60] reported that an extract from lingonberry leaves showed no antimicrobial activity against *E. coli*. Moreover, Tian et al. [61] reported that *E. coli* presented low sensitivity to an extract of berry plants and no inhibitory effect was noticed in extracts derived from bilberry, chokeberry and nettle leaves. Additionally, Silva et al. [62] observed that *E. coli* and *P. aeruginosa* were resistant to a blueberry leaf extract, and, more precisely, they did not identify any inhibition against these strains. Several previous studies reported the antimicrobial activity of other berry leaves against *E. coli* and *P. aeruginosa*. Bilberry and lingonberry leaf extracts displayed antimicrobial activity against *E. coli* and *P. aeruginosa*; these strains were the most resistant bacteria [5]. Similarly, in the research of Gil-Martínez et al. [46], *E. coli*, *P. aeruginosa*, *S. enterica* and *Shigella sonnei* were more resistant to bilberry leaf extract than other bacterial strains.

Regarding the two fungi tested in this study, none of the extracts had any effect against *C. zeylanoides*. Moreover, towards *C. parapsilosis*, only the extracts obtained by alternative extraction methods had an antimicrobial activity, with a MIC of 8.88 mg/mL.

The antimicrobial properties of plant-based extracts are often related to several constituents, including phenolic acids, flavonoids, tannins, alkaloids, terpenoids and lactones. Previous research has demonstrated that the phenolic compounds found in plants play an essential role in their antimicrobial properties. The efficacy of these properties is influenced by the particular mechanism of action of polyphenols, the amount of phenolic compounds, and the techniques used for extraction [63–65]. Moreover, a variety of mechanisms of action, including cytoplasmic membrane destabilization, plasma membrane permeabilization, the suppression of external microbial enzymes, direct effects on the metabolism of microbial cells and the deprivation of a substrate essential to microbial growth, are involved in the inhibition of the proliferation of bacteria [66]. Phenolic compounds are thought to affect the cytoplasmic membrane as their primary antibacterial mechanism. However, the external lipid membrane of Gram-negative bacteria may serve as an adjuvant protective barrier, which could explain why phenolic compounds are ineffective against them [20,62]. The bacteria can be more sensitive or more resistant to the action of the plant extracts.

Growth curves were created, after determining the minimum inhibitory concentrations, using a Microbiology Reader LogPhase 600 (Agilent BioTek, Santa Clara, CA, USA) to improve our comprehension of the extracts' impact on the inhibited microorganisms. The growth curves were created only for two of the microorganisms tested, because they were the only ones sensitive to all the extracts obtained from the bog bilberry leaves. Figure 1 shows the growth curves realized for the *S. aureus* strain using different concentrations of the four extracts obtained from bog bilberry leaves. Visual observation was employed first in our study to determine all MICs. However, it is not more precise than the spectrophotometric method used to create the growth curves. When microbial loads are low, cellular growth can occasionally produce turbidity invisible to the human eye, but it can be identified spectrophotometrically [30]. This situation was observed in our study. Using a resazurin microtiter plate-based antibacterial assay, the MIC for all the extracts of bog bilberry leaves was 17.75 mg/mL.

On the other hand, as can be seen in Figure 1 with the use of a Microbiology Reader LogPhase 600 (BioTek), the MIC for the extracts was lower, 8.88 mg/mL. Additionally, it can be noticed that for lower concentrations the growth of the *S. aureus* was inhibited for a period of time. Afterward, the bacteria started to grow.

Figure 2 shows the growth curves realized for the *S. enterica* strain using different concentrations of the four extracts obtained from bog bilberry leaves. Similar to the *S. aureus* strain, for all extracts, in the case of *S. enterica* strain, the MIC was lower when using the Microbiology Reader LogPhase 600.

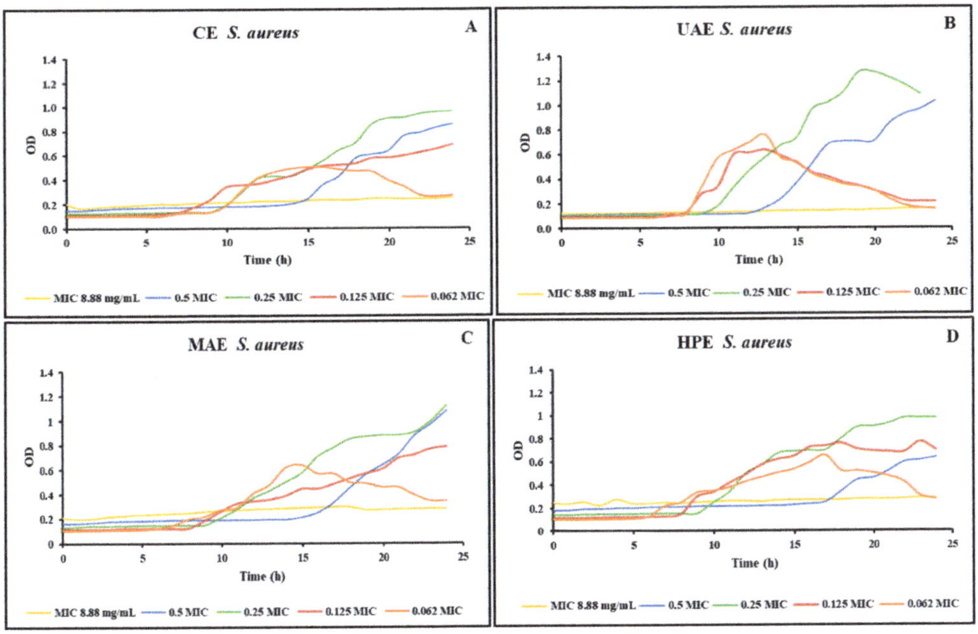

Figure 1. *S. aureus* growth curves for the extracts obtained by (**A**) CE, (**B**) UAE, (**C**) MAE, (**D**) HPE.

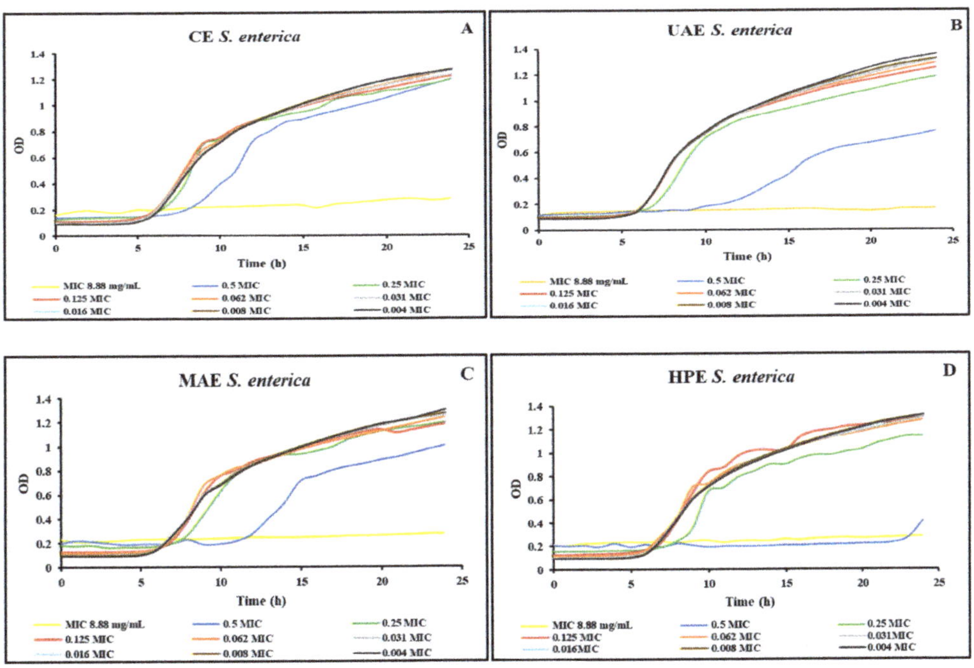

Figure 2. *S. enterica* growth curves for the extracts obtained by (**A**) CE, (**B**) UAE, (**C**) MAE, (**D**) HPE.

4. Conclusions

In conclusion, our study identified a rich profile of phenolic compounds in the leaves of the Romanian bog bilberry, belonging to hydroxycinnamic acids, flavanols and flavonols. Remarkably, all 17 phenolic compounds were consistently detected across all four extraction methods. The predominant phenolic acids were chlorogenic acid, neochlorogenic acid and caffeic acid, while for the flavonols class, quercetin-glucuronide was the most abundant; all in the highest quantities in the HPE, UAE and MAE extracts, and lowest in the CE-derived extract. Hydroxycinnamic acids, especially chlorogenic acid, were the most prevalent subclass across all the extraction methods, with their highest levels in alternative extraction-derived samples.

Quantitatively, HPE, UAE and MAE consistently yielded a significant phenolic and flavonoid content, along with a high antioxidant capacity, reinforcing the potential benefits of the alternative extraction techniques in comparison to conventional methods. However, antimicrobial activity was observed selectively against Gram-positive bacteria and *S. enterica*, underlining the complex relationship between phenolic composition and biological activity.

This comprehensive analysis provides valuable insights into the phenolic composition, extraction efficiency and bioactivity of *Vaccinium uliginosum* L., offering perspectives for future works (e.g., identification of the phenolic compounds/phenolic class responsible for the antibacterial activity) and supporting the potential uses and applications of these extracts in real food systems, such as functional foods and pharmaceuticals.

Author Contributions: Conceptualization, B.E.Ș., C.S., L.F.C. and L.M.; methodology, B.E.Ș., S.A.N., B.E.T. and D.G.; software, B.E.T., S.A.N. and F.R.; validation, B.E.Ș., C.S., L.F.C. and L.M.; formal analysis, S.A.S., A.C.F., S.A.N., F.R. and G.A.M.; resources, S.A.S., A.C.F., G.A.M. and D.G.; data curation, B.E.T., A.C.F., S.A.S. and F.R.; writing—original draft preparation, B.E.Ș., S.A.N., B.E.T. and G.A.M.; writing—review and editing, C.S., L.F.C., L.M. and D.C.V.; supervision, C.S., L.F.C. and L.M.; project administration, B.E.Ș. and D.C.V.; funding acquisition, B.E.Ș. and D.C.V. All authors have read and agreed to the published version of the manuscript.

Funding: This research has received funding from UEFISCDI-MCDI, project number PD 7/2022, PN-III-P1-1.1-PD-2021-0444, and from the European Union's Horizon Europe Framework Programme (HORIZON) under the Marie Skłodowska-Curie grant agreement No. 101086261–FEEDACTIV.

Institutional Review Board Statement: Not applicable.

Informed Consent Statement: Not applicable.

Data Availability Statement: The data presented in this study are available on request from the corresponding author.

Conflicts of Interest: The authors declare no conflicts of interest. The funders had no role in the design of the study; in the collection, analyses, or interpretation of data; in the writing of the manuscript; or in the decision to publish the results.

References

1. Samtiya, M.; Aluko, R.E.; Dhewa, T.; Moreno-Rojas, J.M. Potential health benefits of plant food-derived bioactive components: An overview. *Foods* **2021**, *10*, 839. [CrossRef]
2. Wilson, D.W.; Nash, P.; Singh, H.; Griffiths, K.; Singh, R.; De Meester, F.; Horiuchi, R.; Takahashi, T. The role of food antioxidants, benefits of functional foods, and influence of feeding habits on the health of the older person: An overview. *Antioxidants* **2017**, *6*, 81. [CrossRef] [PubMed]
3. Tian, Y.; Liimatainen, J.; Alanne, A.L.; Lindstedt, A.; Liu, P.; Sinkkonen, J.; Kallio, H.; Yang, B. Phenolic compounds extracted by acidic aqueous ethanol from berries and leaves of different berry plants. *Food Chem.* **2017**, *220*, 266–281. [CrossRef] [PubMed]
4. Bujor, O.C.; Ginies, C.; Popa, V.I.; Dufour, C. Phenolic compounds and antioxidant activity of lingonberry (*Vaccinium vitis-idaea* L.) leaf, stem and fruit at different harvest periods. *Food Chem.* **2018**, *252*, 356–365. [CrossRef]
5. Ștefănescu, B.E.; Călinoiu, L.F.; Ranga, F.; Fetea, F.; Mocan, A.; Vodnar, D.C.; Crișan, G. Chemical composition and biological activities of the nord-west romanian wild bilberry (*Vaccinium myrtillus* L.) and lingonberry (*Vaccinium vitis-idaea* L.) leaves. *Antioxidants* **2020**, *9*, 495. [CrossRef] [PubMed]

6. Kopystecka, A.; Kozioł, I.; Radomska, D.; Bielawski, K.; Bielawska, A.; Wujec, M. *Vaccinium uliginosum* and *Vaccinium myrtillus*—Two Species—One Used as a Functional Food. *Nutrients* **2023**, *15*, 4119. [CrossRef] [PubMed]
7. Ştefanescu, B.E.; Szabo, K.; Mocan, A.; Crisan, G. Phenolic compounds from five ericaceae species leaves and their related bioavailability and health benefits. *Molecules* **2019**, *24*, 2046. [CrossRef]
8. Lätti, A.K.; Jaakola, L.; Riihinen, K.R.; Kainulainen, P.S. Anthocyanin and flavonol variation in bog bilberries (*Vaccinium uliginosum* L.) in Finland. *J. Agric. Food Chem.* **2010**, *58*, 427–433. [CrossRef]
9. Stanoeva, J.P.; Stefova, M.; Andonovska, K.B.; Vankova, A.; Stafilov, T. Phenolics and mineral content in bilberry and bog bilberry from Macedonia. *Int. J. Food Prop.* **2017**, *20*, S863–S883. [CrossRef]
10. Dobroslavić, E.; Elez Garofulić, I.; Šeparović, J.; Zorić, Z.; Pedisić, S.; Dragović-Uzelac, V. Pressurized Liquid Extraction as a Novel Technique for the Isolation of *Laurus nobilis* L. Leaf Polyphenols. *Molecules* **2022**, *27*, 5099. [CrossRef]
11. Lončarić, A.; Celeiro, M.; Jozinović, A.; Jelinić, J.; Kovač, T.; Jokić, S.; Babić, J.; Moslavac, T.; Zavadlav, S.; Lores, M. Green extraction methods for extraction of polyphenolic compounds from blueberry pomace. *Foods* **2020**, *9*, 1521. [CrossRef] [PubMed]
12. Solaberrieta, I.; Mellinas, C.; Jiménez, A.; Garrigós, M.C. Recovery of Antioxidants from Tomato Seed Industrial Wastes by Microwave-Assisted and Ultrasound-Assisted Extraction. *Foods* **2022**, *11*, 3068. [CrossRef] [PubMed]
13. Vieira, E.F.; Souza, S.; Moreira, M.M.; Cruz, R.; da Silva, A.B.; Casal, S.; Delerue-Matos, C. Valorization of Phenolic and Carotenoid Compounds of *Sechium edule* (Jacq. Swartz) Leaves: Comparison between Conventional, Ultrasound- and Microwave-Assisted Extraction Approaches. *Molecules* **2022**, *27*, 7193. [CrossRef] [PubMed]
14. Quitério, E.; Grosso, C.; Ferraz, R.; Delerue-Matos, C.; Soares, C. A Critical Comparison of the Advanced Extraction Techniques Applied to Obtain Health-Promoting Compounds from Seaweeds. *Mar. Drugs* **2022**, *20*, 677. [CrossRef]
15. Watrelot, A.A.; Bouska, L. Optimization of the ultrasound-assisted extraction of polyphenols from Aronia and grapes. *Food Chem.* **2022**, *386*, 132703. [CrossRef] [PubMed]
16. Abi-Khattar, A.M.; Boussetta, N.; Rajha, H.N.; Abdel-Massih, R.M.; Louka, N.; Maroun, R.G.; Vorobiev, E.; Debs, E. Mechanical damage and thermal effect induced by ultrasonic treatment in olive leaf tissue. Impact on polyphenols recovery. *Ultrason. Sonochem.* **2022**, *82*, 105895. [CrossRef]
17. Solaberrieta, I.; Jiménez, A.; Garrigós, M.C. Valorisation of Aloe Vera Skin By-Products to Obtain Bioactive Compounds by Microwave-Assisted Extraction: Antioxidant Activity and Chemical Composition. *SSRN Electron. J.* **2022**, *11*, 1058. [CrossRef]
18. Alexandre, E.M.C.; Castro, L.M.G.; Moreira, S.A.; Pintado, M.; Saraiva, J.A. Comparison of Emerging Technologies to Extract High-Added Value Compounds from Fruit Residues: Pressure- and Electro-Based Technologies. *Food Eng. Rev.* **2017**, *9*, 190–212. [CrossRef]
19. Pais, A.C.S.; Pinto, C.A.; Ramos, P.A.B.; Pinto, R.J.B.; Rosa, D.; Duarte, M.F.; Abreu, M.H.; Rocha, S.M.; Saraiva, J.A.; Silvestre, A.J.D.; et al. High pressure extraction of bioactive diterpenes from the macroalgae: *Bifurcaria bifurcata*: An efficient and environmentally friendly approach. *RSC Adv.* **2019**, *9*, 39893–39903. [CrossRef]
20. Ștefănescu, B.E.; Călinoiu, L.F.; Ranga, F.; Fetea, F.; Mocan, A.; Vodnar, D.C.; Crișan, G. The chemical and biological profiles of leaves from commercial blueberry varieties. *Plants* **2020**, *9*, 1193. [CrossRef]
21. Ștefănescu, B.E.; Nemes, S.A.; Teleky, B.E.; Călinoiu, L.F.; Mitrea, L.; Martău, G.A.; Szabo, K.; Mihai, M.; Vodnar, D.C.; Crișan, G. Microencapsulation and Bioaccessibility of Phenolic Compounds of *Vaccinium* Leaf Extracts. *Antioxidants* **2022**, *11*, 674. [CrossRef] [PubMed]
22. Dahmoune, F.; Boulekbache, L.; Moussi, K.; Aoun, O.; Spigno, G.; Madani, K. Valorization of *Citrus limon* residues for the recovery of antioxidants: Evaluation and optimization of microwave and ultrasound application to solvent extraction. *Ind. Crops Prod.* **2013**, *50*, 77–87. [CrossRef]
23. Nisca, A.; Ștefănescu, R.; Stegăruș, D.I.; Mare, A.D.; Farczadi, L.; Tanase, C. Phytochemical profile and biological effects of spruce (*Picea abies*) bark subjected to ultrasound assisted and microwave-assisted extractions. *Plants* **2021**, *10*, 870. [CrossRef] [PubMed]
24. Ben Hamissa, A.M.; Seffen, M.; Aliakbarian, B.; Casazza, A.A.; Perego, P.; Converti, A. Phenolics extraction from *Agave americana* (L.) leaves using high-temperature, high-pressure reactor. *Food Bioprod. Process.* **2012**, *90*, 17–21. [CrossRef]
25. Dulf, F.V.; Vodnar, D.C.; Dulf, E.H.; Toşa, M.I. Total Phenolic Contents, Antioxidant Activities, and Lipid Fractions from Berry Pomaces Obtained by Solid-State Fermentation of Two *Sambucus* Species with *Aspergillus niger*. *J. Agric. Food Chem.* **2015**, *63*, 3489–3500. [CrossRef]
26. Aryal, S.; Baniya, M.K.; Danekhu, K.; Kunwar, P.; Gurung, R.; Koirala, N. Total Phenolic content, Flavonoid content and antioxidant potential of wild vegetables from western Nepal. *Plants* **2019**, *8*, 96. [CrossRef]
27. Zhishen, J.; Mengcheng, T.; Jianming, W. The determination of flavonoid contents in mulberry and their scavenging effects on superoxide radicals. *Food Chem.* **1999**, *64*, 555–559. [CrossRef]
28. Diaconeasa, Z.; Leopold, L.; Rugină, D.; Ayvaz, H.; Socaciu, C. Antiproliferative and antioxidant properties of anthocyanin rich extracts from blueberry and blackcurrant juice. *Int. J. Mol. Sci.* **2015**, *16*, 2352–2365. [CrossRef]
29. Semeniuc, C.A.; Pop, C.R.; Rotar, A.M. Antibacterial activity and interactions of plant essential oil combinations against Gram-positive and Gram-negative bacteria. *J. Food Drug Anal.* **2017**, *25*, 403–408. [CrossRef]
30. Alexandre, E.M.C.; Silva, S.; Santos, S.A.O.; Silvestre, A.J.D.; Duarte, M.F.; Saraiva, J.A.; Pintado, M. Antimicrobial activity of pomegranate peel extracts performed by high pressure and enzymatic assisted extraction. *Food Res. Int.* **2019**, *115*, 167–176. [CrossRef]

31. Dobroslavić, E.; Garofulić, I.E.; Zorić, Z.; Pedisić, S.; Dragović-Uzelac, V. Polyphenolic characterization and antioxidant capacity of *Laurus nobilis* L. Leaf extracts obtained by green and conventional extraction techniques. *Processes* **2021**, *9*, 1840. [CrossRef]
32. Setyaningsih, W.; Saputro, I.E.; Palma, M.; Barroso, C.G. Stability of 40 phenolic compounds during ultrasound-assisted extractions (UAE). *AIP Conf. Proc.* **2016**, *1755*, 080009. [CrossRef]
33. Bidel, L.P.R.; Meyer, S.; Goulas, Y.; Cadot, Y.; Cerovic, Z.G. Responses of epidermal phenolic compounds to light acclimation: In vivo qualitative and quantitative assessment using chlorophyll fluorescence excitation spectra in leaves of three woody species. *J. Photochem. Photobiol. B Biol.* **2007**, *88*, 163–179. [CrossRef]
34. Jaakola, L.; Määttä-Riihinen, K.; Kärenlampi, S.; Hohtola, A. Activation of flavonoid biosynthesis by solar radiation in bilberry (*Vaccinium myrtillus* L.) leaves. *Planta* **2004**, *218*, 721–728. [CrossRef] [PubMed]
35. Uleberg, E.; Rohloff, J.; Jaakola, L.; Trôst, K.; Junttila, O.; Häggman, H.; Martinussen, I. Effects of temperature and photoperiod on yield and chemical composition of northern and southern clones of bilberry (*Vaccinium myrtillus* L.). *J. Agric. Food Chem.* **2012**, *60*, 10406–10414. [CrossRef]
36. Bujor, O.C.; Le Bourvellec, C.; Volf, I.; Popa, V.I.; Dufour, C. Seasonal variations of the phenolic constituents in bilberry (*Vaccinium myrtillus* L.) leaves, stems and fruits, and their antioxidant activity. *Food Chem.* **2016**, *213*, 58–68. [CrossRef]
37. Caldas, T.W.; Mazza, K.E.L.; Teles, A.S.C.; Mattos, G.N.; Brígida, A.I.S.; Conte-Junior, C.A.; Borguini, R.G.; Godoy, R.L.O.; Cabral, L.M.C.; Tonon, R.V. Phenolic compounds recovery from grape skin using conventional and non-conventional extraction methods. *Ind. Crops Prod.* **2018**, *111*, 86–91. [CrossRef]
38. Mróz, M.; Malinowska-Pańczyk, E.; Bartoszek, A.; Kusznierewicz, B. Comparative Study on Assisted Solvent Extraction Techniques for the Extraction of Biologically Active Compounds from *Sideritis raeseri* and *Sideritis scardica*. *Molecules* **2023**, *28*, 4207. [CrossRef]
39. Brezoiu, A.M.; Deaconu, M.; Mitran, R.A.; Prelipcean, A.M.; Matei, C.; Berger, D. Optimisation of Polyphenols Extraction from Wild Bilberry Leaves—Antimicrobial Properties and Stability Studies. *Molecules* **2023**, *28*, 5795. [CrossRef]
40. Páscoa, R.N.M.J.; Gomes, M.J.; Sousa, C. Antioxidant activity of blueberry (*Vaccinium* spp.) cultivar leaves: Differences across the vegetative stage and the application of near infrared spectroscopy. *Molecules* **2019**, *24*, 3900. [CrossRef]
41. Gao, S.H.; Zhao, T.R.; Liu, Y.P.; Wang, Y.F.; Cheng, G.G.; Cao, J.X. Phenolic constituents, antioxidant activity and neuroprotective effects of ethanol extracts of fruits, leaves and flower buds from *Vaccinium dunalianum* Wight. *Food Chem.* **2022**, *374*, 131752. [CrossRef] [PubMed]
42. Cheng, M.; He, J.; Wang, H.; Li, C.; Wu, G.; Zhu, K.; Chen, X.; Zhang, Y.; Tan, L. Comparison of microwave, ultrasound and ultrasound-microwave assisted solvent extraction methods on phenolic profile and antioxidant activity of extracts from jackfruit (*Artocarpus heterophyllus* Lam.) pulp. *Lwt* **2023**, *173*, 114395. [CrossRef]
43. Routray, W.; Orsat, V. MAE of phenolic compounds from blueberry leaves and comparison with other extraction methods. *Ind. Crops Prod.* **2014**, *58*, 36–45. [CrossRef]
44. Alexandre, E.M.C.; Coelho, M.C.; Ozcan, K.; Pinto, C.A.; Teixeira, J.A.; Saraiva, J.A.; Pintado, M. Emergent technologies for the extraction of antioxidants from prickly pear peel and their antimicrobial activity. *Foods* **2021**, *10*, 570. [CrossRef] [PubMed]
45. Martz, F.; Jaakola, L.; Julkunen-Tiitto, R.; Stark, S. Phenolic Composition and Antioxidant Capacity of Bilberry (*Vaccinium myrtillus*) Leaves in Northern Europe Following Foliar Development and Along Environmental Gradients. *J. Chem. Ecol.* **2010**, *36*, 1017–1028. [CrossRef] [PubMed]
46. Gil-Martínez, L.; Aznar-Ramos, M.J.; del Carmen Razola-Diaz, M.; Mut-Salud, N.; Falcón-Piñeiro, A.; Baños, A.; Guillamón, E.; Gómez-Caravaca, A.M.; Verardo, V. Establishment of a Sonotrode Extraction Method and Evaluation of the Antioxidant, Antimicrobial and Anticancer Potential of an Optimized *Vaccinium myrtillus* L. Leaves Extract as Functional Ingredient. *Foods* **2023**, *12*, 1688. [CrossRef]
47. Wang, Y.; Zhu, J.; Meng, X.; Liu, S.; Mu, J.; Ning, C. Comparison of polyphenol, anthocyanin and antioxidant capacity in four varieties of *Lonicera caerulea* berry extracts. *Food Chem.* **2016**, *197*, 522–529. [CrossRef]
48. Bunea, A.; Rugină, D.; Sconţa, Z.; Pop, R.M.; Pintea, A.; Socaciu, C.; Tăbăran, F.; Grootaert, C.; Struijs, K.; VanCamp, J. Anthocyanin determination in blueberry extracts from various cultivars and their antiproliferative and apoptotic properties in B16-F10 metastatic murine melanoma cells. *Phytochemistry* **2013**, *95*, 436–444. [CrossRef]
49. Xu, J.G.; Hu, Q.P.; Liu, Y. Antioxidant and DNA-protective activities of chlorogenic acid isomers. *J. Agric. Food Chem.* **2012**, *60*, 11625–11630. [CrossRef]
50. Lesjak, M.; Beara, I.; Simin, N.; Pintać, D.; Majkić, T.; Bekvalac, K.; Orčić, D.; Mimica-Dukić, N. Antioxidant and anti-inflammatory activities of quercetin and its derivatives. *J. Funct. Foods* **2018**, *40*, 68–75. [CrossRef]
51. Soobrattee, M.A.; Neergheen, V.S.; Luximon-Ramma, A.; Aruoma, O.I.; Bahorun, T. Phenolics as potential antioxidant therapeutic agents: Mechanism and actions. *Mutat. Res.—Fundam. Mol. Mech. Mutagen.* **2005**, *579*, 200–213. [CrossRef] [PubMed]
52. Heim, K.E.; Tagliaferro, A.R.; Bobilya, D.J. Flavonoid antioxidants: Chemistry, metabolism and structure-activity relationships. *J. Nutr. Biochem.* **2002**, *13*, 572–584. [CrossRef] [PubMed]
53. Aguilar-Villalva, R.; Molina, G.A.; España-Sánchez, B.L.; Díaz-Peña, L.F.; Elizalde-Mata, A.; Valerio, E.; Azanza-Ricardo, C.; Estevez, M. Antioxidant capacity and antibacterial activity from *Annona cherimola* phytochemicals by ultrasound-assisted extraction and its comparison to conventional methods. *Arab. J. Chem.* **2021**, *14*, 103239. [CrossRef]

54. Saifullah, M.; McCullum, R.; Van Vuong, Q. Optimization of microwave-assisted extraction of polyphenols from lemon myrtle: Comparison of modern and conventional extraction techniques based on bioactivity and total polyphenols in dry extracts. *Processes* **2021**, *9*, 2212. [CrossRef]
55. Boy, F.R.; Casquete, R.; Martínez, A.; Córdoba, M.d.G.; Ruíz-Moyano, S.; Benito, M.J. Antioxidant, antihypertensive and antimicrobial properties of phenolic compounds obtained from native plants by different extraction methods. *Int. J. Environ. Res. Public Health* **2021**, *18*, 2475. [CrossRef] [PubMed]
56. Mašković, P.; Veličković, V.; Mitić, M.; Đurović, S.; Zeković, Z.; Radojković, M.; Cvetanović, A.; Švarc-Gajić, J.; Vujić, J. Summer savory extracts prepared by novel extraction methods resulted in enhanced biological activity. *Ind. Crops Prod.* **2017**, *109*, 875–881. [CrossRef]
57. Veličković, V.; Đurović, S.; Radojković, M.; Cvetanović, A.; Švarc-Gajić, J.; Vujić, J.; Trifunović, S.; Mašković, P.Z. Application of conventional and non-conventional extraction approaches for extraction of *Erica carnea* L.: Chemical profile and biological activity of obtained extracts. *J. Supercrit. Fluids* **2017**, *128*, 331–337. [CrossRef]
58. Gutiérrez-Sánchez, M.D.C.; Aguilar-Zárate, P.; Michel-Michel, M.R.; Ascacio-Valdés, J.A.; Reyes-Munguía, A. The Ultrasound-Assisted Extraction of Polyphenols from Mexican Firecracker (*Hamelia patens* Jacq.): Evaluation of Bioactivities and Identification of Phytochemicals by HPLC-ESI-MS. *Molecules* **2022**, *27*, 8845. [CrossRef]
59. Tanase, C.; Cosarca, S.; Toma, F.; Mare, A.; Cosarca, A.; Man, A.; Miklos, A.; Imre, S. Antibacterial activities of spruce bark (*Picea abies* L.) extract and its components against human pathogens. *Rev. Chim.* **2018**, *69*, 1462–1467. [CrossRef]
60. Vilkickyte, G.; Petrikaite, V.; Pukalskas, A.; Sipailiene, A.; Raudone, L. Exploring *Vaccinium vitis-idaea* L. as a potential source of therapeutic agents: Antimicrobial, antioxidant, and anti-inflammatory activities of extracts and fractions. *J. Ethnopharmacol.* **2022**, *292*, 115207. [CrossRef]
61. Tian, Y.; Puganen, A.; Alakomi, H.L.; Uusitupa, A.; Saarela, M.; Yang, B. Antioxidative and antibacterial activities of aqueous ethanol extracts of berries, leaves, and branches of berry plants. *Food Res. Int.* **2018**, *106*, 291–303. [CrossRef] [PubMed]
62. Silva, S.; Costa, E.M.; Pereira, M.F.; Costa, M.R.; Pintado, M.E. Evaluation of the antimicrobial activity of aqueous extracts from dry *Vaccinium corymbosum* extracts upon food microorganism. *Food Control* **2013**, *34*, 645–650. [CrossRef]
63. Cowan, M.M. Plant products as antimicrobial agents. *Clin. Microbiol. Rev.* **1999**, *12*, 564–582. [CrossRef] [PubMed]
64. Gurjar, M.S.; Ali, S.; Akhtar, M.; Singh, K.S. Efficacy of plant extracts in plant disease management. *Agric. Sci.* **2012**, *03*, 425–433. [CrossRef]
65. Deng, Y.; Yang, G.; Yue, J.; Qian, B.; Liu, Z.; Wang, D.; Zhong, Y.; Zhao, Y. Influences of ripening stages and extracting solvents on the polyphenolic compounds, antimicrobial and antioxidant activities of blueberry leaf extracts. *Food Control* **2014**, *38*, 184–191. [CrossRef]
66. Puupponen-Pimiä, R.; Nohynek, L.; Alakomi, H.L.; Oksman-Caldentey, K.M. Bioactive berry compounds—Novel tools against human pathogens. *Appl. Microbiol. Biotechnol.* **2005**, *67*, 8–18. [CrossRef]

foods

MDPI

Article

Carvacrol Treatment Reduces Decay and Maintains the Postharvest Quality of Red Grape Fruits (*Vitis vinifera* L.) Inoculated with *Alternaria alternata*

Hongying Li [1,†], Jie Ding [2,†], Chunyan Liu [3], Peng Huang [1,4], Yifan Yang [1], Zilu Jin [1] and Wen Qin [1,*]

1 College of Food Science, Sichuan Agricultural University, Ya'an 625014, China; lihying1998@163.com (H.L.); 2016106005@yibinu.edu.cn (P.H.); 18599994741@163.com (Y.Y.); 15697053681@163.com (Z.J.)
2 College of Food Science, Sichuan Tourism University, Chengdu 610100, China; dingjiedream@163.com
3 Chengdu Kuafu Technology Co., Ltd., Chengdu 610100, China; chunyan202103@163.com
4 Department of Quality Management and Inspection and Detection, Yibin University, Yibin 644000, China
* Correspondence: qinwen@sicau.edu.cn; Tel.: +86-139-8161-6637
† These authors contributed equally to this work.

Abstract: In this study, we isolated and identified pathogenic fungi from the naturally occurring fruits of red grapes, studied their biological characteristics, screened fifteen essential oil components to find the best natural antibacterial agent with the strongest inhibitory effect, and then compared the incidence of postharvest diseases and storage potential of red grapes treated with two concentrations (0.5 EC_{50}/EC_{50}) of essential oil components (inoculated with pathogenic fungi) during storage for 12 d at room temperature. In our research, *Alternaria alternata* was the primary pathogenic fungus of red grapes. Specifically, red grapes became infected which caused diseases, regardless of whether they were inoculated with *Alternaria alternata* in an injured or uninjured state. Our findings demonstrated that the following conditions were ideal for *Alternaria alternata* mycelial development and spore germination: BSA medium, D-maltose, ammonium nitrate, 28 °C, pH 6, and exposure to light. For the best *Alternaria alternata* spore production, OA medium, mannitol, urea, 34 °C, pH 9, and dark conditions were advised. Furthermore, with an EC_{50} value of 36.71 μg/mL, carvacrol demonstrated the highest inhibitory impact on *Alternaria alternata* among the 15 components of essential oils. In the meantime, treatment with EC_{50} concentration of carvacrol was found to be more effective than 0.5 EC_{50} concentration for controlling *Alternaria alternata*-induced decay disease of red grapes. The fruits exhibited remarkable improvements in the activity of defense-related enzymes, preservation of the greatest hardness and total soluble solids content, reduction in membrane lipid peroxidation in the peel, and preservation of the structural integrity of peel cells. Consequently, carvacrol was able to prevent the *Alternaria alternata* infestation disease that affects red grapes, and its EC_{50} concentration produced the greatest outcomes.

Keywords: *Alternaria alternata*; isolation and identification; biological characteristics; natural bacteriostatic agent; postharvest disease; antifungal activity

1. Introduction

Alternaria Nees, a member of the *Pleosporaceae* family in the *Dothideomycetes* class of *Ascomycota*, is a highly prevalent fungus that includes saprophytic, parasitic, and endophytic strains. Among these, *Alternaria alternata* (*A. alternata*) is a saprophytic fungus capable of surviving in various substrates, including water, soil, and air. Additionally, certain strains have the ability to re-colonize and grow on host tissues affected by diseases, exacerbating the severity of the infection [1]. *A. alternata*, as a parasitic fungus, has the ability to extensively infect fruits and vegetables during field cultivation, storage, and through direct contact. Significant losses in the nutritional and commercial value of crops occur as a result of this pervasive contamination, both in the field and during the postharvest procedures [2]. Currently, over 95% of reported cases indicate that *Alternaria* has the ability to facultatively parasitize various crops. It is also recognized as a major cause of diseases of blueberries, cherries, maize, tomatoes, and other fruits and vegetables [3–5]. In addition, *Alternaria* is known to produce mycotoxins that can be harmful to humans and animals. Some of these mycotoxins include alternariol (AOH), alternariol methyl ether (AME), and tenuazonic acid (TeA). These toxic secondary metabolites are secreted by pathogenic fungi such as *Alternaria* [6]. At low concentrations, these mycotoxins can disrupt the normal physiological and metabolic functions of the host and even induce apoptosis. They have genotoxic, mutagenic, cytotoxic, and other adverse effects. In addition, there is evidence of synergistic effects between AOH and AME [7]. Grapes (*Vitis vinifera* L.) are one of the most popular fruits in the world, with rich nutritional and medicinal value [8]. Apart from being eaten raw, grapes are also extensively processed to make wine, grape juice, raisins, and other products to satisfy a variety of customer demands [9–11]. Consequently, grapes hold significant economic worth within the worldwide fruit industry [12]. However, grapes are berries, which are susceptible to mechanical damage and pathogen infection while being picked, stored, and transported [13]. Among them, pathogens such as *Alternaria* can quickly invade through the wounds or skin of fruits [14]. Once the conditions are suitable, its mycelium will spread out and generate more conidia, which will lower quality and cause postharvest losses [15]. In order to protect the interests of consumers and the red grape fruit industry, it is therefore essential to gain a full understanding of the biological characteristics of *Alternaria*. With this understanding, effective control measures for preventing *Alternaria*-caused postharvest deterioration can be developed and put into practice.

Synthetic fungicides are widely used in pre- and postharvest production to control plant diseases. Consumer demand for organic fruits and vegetables free of pesticides and chemical residues is rising; nevertheless, long-term usage of improved fungicides can result in fungal strains, chemical residues, and environmental contamination of antifungal agents [16]. In this context, sulphur dioxide (SO_2) fumigation has been shown to be an effective method of controlling the postharvest pathogen *A. alternata*. However, the optimal dosage of SO_2 treatment is often close to causing bleaching damage, fruit cracking, and softening. In addition, potential residues of SO_2 pose a risk to human health, limiting its practical use for prolonged exposure in production practices [17]. In contrast, dielectric-barrier discharge plasma (DBD) [18] and ultraviolet irradiation (UVI) [19] have developed into large-scale environmental protection and physical sterilization methods for controlling postharvest diseases. The safety profiles of these non-thermal sterilizing methods are enhanced. Unfortunately, their high cost and sophisticated nature prevent them from being used in areas with poor infrastructure. Moreover, microbial antagonists are known to be highly plastic in their growth and reproduction stages, sensitive to environmental factors, and difficult to ensure stability, thus resulting in limited therapeutic effects in practical applications. Thus, over the past ten years, more research has been conducted on biocontrol agents (BCAs) based on natural substances, paying particular emphasis to the use of plant extracts or essential oils as fungicides in place of their synthetic counterparts [20].

For example, carvacrol is a monoterpenoid phenol, a light-yellow oily liquid with lipophilicity and good inhibitory effects on various pathogenic bacteria. It has been reported to cause structural damage to cell walls and membranes [21], and it also exhibits strong antibacterial and antifungal properties in the field of food preservation. Indeed, the excellent antibacterial properties of carvacrol make it a suitable substitute for ketanol and carbolic acid in the dental field [22]. In addition, it is reported that compared to other plant essential oil components, carvacrol exhibits superior postharvest disease inhibitory activity. In a recent study by MI et al. [23], it was found that the growth of *A. alternata* was significantly inhibited after treatment with carvacrol, ultimately improving the storage quality parameters of kiwifruit. Due to the harmful effects of high concentrations of carvacrol on fruits, it is not suitable to choose high concentrations in practical applications [24]. However, little is known about the possible outcomes and workings of applying low carvacrol concentrations ($\leq EC_{50}$) to stop *A. alternata*-caused postharvest illnesses. Thus, the goal of this study was to investigate the antifungal potential and control effect of low concentrations of carvacrol on the main pathogenic fungus, *A. alternata*, in red grapes after harvest. In addition, we aimed to study the quality parameters and disease resistance of Streptomyces in vivo and the effects of low-dose carvacrol treatment on red grape infections.

2. Materials and Methods

2.1. Materials and Reagents

Red grape (Krenson) samples were purchased from a local supermarket in Yucheng District, Ya 'an City, Sichuan Province, during the growing season. Fruits that exhibited uniformity in color and size, consistent maturity, and no mechanical damage were chosen, and pathogens were isolated after natural onset and decay.

Potato dextrose agar medium (PDA) was purchased from Beijing Aoboxing Biotechnology Co., Ltd., Beijing, China. Wort agar medium (MEA) was purchased from Qingdao Hi-Tech Industrial Park Haibo Biotechnology Co., Ltd., Qingdao, China.

Carvacrol, thymol, geraniol, citral, L-menthol, menthone, anisaldehyde, linalool, citronellal, trans-2-hexenal, diallyl disulfide, trans-caryophyllene, piperoneone, eugenol, and anethole (Analytical pure) were all purchased from RON Reagent Shanghai Yi En Chemical Technology Co., Ltd., Shanghai, China. All other chemicals and reagents were of analytical grade and were purchased from Chengdu Cologne Chemical Co., Ltd., Chengdu, China.

2.2. Isolation and Identification of Pathogens

2.2.1. Isolation and Purification of Pathogens

Pathogenic fungi of red grapes were injected as described by Zhang et al. [25]. The experiment was conducted on an ultraclean workbench (SW-CJ-1F, Shanghai Bangxi Instrument Technology Co., Ltd., Shanghai, China), and related materials were sterilized using a high-pressure steam sterilization pot (GI54DS, Xiamen Zhihui Instrument Co., Ltd., Xiamen, China). It was then placed in a biochemical incubator (LRH-250F, Shanghai Yiheng Technology Co., Ltd., Shanghai, China) and cultured (28 °C, 48 h). Subsequently, the edge of the colony mycelium was chosen based on the cultivation methods until one colony formed.

2.2.2. Morphological Identification

Fungi were inoculated on potato dextrose agar (PDA) and cultured (28 °C, 5 d). We looked at the colony's size, morphology, color, texture, and other aspects. Water lenses were made so that hyphae and spore morphological characteristics might be seen and photographed using a biomicroscope (N-126, Ninon Corporation, Tokyo, Japan). The species of fungi were preliminarily identified by using a fungal identification manual.

2.2.3. Molecular Biology Assay

After fungi were purified, their genomic DNA was amplified using the universal primers ITS1 and ITS4 of the fungal ribosomal gene transcribed spacer. The amplified material was then transferred to the Chengdu branch of Beijing Qingke Biotechnology Co., Ltd., Beijing, China for sequencing.

2.2.4. Pathogenicity of Pathogens

The purified fungi were reversely infected in accordance with Koch's rule [26]. When disease occurred, pathogenic fungi were isolated and purified again, and compared with the previously obtained fungi.

2.3. Biological Characteristics of Pathogenic Fungi

The growth curves of pathogenic fungi were determined, and the effects of different growth conditions were investigated. The media included potato dextrose agar (PDA), potato sucrose agar (PSA), Chagall medium (Czapek), wort agar medium (MEA), bean sprout juice medium (BSA), oat agar medium (OA), and red grape agar medium (RGA). Carbon and nitrogen sources were also used (carbon sources: sucrose, mannitol, glucose, soluble starch, D-maltose, lactose, and D-fructose. Nitrogen sources: sodium nitrate, ammonium sulfate, ammonium nitrate, potassium chloride, potassium nitrate, L-Alanine, and urea). The temperatures were 4, 10, 16, 22, 28, 34, 40. The pH values were 3, 4, 5, 6, 7, 8, 9. The light conditions included full light, alternating light and dark, and full dark. The fatal temperatures of pathogenic fungus hyphae and spores, as well as colony diameter, spore production, and spore germination, were measured. Each treatment was repeated three times, and the average value was taken to determine the optimal growth conditions of pathogenic fungi.

2.3.1. Growth Curve of Pathogenic Fungi

With a few minor adjustments, spore suspension was made as described by Zhang et al. [27]. The concentration of pathogenic fungi was adjusted with sterile water to 10^5 spores/mL for later use.

A 200 uL spore suspension was cultured in 100 mL of PDB medium for 7 d at 28 °C and 110 rpm in a water bath thermostatic shaker (SHA-BA Shanghai Yiheng Technology Co., Ltd., Shanghai, China), with samples being obtained every 12 h. After centrifuging the samples (6000× g r/min for 10 min) to extract the culture solution, they were dried to a consistent weight in an electric blast drying oven (DHG-9245A Shanghai Yiheng Technology Co., Ltd., Shanghai, China) set at 60 °C. The growth curve was then drawn using the dry weight of hyphae as the ordinate and the culture time of pathogenic fungi as the abscissa.

2.3.2. Culture Medium

Seven different types of culture media were infected with the pathogenic fungal cake (28 °C, 5 d). The total number of spores was counted using a blood cell counting plate, and the colony diameter was measured using the cross method.

2.3.3. Carbon and Nitrogen Sources

Seven carbon and nitrogen sources were employed in place of sucrose and sodium nitrate, and the other procedures were the same as above. Czapek medium was used as the basic medium. The carbon and nitrogen sources were prepared into 1% solution with sterile water and pathogenic fungal spores were collected. A 25 μL spore suspension was coated on the concave slide and cultured (28 °C, 4 h). Three spore germination zones were observed. The spore germination rate was estimated, and the total number of spores should not be less than 150. The length of the germ tube must be greater than half of the diameter of the spore to germinate. The germination rate of spores was calculated according to the following formula:

$$\text{Germination rate of spores} = (n_0/n) \times 100\%$$

where n_0 is the number of germinated spores and n is the total number of spores.

2.3.4. Temperature

Other steps were the same as above, in addition to temperature.

2.3.5. Lethal Temperature

The test tube holding the bacterial cake or spore solution was heated (35, 40, 45, 50, 55, 60, and 65 °C for 10 min) in a digital display thermostatic water bath (HH-4 Shanghai Bangxi Instrument Technology Co., Ltd., Shanghai, China), and then cultured (28 °C, 5 d). According to the highest temperature that fungi could grow, the temperature gradient that fungi could not grow was set at an interval of 1 °C, and then the above process was repeated.

2.3.6. pH

The pH of sterile water and PDA medium was adjusted by using 1 mol/L HCl and 1 mol/L NaOH. The other operation steps were the same as above.

2.3.7. Light

Other steps were the same as above, in addition to lighting options.

2.4. Screening of Natural Bacteriostatic Agents

Pathogenic fungi were inoculated into PDA medium that contained 15 essential oil components at a specific concentration, including carvacrol, thymol, and geraniol (Table A1). By measuring the colony diameter, the growth inhibition rate was estimated using the formula shown below. The logarithm of the concentration of the essential oil's constituents as the abscissa and the inhibition rate as the ordinate were used to calculate the virulence regression equation and the correlation coefficient R. The virulence regression equation was then used to obtain the EC_{50} value as follows:

$$\text{Mycelial growth inhibition rate} = [(d_0 - d)/(d_0 - 5)] \times 100\%$$

where d_0 is the control colony diameter and d is the colony diameter of the experimental group.

2.5. Inhibitory Effect of Essential Oil Components on Pathogenic Fungi In Vitro

The mechanism of action of carvacrol was determined using scanning electron microscopy (SEM) and transmission electron microscopy (TEM) as previously described [28], with minor modifications.

The cultured pathogenic fungi were fixed in 3% glutaraldehyde solution (4 °C, 2 h), washed three times with phosphate-buffered solution (PBS; pH 7.2, 0.1 mol/L), and dehydrated in 30%, 50%, 70%, 80%, 90%, 95%, and 100% alcohol gradient for 15 min. With conductive adhesive, the sample was attached to the sample holder, which was then placed in an ion sputter (e-1045, Hitachi High Tech Nako, Tokyo, Japan) for spraying treatment. The sample was photographed using a scanning electron microscope (jsm-it700 hr, Japan Electronics, Tokyo, Japan).

The tissue was first prefixed with 3% glutaraldehyde, followed by 1% osmium tetroxide postfixation, series acetone dehydration, extended Epon 812 infiltration, and embedding. The semithin sections were stained with methylene blue, while the ultrathin sections were cut with a diamond knife and stained with uranyl acetate and lead citrate. A JEM-1400-FLASH transmission electron microscope was used to examine the sections.

2.6. Inhibitory Effect of Essential Oil Components on Pathogenic Fungi In Vivo and Their Preservation Effect

To ensure uniformity in size and ripeness and the absence of mechanical damage, red grapes were cut off from the stem. Pretreatment was performed as explained by Ding et al. [29]. These sterilized red grapes were used as the negative control group (CK1), and the inoculated fruits were used as the positive control group (CK2), following spraying of the sterilized red grapes' surfaces with *A. alternata* spore suspension. The remaining fruits were divided into two groups and sprayed with 0.5 EC_{50} and EC_{50} concentrations of carvacrol, marked as Q1 and Q2. Each group was prepared with fifteen boxes of red grapes, and each measurement was performed three times or more to replicate the natural shelf-life state while the fruits were kept at room temperature. The disease incidence, the degree of pericarp membrane lipid peroxidation of the peel, and the activities of defense-related enzymes of the peel were evaluated and recorded over 12 d at intervals of 3 d.

2.6.1. Postharvest Disease Incidence Rate

We referred to a previous study method, with minor modifications [30]. Based on the size of the fruit decay area, it was divided into different grades as described by Ding et al. [29]. The decay rate was calculated according to the following formula:

$$\text{Decay rate} = [(\text{i} \times \text{n}_\text{i})/4\text{n}] \times 100\%$$

where i is the level of decay and the number of fruits at that level of decay, and n is the total number of fruits per box of red grapes.

2.6.2. Postharvest Changes in Physical and Chemical Properties

Each group consisted of three red grape groups that were fixed at random. The fixed group was weighed at each sampling and the findings were recorded. The weight at 0 d was the initial weight. The change in red grape hardness was measured using a fruit hardness tester (GY-4, Yueqing Edbao Instrument Co., Ltd., Yueqing, China) based on a previous report by Xu et al. [31], with some modifications. The total soluble solids (TSS) content of red grapes was determined using a digital refractometer (LH-B55, Hangzhou Luheng Biotechnology Co., Ltd., Hangzhou, China), and the method for determining titratable acid (TA) was acid–base titration, which had been slightly modified according to a previous report [32]. The weight loss rate was calculated according to the following formula:

$$\text{Weight loss rate} = [(\text{m}_0 - \text{m}_\text{t})/\text{m}_0] \times 100\%$$

where m_0 is the initial weight of red grapes and m_t is the weight of red grapes on t d.

2.6.3. Degree of Membrane Lipid Peroxidation in Fruit Peel

The method used to determine the content of malondialdehyde in fruits was based on a previous study [33], with some modifications. This method required the use of a UV-Vis spectrophotometer (UV BlueStar A, Beijing Lebertech Instrument Co., Ltd., Beijing, China). The relative conductivity of the red grapes' peel was measured using a conductivity meter (DDS-11A, Shanghai Youyou Instrument Co., Ltd., Shanghai, China), which was determined based on a previous report by Han et al. [34], with minor modifications.

2.6.4. Defense-Related Enzyme Activity

Red grapes (5 g) were collected and homogenized in ice-cold PBS (5 mL, pH 7.2, 0.1 mol/L). Subsequently, the homogenates were centrifuged (4 °C, 10,000 r/min, 15 min). The resultant supernatants were then used as enzyme extracts to measure the activities of defense-related enzymes using plant-based ELISA kits from Chengdu Kuafu Technology Co., Ltd., which included superoxide dismutase (SOD), peroxidase (POD), catalase (CAT), and phenylalanine ammonia lyase (PAL).

2.7. Statistical Analysis

The final results were expressed as the means ± standard deviations of three independent replicates. The phylogenetic tree was drawn using MEGA-X. Excel 2022 was used to calculate data statistics. Means were compared using one-way analysis of variance and Duncan's multi-range test using SPSS 25. $p < 0.05$ was considered significant. The Origin 2021 software was used for statistical analyses.

3. Results and Discussion

3.1. Isolation and Identification of Pathogens

The fungal group *Alternaria*, which exhibits high adaptability and a wide range of species, is a significant cause of diseases in fruits and vegetables [35]. As shown in Figure 1, the colonies of strain *LQ* on PDA medium had an almost circular shape. The hyphae were initially greyish white and radially growing, gradually deepening to greyish green or ink green. The colonies had flat and dense villi with clean edges, developed aerial hyphae, and exhibited a growth rate of 6.01 mm/d (Figure 1a). Over time, the colonies changed color from pale yellow to blackish brown (Figure 1b). Microscopic examination revealed largely unbranched mycelia with sizes ranging from 1.8 to 4.7 μm and septa (Figure 1c). The conidial chains were upright or curved and the conidia appeared yellowish brown or brown with different shapes and sizes (11.83–42.66 μm × 5.69–16.12 μm) (Figure 1c). The conidia's surface was smooth or drum-shaped, with 0–2 longitudinal or oblique septa and 2–5 transverse septa that were constricted at the septa. The septa in the middle part of the spore body were thicker and dark brown (Figure 1d), with most having a light brown columnar short beak. The long beak had a septum and measured 2.40–25.24 μm × 2.80–5.30 μm (Figure 1c). Based on the colony characteristics and reference to fungal identification manuals, strain *LQ* was tentatively identified as *Alternaria*.

Figure 1. Growth and morphological characteristics of *LQ* and diseases of red grapes caused by reverse inoculation: (**a**) colony in PDA medium (front); (**b**) colony in PDA medium (reverse); (**c**) microscopic images of hyphae and spores (40×); (**d**) observation of hyphal and spore morphology; (**e**) inoculation of pathogenic fungi in injured fruits; (**f**) inoculation of pathogenic fungi without injury; (**g**) injured control; and (**h**) no-injury control.

The pathogenicity test confirmed that the red grapes in the control group showed no disease response. However, the red grapes in the experimental group, whether injured or uninjured, were all infected with the fungus and produced a significant number of spores. In addition to exhibiting mycelial development and spore germination on the uninoculated side, the red grapes in the wounded, inoculated treatment group also had disease signs, which was consistent with the natural symptoms of red grapes. The fungi were reisolated and purified and found to match the characteristics of the inoculated strain. This agreement with Koch's postulates indicated that this fungus was the main pathogen responsible for red grape storage diseases.

It was found in this study that *A. alternata* was the predominant pathogenic fungus of red grapes, which was compatible with the findings of Ghuffar S in the Rawalpindi district of Punjab province, Pakistan [36]. Furthermore, Li Z isolated *A. alternata* from the leaves of the Amur grape *Vitis amurensis* [37]. Together, these data demonstrated that *A. alternata* was the primary pathogenic fungus of grapes, and it was possible that there were no geographical variations.

Sequencing revealed a fragment size of 550 bp in strain *LQ* (Figure A1a). With 100% coverage and homology, the ITS sequences of pathogenic fungi with a higher homology were all recognized as *Alternaria species* through further homology alignment with known sequences (Figure A1b) in the NCBI database. The phylogenetic tree is shown in Figure A1c. Sequences from different *Alternaria species* formed a single large clade, with the branch confidence scores mostly above 70% and often above 90%. *LQ* clustered closely with two *A. alternata* sequences, with high (100%) branch confidence and close genetic distance. This cluster of branches was relatively independent of those of other *Alternaria species*, indicating that *LQ* could be confidently identified as *A. alternata*.

3.2. Biological Characteristics of LQ

The mycelium of this strain, as depicted in Figure A2f, grew slowly throughout the course of 0–24 h without significant change in weight, suggesting a lag phase. Rapid mycelial growth then transpired between 24 and 72 h, signifying a logarithmic growth phase. During 72–108 h, a little rise in mycelial weight was noted. Mycelial weight did not significantly alter between 108 and 132 h, suggesting that the strain had stabilized. The strain's weight steadily dropped starting at 132 h, signifying the start of a decline phase.

Significant differences ($p < 0.05$) were observed in mycelial growth and spore production of strain *LQ* under different nutrient and environmental conditions, as shown in Figure 2. *LQ* showed the highest growth rate in BSA medium (60.29 $\pm$ 0.48 mm). D-maltose and ammonium nitrate supported superior growth compared to other groups. Similarly, potassium nitrate and L-alanine also promoted accelerated growth with no significant difference between them. Optimal growth occurred at 28 °C, while growth slowed significantly above 34 °C and was absent at 4 °C and 40 °C. *LQ* preferred neutral or slightly acidic conditions with pH of 6 for maximum growth, while alkaline conditions inhibited it. Mycelial growth rate was higher in the light/dark alternation and full-light treatments compared to full darkness, with no significant difference between them. Thus, the ideal conditions for *LQ* mycelial growth were BSA medium, D-maltose, ammonium nitrate, 28 °C, pH 6, and exposure to light, and the optimum conditions for sporulation were consistent with this (Figure A2). The hyphae and spores of *LQ* were effectively sterilized at 56 °C for 10 min (Table A2).

LQ did not produce spores in Czapek and MEA media. Of the media tested, OA was the most favorable for spore production. When mannitol and urea were used as carbon and nitrogen sources, *LQ* showed the highest spore production. However, *LQ* did not produce spores in the medium containing ammonium sulphate and L-alanine as nitrogen sources. The temperature was correlated with spore production, which peaked at 34 °C. At 4 and 40 °C, no spores were generated. In contrast to mycelial growth, alkaline and dark conditions favored *LQ* spore production. At pH 9, spore production reached $(2.117 \pm 0.0764) \times 10^6$ spores/mL. Therefore, OA medium, mannitol, urea, 34 °C, pH 9, and dark conditions were recommended for optimal *LQ* spore production.

Remarkably, we discovered that there was no positive correlation between the strain's capacity for mycelial development and sporulation. The specific results were different from Wang et al. [38], which may be because different choices were made for culture medium types and environmental factors. Due to the differences in the utilization of nitrogen sources by *LQ*, it may be safer to choose ammonium sulfate as a nitrogen fertilizer compared to nitrate and urea fertilizers.

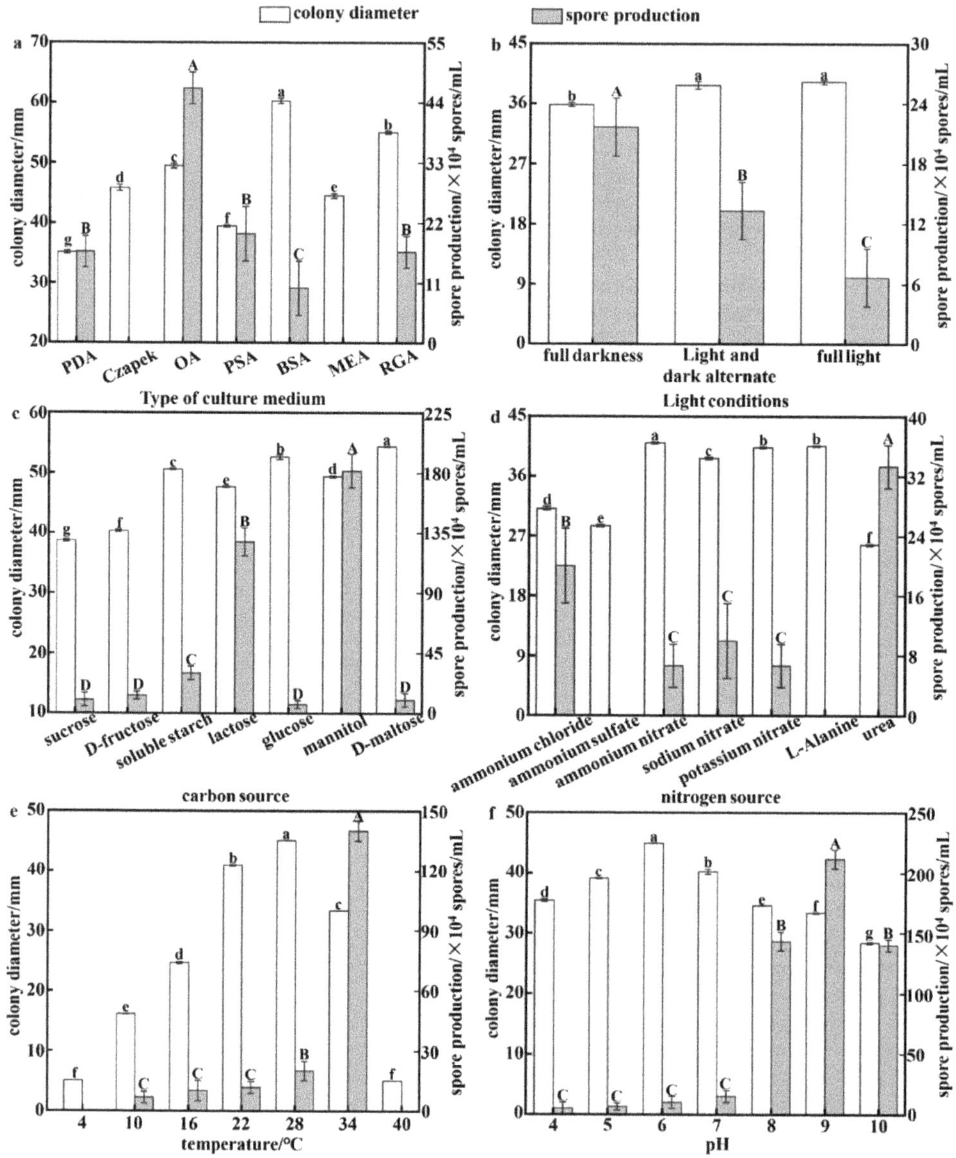

Figure 2. Colony diameter and spore production of *LQ* under different growth conditions: (**a**) culture media; (**b**) carbon sources; (**c**) nitrogen sources; (**d**) temperature; (**e**) pH; and (**f**) lighting conditions. Different minuscule letters indicate significant differences in colony diameter ($p < 0.05$). Different capital letters indicate significant differences in spore production ($p < 0.05$).

3.3. Effect of Carvacrol on LQ In Vitro

The growth of *LQ* was inhibited to varying degrees by 15 essential oil ingredients, including carvacrol, thymol, and geraniol, with the inhibition proportional to the concentration (Figure A3). As shown in Table A3, the toxicity regression equations for these components showed good linear correlation (R > 0.97). Carvacrol exhibited the strongest antibacterial effect, with an EC_{50} of 36.71 μg/mL, surpassing the other 14 ingredients.

Thymol, geraniol, citral, and L-menthol followed, with EC_{50} values ranging from 40 to 90 μg/mL. Consequently, carvacrol was selected for subsequent experiments as an inhibitor against *LQ*. Carvacrol is a lipophilic, light-yellow oily liquid monoterpene phenol that has good pathogen inhibitory activity. It is present in a variety of oregano plants, including sage, thyme, and oregano, of which it is most prevalent in oregano and the primary active component of oregano essential oil [39]. As a compound with a variety of biological activities, it has broad application prospects. The World Health Organization (WHO) reported that when the residue of carvacrol in food was less than 50.0 mg/kg, it had no harm to human health and could be used for postharvest preservation of fruits [40].

Damage to the hyphae and spores of *LQ* increased with carvacrol concentration (Figure 3a). The control group showed regular, plump, and smooth mycelia with intact cell walls and germinated spores. In contrast, carvacrol-treated mycelia and spores showed shrinkage, which became more pronounced with increasing concentration. At EC_{50}, wrinkles and craters similar to meteorite craters appeared and small cracks appeared on the mycelial surface. Above EC_{50}, hyphae and spores were completely destroyed and spores did not germinate. Specifically, at 2 EC_{50}, hyphae shrank and bent, and the spore surface suffered extensive damage with disrupted cell walls. As a result, higher carvacrol concentrations harmed *LQ* hyphae and spores more severely and prevented them from growing normally.

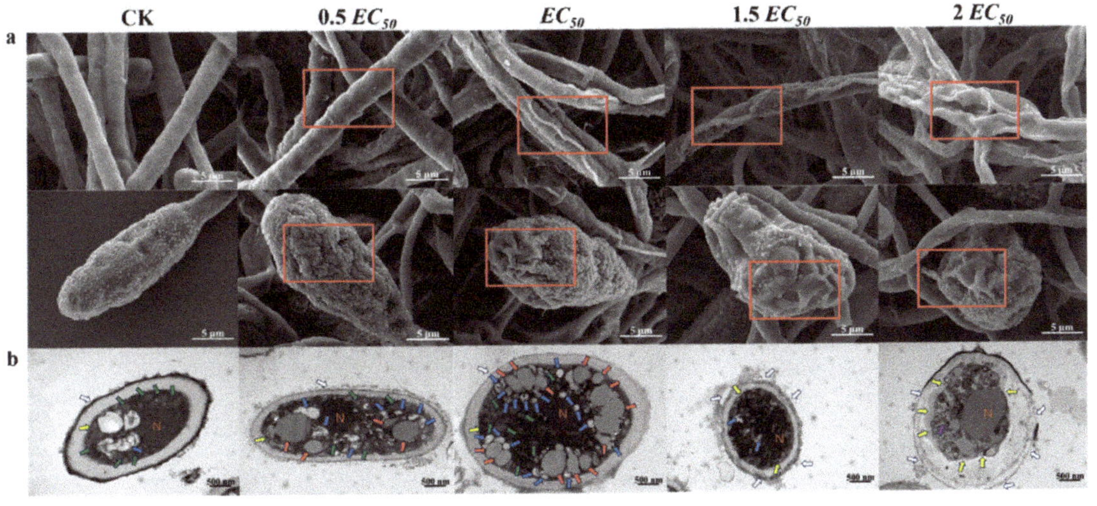

Figure 3. SEM and TEM results: (**a**) surface morphology of hyphae and spores observed under SEM, where the red box indicates the twisted morphology of hyphae and the wrinkled and damaged state of spores; (**b**) structure of cells observed under TEM, where the white arrows indicate cell walls, the yellow arrows show cell membranes, the red arrows show mitochondria, the green arrows show lipid droplets, the blue arrows show autophagosomes, the purple arrows indicate the loss of cytoplasmic material, and N indicates the cell nucleus.

The TEM observations (Figure 3b) were highly consistent with the SEM results. The control group showed intact cell walls and membranes, uniformly distributed plasma membranes, abundant cytoplasm and contents, and visible organelle structures. In the 0.5 EC_{50} and EC_{50} treatment groups, *LQ* showed cell swelling, distorted mitochondria, and increased lipid droplets and autophagosomes, while the cell walls and membranes of pathogenic fungi remained largely intact. Significant damage to the cell wall and membrane, the loss of cell contents, a decrease in the number of lipid droplets, an unorganized internal matrix, and a distortion in the shape of organelles were the outcomes of concentrations higher than EC_{50}. It was hypothesized that at lower concentrations ($\leq EC_{50}$), carvacrol inhibited *LQ* growth by causing damage to mitochondria and causing aberrant formation of lipid droplets, while at higher concentrations (>EC_{50}), carvacrol was likely to cause irreparable cell damage by rupturing the integrity of the *LQ* cell wall and membrane.

Through in vitro experiments, we found that carvacrol could inhibit the growth of mycelium and spore germination of *A. alternata*, and there was a certain dose relationship. On the one hand, carvacrol inhibited important enzyme activities linked to energy metabolism in pathogenic fungal cells, which might have further harmed mitochondria by altering the mitochondrial shape and causing an aberrant increase in lipid droplets. Insufficient energy supply to the cells could lead to the activation of autophagosomes, which are important molecules for maintaining cell homeostasis [41]. On the other hand, carvacrol might inhibit important enzymes involved in the synthesis of cell walls and membranes, which hinders the production of substances like chitin and ergosterol. Additionally, accumulated ROS resulted in membrane lipid peroxidation, which damaged the integrity of cell walls and membranes and caused pathogenic fungal cells to cease growing and die [42].

3.4. Effect of Carvacrol on the Physical and Chemical Properties of LQ-Inoculated Red Grapes

According to the results shown in Figure 4a, carvacrol effectively inhibited fruit decay caused by *LQ* at different concentrations. The positive control group (CK2) showed a significant increase in decay rate from day 0, which was significantly higher than the other treatment groups ($p < 0.05$). On 12 d, the degradation rate reached 63.37 ± 1.75%. The degradation rate in the CK1, Q1, and Q2 groups significantly increased starting on 6 d, and by 9 d, notable changes were seen. On 12 d, the degradation rates were 11.56 ± 0.89%, 29.29 ± 1.29%, and 14.73 ± 0.69%, respectively. The Q2 group showed stronger inhibitory effects and closely resembled the CK1 group, with a significant difference observed only on 12 d (Figure 4f).

As shown in Figure 4b, carvacrol effectively slowed the weight loss caused by *LQ*, in line with the degradation trend. With the highest rate of 12.27 ± 0.38%, CK2 showed increased water loss and faster respiration as a result of *LQ*. On 9 d, both essential oil treatment groups and CK1 showed similar patterns with no significant differences ($p > 0.05$). In general, carvacrol slowed fruit respiration and reduced water loss, with Q2 giving the most favorable results.

As shown in Figure 4c, carvacrol considerably inhibited the deterioration of red grapes and preserved their hardness during storage, while *LQ* caused the grapes to soften more during storage. CK2 showed the most severe softening and the most rapid decrease in hardness. Fruits treated with carvacrol showed a milder decrease. On 12 d, the hardness of the four groups was 3.71 ± 0.07, 0.6 ± 0.13, 3.12 ± 0.09, and 3.68 ± 0.09 N, respectively. Compared to 0 d, the hardness decreased by 26.97%, 87.93%, 37.35%, and 25.35%, respectively. Q2 and CK1 showed similar trends without significant differences until the last day ($p > 0.05$). In conclusion, carvacrol reduced fruit softening and retained firmness, with Q2 showing greater efficacy than Q1.

TSS in all four treatment groups had a similar trend, rising at first and then dropping over time (Figure 4d). This is most likely because TSS functions as a structural carbohydrate that is mostly hydrolyzed during fruit metabolism [43]. CK2 showed a sharp increase to 20.2 ± 0.02%, which decreased after the third day. CK1 and Q1 reached their peak on 6 d with contents of 18.61 ± 0.04% and 18.16 ± 0.08%, respectively. Notably, Q2 reached its peak on 9 d (18.89 ± 0.04%). On 12 d, the TSS of Q2 was significantly higher than that of CK1, indicating the beneficial effect of higher carvacrol concentration on red grape preservation. Overall, carvacrol inhibited the metabolism of *LQ*-infected red grapes, with Q2 showing the best results.

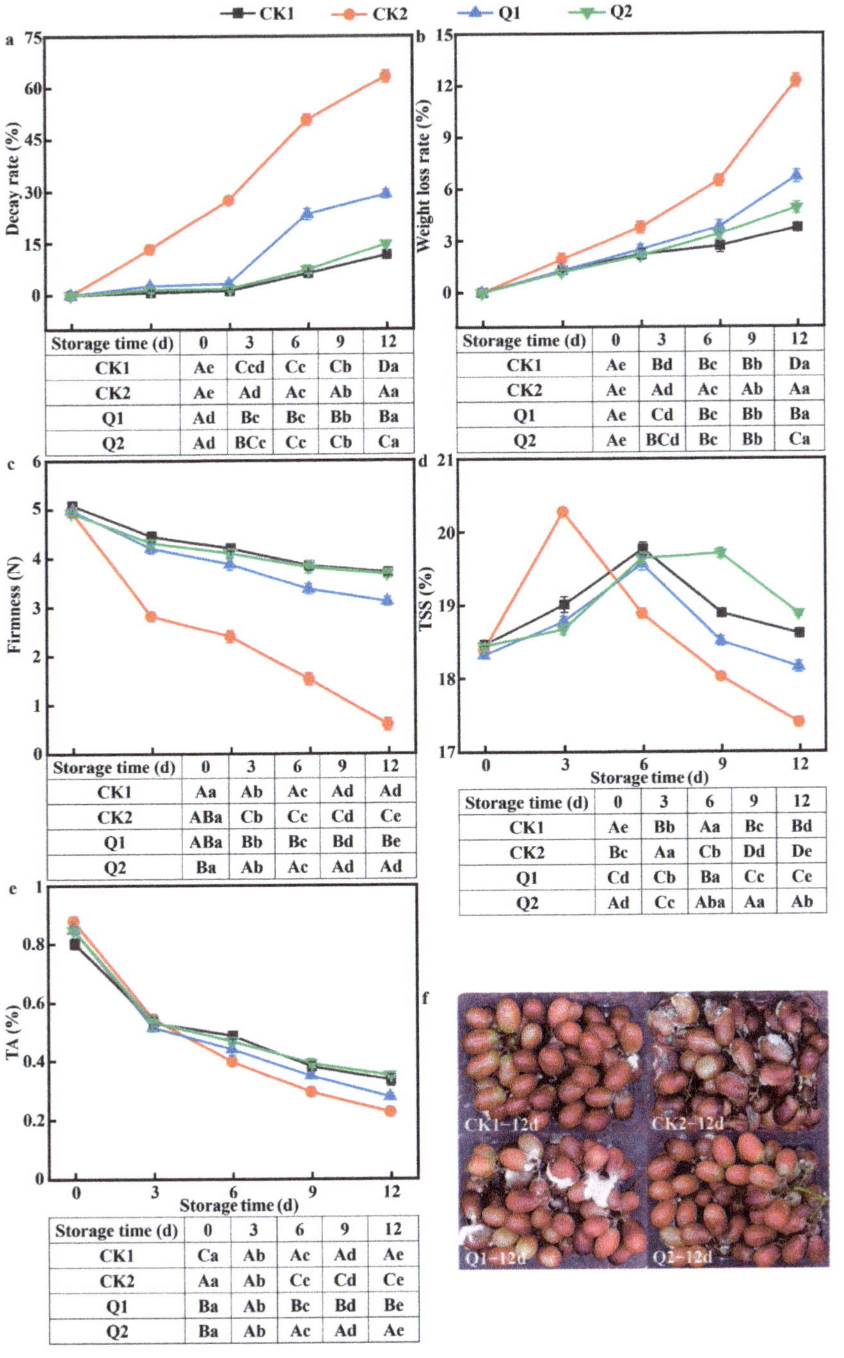

(a)

Storage time (d)	0	3	6	9	12
CK1	Ae	Ccd	Cc	Cb	Da
CK2	Ae	Ad	Ac	Ab	Aa
Q1	Ad	Bc	Bc	Bb	Ba
Q2	Ad	BCc	Cc	Cb	Ca

(b)

Storage time (d)	0	3	6	9	12
CK1	Ae	Bd	Bc	Bb	Da
CK2	Ae	Ad	Ac	Ab	Aa
Q1	Ae	Cd	Bc	Bb	Ba
Q2	Ae	BCd	Bc	Bb	Ca

(c)

Storage time (d)	0	3	6	9	12
CK1	Aa	Ab	Ac	Ad	Ad
CK2	ABa	Cb	Cc	Cd	Ce
Q1	ABa	Bb	Bc	Bd	Be
Q2	Ba	Ab	Ac	Ad	Ad

(d)

Storage time (d)	0	3	6	9	12
CK1	Ae	Bb	Aa	Bc	Bd
CK2	Bc	Aa	Cb	Dd	De
Q1	Cd	Cb	Ba	Cc	Ce
Q2	Ad	Cc	Aba	Aa	Ab

(e)

Storage time (d)	0	3	6	9	12
CK1	Ca	Ab	Ac	Ad	Ae
CK2	Aa	Ab	Cc	Cd	Ce
Q1	Ba	Ab	Bc	Bd	Be
Q2	Ba	Ab	Ac	Ad	Ae

Figure 4. Effect of carvacrol on the quality of red grapes during storage with different treatments: (**a**) decay rate; (**b**) weight loss rate; (**c**) hardness; (**d**) TSS; (**e**) TA; and (**f**) decay on the last day of storage. Different minuscule letters indicate significant differences in colony diameter ($p < 0.05$). Different capital letters indicate significant differences in spore production ($p < 0.05$).

TA of red grapes dropped during storage (Figure 4e), most likely as a result of additional materials and organic acids being consumed during fruit respiration. In the later stages of storage, CK2 had a significantly lower TA than the other three groups ($p < 0.05$), while the essential oil-treated group showed a mitigated decline. On 12 d, the TA of CK2 was 0.23 ± 0.02%, a decrease of 74.22% compared to 0 d, while the other three groups (CK1/Q1/Q2) showed decreases of 58.29%, 67.29%, and 58.52%, respectively. Overall, this suggests that carvacrol treatment could delay the rate of titratable acid consumption, with Q2 showing a slower metabolic rate in red grapes infected with *LQ*.

Red grapes naturally have life, and during storage, they engage in a variety of biological processes like transpiration and respiration, which cause the fruit to soften after dehydration. Pectin hydrolysis and depolymerization immediately contribute to cell damage and cell wall rupture, which in turn cause fruit softening and hardness loss [44]. It has been reported that carvacrol-containing biofilms effectively preserve tomatoes and persimmons. After twenty days, tomatoes' inhibition rate reached 90%, while by two months, persimmons showed no damaged fruit [45]. In addition, carvacrol could significantly maintain the hardness of fresh-cut apples, prevent their softening, and significantly alleviate the decrease in TSS and TA contents [46]. This was consistent with the research results. All these indicated that carvacrol had broad-spectrum antibacterial activity, which could reduce the aging of fruits, slow down the deterioration of quality, prolong the storage time of fruits, and produce effective fresh-keeping effects. However, the preservation effect may vary with the type, processing method, and concentration of fruits and vegetables.

3.5. Effect of Carvacrol on Membrane Lipid Peroxidation of LQ-Inoculated Red Grapes

There were variations in the degree of membrane lipid peroxidation of red grape peel in each group over the course of storage (Figure 5). It was observed that the change trend of MDA and relative conductivity increased. Additionally, the CK2 group had a significantly higher degree of membrane lipid peroxidation than the other three groups ($p < 0.05$). The final byproduct of membrane lipid peroxidation is the content of MDA, which is widely used to assess the level of senescence damage to fruit cells in conjunction with relative conductivity [34]. Thus, this case showed that when red grapes were infected with *LQ*, the degree of membrane lipid peroxidation rose, resulting in ongoing accumulation of MDA, damage to the cell membrane, electrolyte leakage, and concurrently elevated relative conductivity. On 12 d, the MDA content of CK2 reached 19.72 ± 0.41 μmol/g, while Q1 and Q2 had a lower concentration with a significant difference ($p < 0.05$). According to this study, carvacrol may be able to stop red grape peel's membrane lipids from peroxiding. Overall, carvacrol can effectively delay fruit senescence, preserve the integrity of cell membrane, and prevent oxidative damage to pericarp cell membrane lipids caused by an infection with *LQ*, with Q2 showing greater efficacy.

Active oxygen buildup in red grapes during storage caused the membrane system to be attacked and the membrane structure to be degraded, which increased the fruit's MDA level, caused electrolyte leakage, and accelerated fruit aging. The reason why carvacrol could alleviate this situation may be because it had an inhibitory effect on *A. alternata*, which initially inactivated some pathogenic fungi and prevented their infection and colonization. This was consistent with the results of Chen et al. [47].

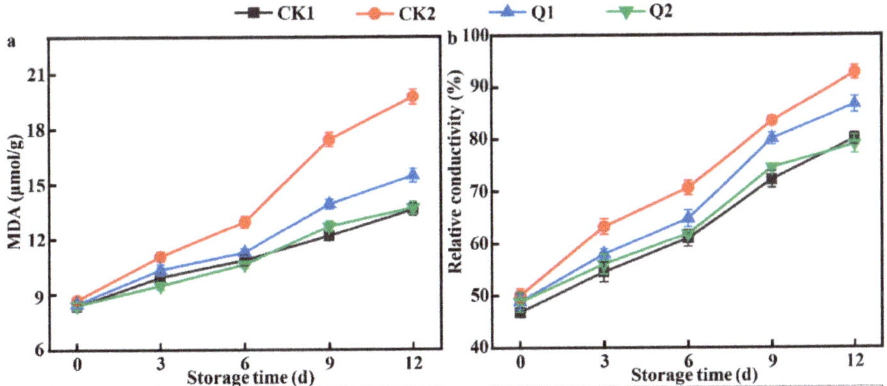

Storage time (d)	0	3	6	9	12
CK1	Ae	Bd	BCc	Cb	Ca
CK2	Ae	Ad	Ac	Ab	Aa
Q1	Ae	Bd	Bc	Bb	Ba
Q2	Ae	Cd	Cc	Cb	Ca

Storage time (d)	0	3	6	9	12
CK1	Be	Cd	Cc	Db	Ca
CK2	Ae	Ad	Ac	Ab	Aa
Q1	ABe	Bd	Bc	Bb	Ba
Q2	ABe	BCd	Cc	Cb	Ca

Figure 5. Effects of carvacrol on the quality of *LQ*-inoculated red grapes during storage: (**a**) MDA and (**b**) relative conductivity. Different minuscule letters indicate significant differences in colony diameter ($p < 0.05$). Different capital letters indicate significant differences in spore production ($p < 0.05$).

3.6. Effect of Carvacrol on Defense-Related Enzyme Activities of LQ-Inoculated Red Grapes

All defense enzymes' activity increased during storage at first before declining once more (Figure 6). The peak value for the CK2 group was higher than the CK1 group, occurred 3–6 d earlier, and then dropped off quickly. The enzyme activities decreased in the CK1 group on 12 d. When carvacrol was introduced, it interacted with *LQ* and changed the defense enzymes' activity in a statistically significant way ($p < 0.05$). The overall enzyme activities of the Q1 and Q2 groups were significantly better than those of the CK1 group and persisted at a higher level until the last day. Particularly, the enzyme activities of the Q2 and CK1 groups peaked after 9 d of storage, while SOD, POD, CAT, and PAL activities were 1.25, 1.14, 1.25, and 1.2 times those of CK1, respectively. In comparison to the CK2 group, the peak times of the CAT and PAL enzyme activities in the Q1 group were delayed by 3 d, while the Q2 group saw a 9 d delay in the peak time of each enzyme activity. Overall, carvacrol increased disease resistance, stimulated the activities of antioxidant defense enzymes, and postponed the aging of red grapes, with Q2 demonstrating greater preservation impact.

ROS are highly active and toxic, which could cause damage to lipids, proteins, DNA, and other substances, which would result in oxidative stress. SOD could convert O^{2-} to H_2O_2, and excessive H_2O_2 could promote the peroxidation of membrane lipids. Two ROS scavengers, POD and CAT, work together to get rid of them and protect plant cells [48]. This study showed that carvacrol raised the SOD, POD, CAT, and PAL activities in red grapes during storage. Carvacrol is a naturally occurring monoterpene phenol that includes both hydrophobic benzene rings and hydrophilic phenolic hydroxyl groups, which may be the reason for this. It may, therefore, readily pass through epidermal cells that comprise the cell wall, boost the production of ROS, change the membrane's permeability, stimulate an antioxidant response, and strengthen the disease resistance of red grapes. In summary, carvacrol has a significant inhibitory effect on *A. alternata* [49].

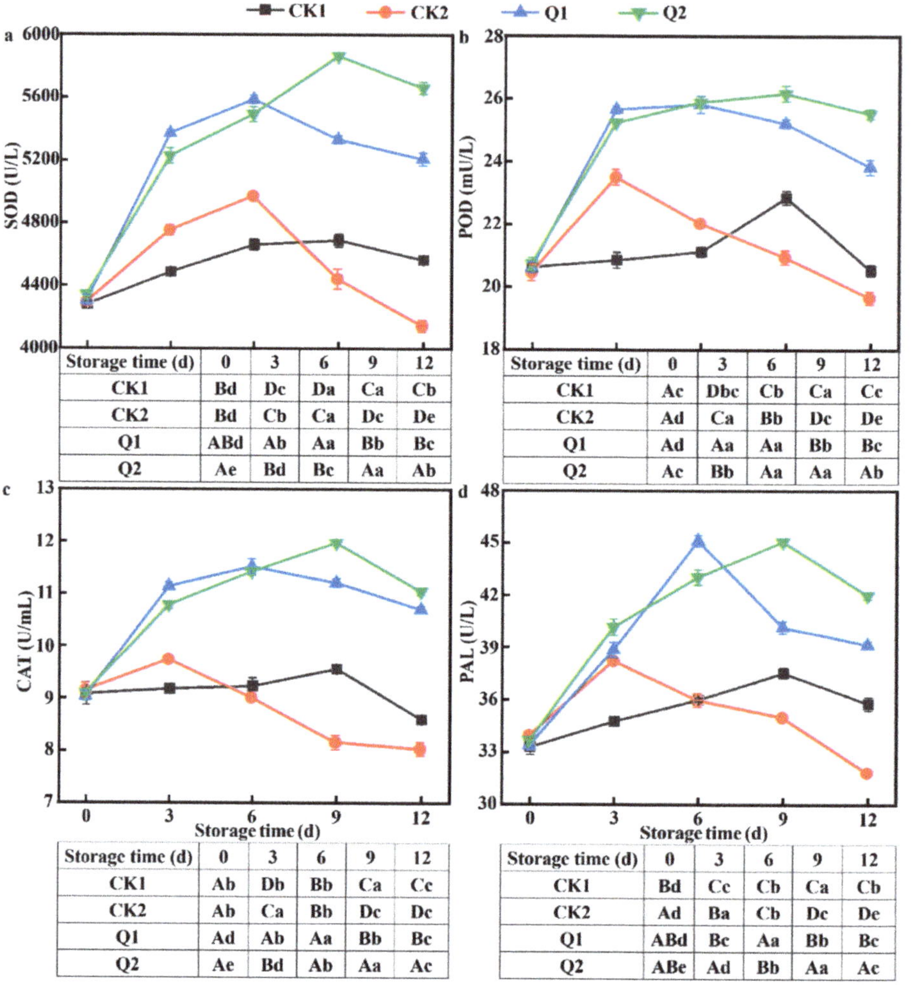

(a)

Storage time (d)	0	3	6	9	12
CK1	Bd	Dc	Da	Ca	Cb
CK2	Bd	Cb	Ca	Dc	De
Q1	ABd	Ab	Aa	Bb	Bc
Q2	Ae	Bd	Bc	Aa	Ab

(b)

Storage time (d)	0	3	6	9	12
CK1	Ac	Dbc	Cb	Ca	Cc
CK2	Ad	Ca	Bb	Dc	De
Q1	Ad	Aa	Aa	Bb	Bc
Q2	Ac	Bb	Aa	Aa	Ab

(c)

Storage time (d)	0	3	6	9	12
CK1	Ab	Db	Bb	Ca	Cc
CK2	Ab	Ca	Bb	Dc	Dc
Q1	Ad	Ab	Aa	Bb	Bc
Q2	Ae	Bd	Ab	Aa	Ac

(d)

Storage time (d)	0	3	6	9	12
CK1	Bd	Cc	Cb	Ca	Cb
CK2	Ad	Ba	Cb	Dc	De
Q1	ABd	Bc	Aa	Bb	Bc
Q2	ABe	Ad	Bb	Aa	Ac

Figure 6. Effect of carvacrol on defense-related enzyme activities of red grapes during storage with different treatments: (**a**) SOD; (**b**) POD; (**c**) CAT; and (**d**) PAL. Different minuscule letters indicate significant differences in colony diameter (p < 0.05). Different capital letters indicate significant differences in spore production (p < 0.05).

4. Conclusions

A. alternata has strong pathogenicity under growth conditions that are not harsh and strong vitality. It has disease effects on red grapes and affects the quality and economic benefits of red grapes. In this study, it was discovered that carvacrol exhibited a broad-spectrum antifungal action in vitro against *A. alternata*, the main postharvest fungal pathogen of red grapes. More specifically, upon treatment with carvacrol at a low concentration ($<EC_{50}$), fungal hyphae and spores began to distort. At EC_{50}, the complete mitochondrial morphology of *A. alternata* was destroyed, and lipid droplets were abnormal, which led to the disorder of cell energy metabolism and ultimately affected the normal growth of this strain. However, when exposed to high concentrations of carvacrol ($>EC_{50}$), spores shrank or even broke, hyphae were completely deformed, the cell wall and membrane integrity was lost, and the cytoplasm was exhausted, which resulted in cell death and growth restriction. At the appropriate concentration of carvacrol (0.5 EC_{50}/EC_{50}), red grapes retained a higher firmness and TSS, TA, and other quality parameters. Overall, carvacrol is capable of preventing *Alternaria alternata* from growing, and it could be utilized as a control measure for preventing *Alternaria alternata*-caused red grape decay diseases.

Author Contributions: H.L. and J.D. designed the study; H.L., J.D., C.L., P.H., Y.Y. and Z.J. performed the experiments; H.L. and J.D. drafted the work; H.L. and J.D. wrote and revised the manuscript; W.Q. revised the final version to be published. All authors have read and agreed to the published version of the manuscript.

Funding: This work was supported by the Sichuan Science and Technology Program (Grant No. 23ZDYF3041) and the project of Chongqing Science and Technology Bureau (cstc2021jscx-cylhX0015).

Data Availability Statement: The data presented in this study are available on request from the corresponding author.

Acknowledgments: We are very grateful to Jie Ding, Chunyan Liu, and Wen Qin for their support for this study.

Conflicts of Interest: Author Chunyan Liu was employed by the company Chengdu Kuafu Technology Co., Ltd. The remaining authors declare that the research was conducted in the absence of any commercial or financial relationships that could be construed as a potential conflict of interest.

Appendix A

Table A1. Types and concentrations of essential oil components.

Essential Oil Name	Concentration/μg·mL^{-1}
Carvacrol	1, 5, 10, 25, 50, 75, 100
Thymol	1, 5, 10, 25, 50, 75, 100
Geraniol	1, 25, 50, 100, 125, 150, 200
Citral	1, 25, 50, 100, 125, 150, 200
L-menthol	1, 10, 25, 50, 100, 150, 200
Menthone	10, 50, 100, 150, 200, 250, 300
Anisaldehyde	10, 50, 100, 150, 200, 250, 300
Linalool	10, 50, 100, 150, 200, 250, 300
Citronellal	10, 50, 100, 150, 200, 250, 300
Trans-2-hexenal	10, 50, 100, 200, 250, 300, 350
Diallyl disulfide	5, 10, 25, 50, 100, 150, 200
Trans caryophyllene	10, 50, 100, 200, 250, 300, 350
Piperitone	10, 50, 100, 200, 250, 300, 350
Eugenol	5, 10, 25, 50, 100, 150, 200
Anethole	5, 10, 25, 50, 100, 150, 200

Table A2. Determination of lethal temperature for *LQ* hyphae and spores.

Temperature	50	51	52	53	54	55	56	57	58	59	60
hyphal growth	+	+	+	+	+	+	−	−	−	−	−
spore germination	+	+	+	+	+	+	−	−	−	−	−

Note: "+" means hyphal growth and spore germination, and "−" means sterile hyphal growth and no spore germination.

Table A3. Indoor toxicity determination of 15 essential oil components on *LQ*.

Essential Oil Name	Toxicity Regression Equation	Correlation Coefficient (R)	EC_{50}/μg·mL^{-1}
Carvacrol	y = 1.6594x + 2.4033	0.9859	36.71
Thymol	y = 1.7174x + 2.2044	0.9811	42.44
Geraniol	y = 1.5884x + 1.9015	0.9809	89.26
Citral	y = 1.4716x + 2.2463	0.9802	74.33
L-menthol	y = 1.8481x + 1.5036	0.9831	77.96

Table A3. *Cont.*

Essential Oil Name	Toxicity Regression Equation	Correlation Coefficient (R)	$EC_{50}/\mu g \cdot mL^{-1}$
Menthone	y = 2.1883x + 0.4048	0.9827	125.86
Anisaldehyde	y = 1.8525x + 1.0666	0.9745	132.83
Linalool	y = 1.9732x + 0.8564	0.9704	125.88
Citronellal	y = 1.9424x + 1.0141	0.9738	112.73
Trans-2-hexenal	y = 1.4661x + 1.7086	0.9736	175.81
Diallyl disulfide	y = 1.4641x + 1.6651	0.9820	189.55
Trans caryophyllene	y = 2.2075x + 0.4428	0.9809	116.00
Piperitone	y = 2.1541x + 0.5331	0.9804	118.49
Eugenol	y = 1.4085x + 1.8369	0.9850	176.06
Anethole	y = 1.4836x + 1.7170	0.9925	163.25

Appendix B

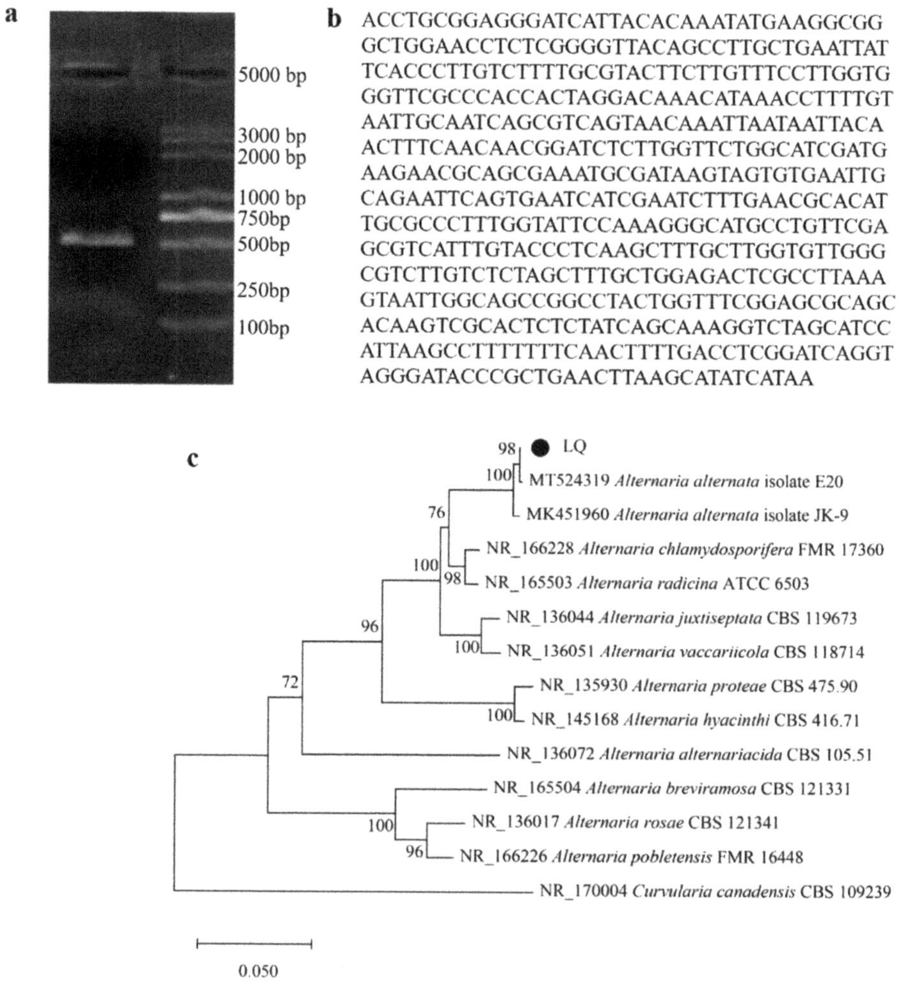

Figure A1. Molecular biological identification results of *LQ*: (**a**) electropherograms; (**b**) DNA sequence of the ITS region of *LQ*; and (**c**) phylogenetic tree based on rDNA sequence analysis of pathogenic fungus *LQ*.

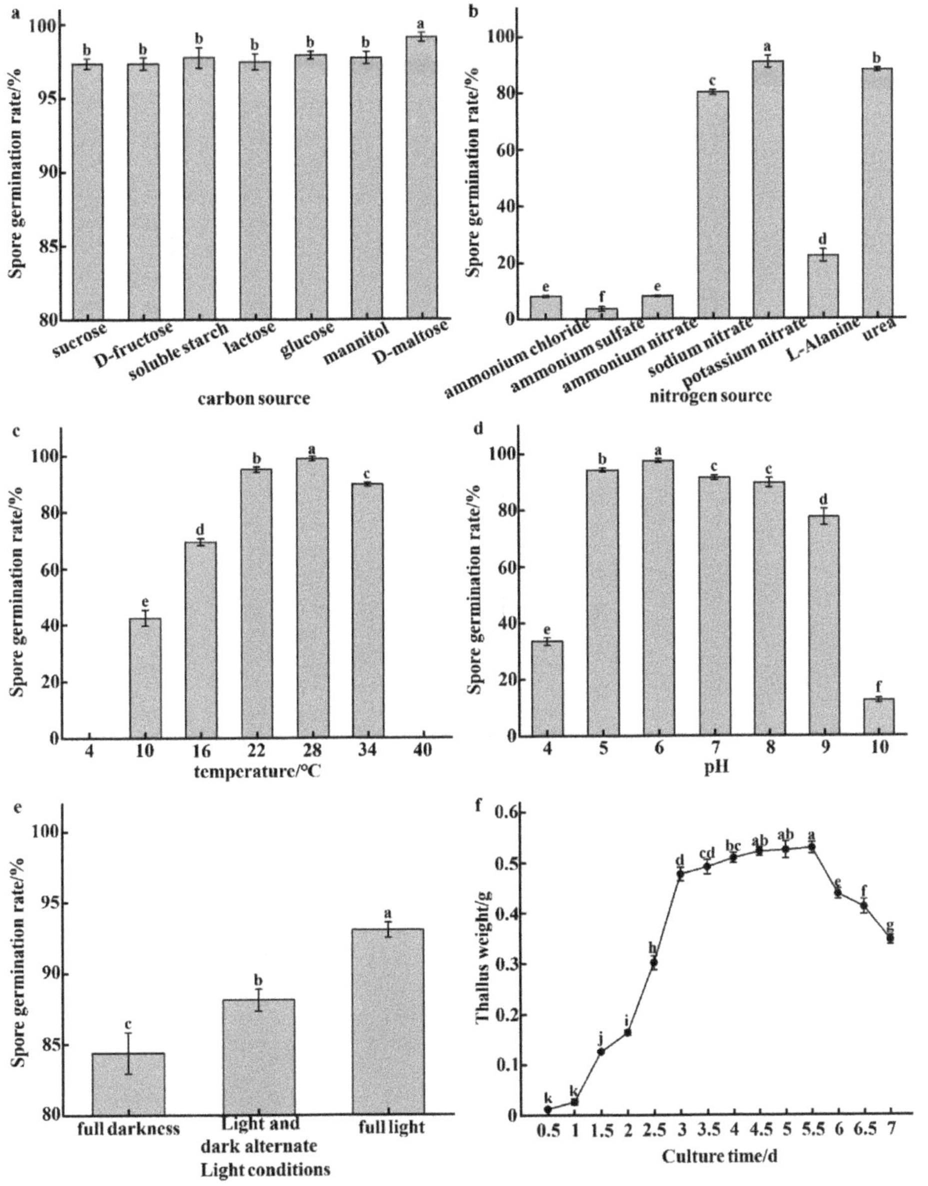

Figure A2. Spore germination rate of *LQ* under different growth conditions and the growth curve of *LQ*: (**a**) carbon sources; (**b**) nitrogen sources; (**c**) temperature; (**d**) pH; (**e**) lighting conditions; and (**f**) growth curve. Different letters indicate significant difference between the groups ($p < 0.05$).

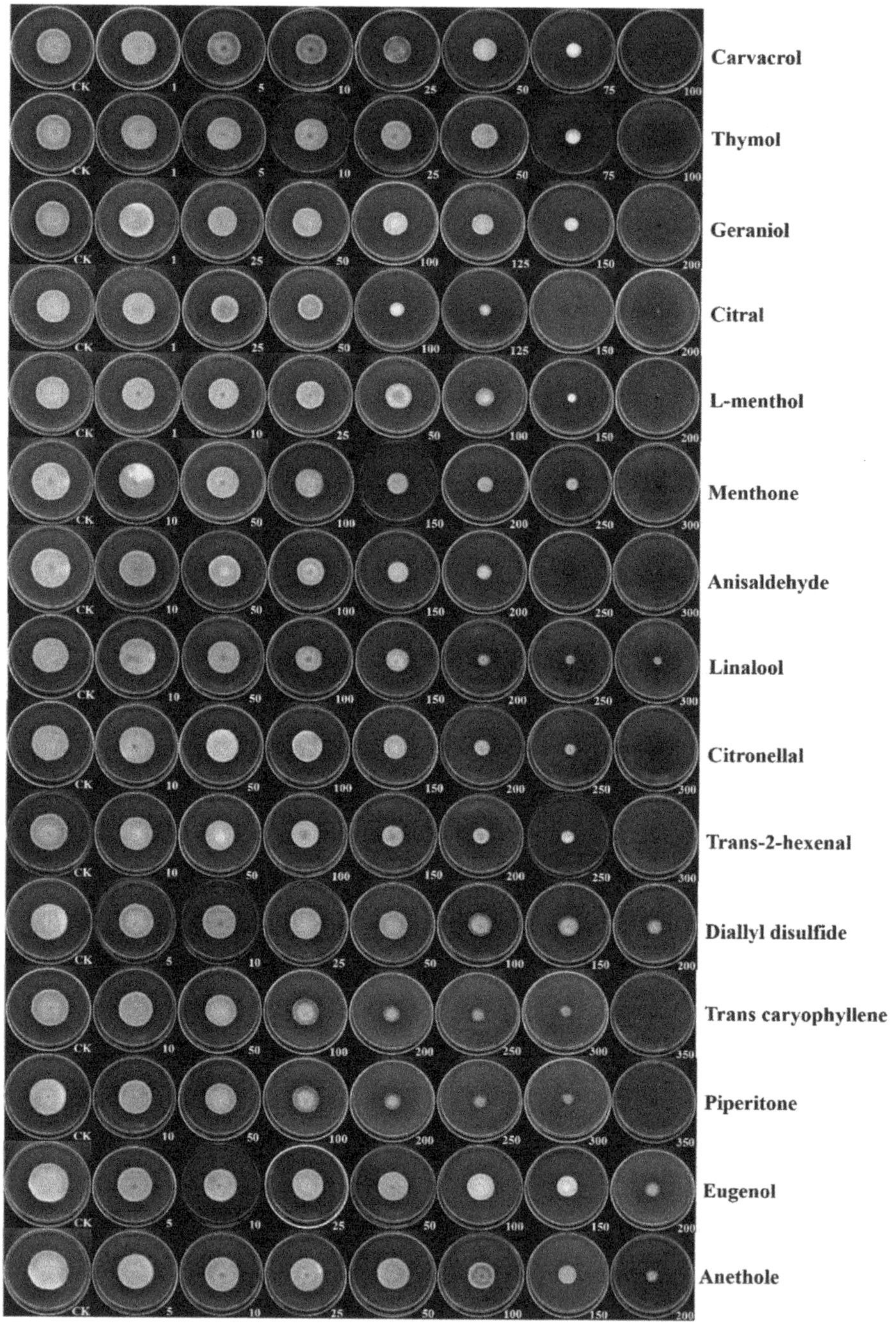

Figure A3. The inhibitory effect of 15 different concentrations of essential oil components on *Alternaria alternata*. The unit of concentration in the figure is μg/mL.

References

1. Zhao, A.; Zhang, Y.; Li, F.; Chen, L.; Huang, X. Analysis of the Antibacterial Properties of Compound Essential Oil and the Main Antibacterial Components of Unilateral Essential Oils. *Molecules* **2023**, *28*, 6304. [CrossRef]
2. Pavicich, M.A.; Nielsen, K.F.; Patriarca, A. Morphological and Chemical Characterization of Alternaria Populations from Apple Fruit. *Int. J. Food Microbiol.* **2022**, *379*, 109842. [CrossRef]
3. Jiang, X.Q.; Ding, J.; Chu, X.Y.; Zhao, Y.P.; Chen, M.R.; Yan, J.; Dai, J.W.; Qin, W.; Liu, Y.W. Effects of H.Uvarum Combined with KGM on Postharvest Diseases of Blueberry. *Food Biosci.* **2023**, *53*, 102730. [CrossRef]
4. Pan, L.Y.; Zhou, J.; Sun, Y.; Qiao, B.X.; Wan, T.; Guo, R.Q.; Zhang, J.; Shan, D.Q.; Cai, Y.L. Comparative Transcriptome and Metabolome Analyses of Cherry Leaves Spot Disease Caused by Alternaria Alternata. *Front. Plant Sci.* **2023**, *14*, 1129515. [CrossRef]
5. Haro, M.L.M.; Cabrera, G.; Pinto, V.F.; Patriarca, A. Alternaria Toxins in Tomato Products from the Argentinean Market. *Food Control* **2023**, *147*, 109607. [CrossRef]
6. Bacha, S.A.S.; Li, Y.P.; Nie, J.Y.; Xu, G.F.; Han, L.X.; Farooq, S. Comprehensive Review on Patulin and Alternaria Toxins in Fruit and Derived Products. *Front. Plant Sci.* **2023**, *14*, 1139757. [CrossRef]
7. Han, X.M.; Xu, W.J.; Wang, L.X.Y.; Zhang, R.N.; Ye, J.; Zhang, J.; Xu, J.; Wu, Y. Natural Occurrence of Alternaria Toxins in Citrus-Based Products Collected from China in 2021. *Toxins* **2023**, *15*, 325. [CrossRef]
8. Liu, Y.P.; Ren, G.; Zhao, X.T.; Wang, W.; Di, J.B.; Wang, Y.; Lin, W.Y. Selenium-Chitosan Treatment Affects Amino Acid Content and Volatile Components of Red Globe Grape during Storage. *J. Food Process. Preserv.* **2023**, *2023*, 3848092. [CrossRef]
9. Buljeta, I.; Pichler, A.; Simunovic, J.; Kopjar, M. Beneficial Effects of Red Wine Polyphenols on Human Health: Comprehensive Review. *Curr. Issues Mol. Biol.* **2023**, *45*, 782–798. [CrossRef]
10. Lee, M.K.; Jung, M.Y. A highly Efficient Strategy for Simultaneous Analysis of Eleven Monomeric Hydroxylated-Stilbenes in Grape Wines and Juices by a Salting-Out Assisted Liquid-Liquid (SALLE) Extraction Combined with a High-Performance Liquid Chromatography-Tandem Mass Spectrometry (HPLC-MS/MS). *Food Chem.* **2023**, *426*, 136661. [CrossRef]
11. Qin, L.W.; Xie, H.; Xiang, N.; Wang, M.; Han, S.A.; Pan, M.Q.; Guo, X.B.; Zhang, W. Dynamic Changes in Anthocyanin Accumulation and Cellular Antioxidant Activities in Two Varieties of Grape Berries during Fruit Maturation under Different Climates. *Molecules* **2022**, *27*, 384. [CrossRef]
12. Xia, K.; Zhang, C.C.; Zhang, X.; Cao, J.D.; He, L.; Liu, C.H. Control of Grey Mould by Sodium Diacetate Treatments and Its Effects on Postharvest Quality of 'Red Globe' Grapes. *Physiol. Mol. Plant Pathol.* **2023**, *125*, 102014. [CrossRef]
13. Buzón-Durán, L.; Sánchez-Hernández, E.; Sánchez-Báscones, M.; García-González, M.C.; Hernández-Navarro, S.; Correa-Guimaraes, A.; Martín-Ramos, P. A Coating Based on Bioactive Compounds from *Streptomyces* spp. and Chitosan Oligomers to Control *Botrytis cinerea* Preserves the Quality and Improves the Shelf Life of Table Grapes. *Plants* **2023**, *12*, 577. [CrossRef]
14. Shi, J.X.; Huang, D.D.; Du, Y.J.; Zhu, S.H.; Hussain, Z.; Haider, M.S.; Anwar, R. Effects of Exogenous Nitric Oxide Treatment on Grape Berries Against *Botrytis cinerea* and *Alternaria alternata* Related Enzymes and Metabolites. *Plant Dis.* **2023**, *107*, 1510–1521. [CrossRef]
15. Huang, R.; Shen, L.; Yu, H.F.; Jiang, J.; Qin, Y.; Liu, Y.L.; Zhang, J.; Song, Y.Y. Evaluation of Rain-Shelter Cultivation Mode Effects on Microbial Diversity during Cabernet Sauvignon (*Vitis vinifera* L.) Maturation in Jingyang, Shaanxi, China. *Food Res. Int.* **2022**, *156*, 111165. [CrossRef]
16. Wang, F.; Saito, S.; Xiao, C.L. Fungicide Resistance of *Alternaria alternata* and A. arborescens Isolates from Mandarin Fruit and Its Influence on Control of Postharvest Alternaria Rot. *Plant Dis.* **2023**, *107*, 1538–1543. [CrossRef]
17. Jiang, Y.Q.; Zhang, L.; Li, X.H.; Chen, L.; Yuan, J.W.; Wang, H.F.; Li, W.H.; Duan, L.H.; Jiang, Y.B.; Tang, Y. Preharvest Fungicide Treatments Reduce the Effective SO_2 Threshold of Postharvest Fumigation to Control Pathogens and Maintain Quality of "red globe" (*Vitis vinifera*) Grapes. *J. Food Saf.* **2023**, *43*, e13047. [CrossRef]
18. Du, Y.H.; Mi, S.N.; Wang, H.H.; Yang, F.W.; Yu, H.; Xie, Y.F.; Guo, Y.H.; Cheng, Y.L.; Yao, W.R. Inactivation Mechanism of *Alternaria alternata* by Dielectric Barrier Discharge Plasma and Its Quality Control on Fresh Wolfberries. *Food Control* **2023**, *148*, 109620. [CrossRef]
19. Wang, J.; Zhang, F.L.; Yao, T.; Li, Y.; Wei, N. Risk Assessment of Mycotoxins, the Identification and Environmental Influence on Toxin-Producing Ability of *Alternaria alternate* in the Main Tibetan Plateau Triticeae Crops. *Front. Microbiol.* **2023**, *13*, 1115592. [CrossRef]
20. Radi, M.; Ahmadi, H.; Amiri, S. Effect of Cinnamon Essential Oil-Loaded Nanostructured Lipid Carriers (NLC) Against Penicillium Citrinum and Penicillium Expansum Involved in Tangerine Decay. *Food Bioprocess Technol.* **2022**, *15*, 306–318. [CrossRef]
21. Zhao, L.A.K.; Wang, J.J.; Zhang, H.Y.; Wang, P.; Wang, C.; Zhou, Y.L.; Li, H.H.; Yu, S.K.; Wu, R.A. Inhibitory Effect of Carvacrol against *Alternaria alternata* Causing Goji Fruit Rot by Disrupting the Integrity and Composition of Cell Wall. *Front. Microbiol.* **2023**, *14*, 1139749. [CrossRef] [PubMed]
22. Maczka, W.; Twardawska, M.; Grabarczyk, M.; Winska, K. Carvacrol-A Natural Phenolic Compound with Antimicrobial Properties. *Antibiotics* **2023**, *12*, 824. [CrossRef]
23. Mi, T.T.; Luo, D.L.; Li, J.K.; Qu, G.F.; Sun, Y.Z.; Cao, S. Carvacrol Exhibits Direct Antifungal Activity against Stem-End Rot Disease and Induces Disease Resistance to Stem-End Rot Disease in Kiwifruit. *Physiol. Mol. Plant Pathol.* **2023**, *127*, 102065. [CrossRef]

24. Zhang, Z.Y.; Tan, Y.B.; McClements, D.J. Investigate the Adverse Effects of Foliarly Applied Antimicrobial Nanoemulsion (Carvacrol) on Spinach. *Lwt-Food Sci. Technol.* **2021**, *141*, 110936. [CrossRef]
25. Zhang, Z.W.; Yang, X.Y.; Zheng, X.J.; Fu, Y.F.; Lan, T.; Tang, X.Y.; Wang, C.Q.; Chen, G.D.; Zeng, J.; Yuan, S. Vitamin E Is Superior to Vitamin C in Delaying Seedling Senescence and Improving Resistance in Arabidopsis Deficient in Macro-Elements. *Int. J. Mol. Sci.* **2020**, *21*, 7429. [CrossRef]
26. Yang, Y.L.; Fang, B.P.; Feng, S.J.; Wang, Z.Y.; Luo, Z.X.; Yao, Z.F.; Zou, H.D.; Huang, L.F. Isolation and Identification of Trichoderma asperellum, the Novel Causal Agent of Green Mold Disease in Sweetpotato. *Plant Dis.* **2021**, *105*, 1711–1718. [CrossRef] [PubMed]
27. Zhang, S.W.; Zheng, Q.; Xu, B.L.; Liu, J. Identification of the Fungal Pathogens of Postharvest Disease on Peach Fruits and the Control Mechanisms of Bacillus subtilis JK-14. *Toxins* **2019**, *11*, 322. [CrossRef]
28. Li, T.T.; Jian, Q.J.; Chen, F.; Wang, Y.; Gong, L.; Duan, X.W.; Yang, B.; Jiang, Y.M. Influence of Butylated Hydroxyanisole on the Growth, Hyphal Morphology, and the Biosynthesis of Fumonisins in Fusarium proliferaturn. *Front. Microbiol.* **2016**, *7*, 1038. [CrossRef]
29. Ding, J.; Liu, C.Y.; Huang, P.; Zhang, Y.W.; Hu, X.J.; Li, H.Y.; Liu, Y.; Chen, L.W.; Liu, Y.W.; Qin, W. Effects of Thymol Concentration on Postharvest Diseases and Quality of Blueberry Fruit. *Food Chem.* **2023**, *402*, 134227. [CrossRef]
30. Kong, J.; Xie, Y.F.; Guo, Y.H.; Cheng, Y.L.; Qian, H.; Yao, W.R. Biocontrol of Postharvest Fungal Decay of Tomatoes with a Combination of Thymol and Salicylic Acid Screening from 11 Natural Agents. *Lwt-Food Sci. Technol.* **2016**, *72*, 215–222. [CrossRef]
31. Xu, M.; Zhou, W.Q.; Geng, W.J.; Zhao, S.R.; Pan, Y.; Fan, G.Q.; Zhang, S.K.; Wang, Y.T.; Liao, K. Transcriptome Analysis Insight into Ethylene Metabolism and Pectinase Activity of Apricot (*Prunus armeniaca* L.) Development and Ripening. *Sci. Rep.* **2021**, *11*, 13569. [CrossRef] [PubMed]
32. Lan, W.T.; Li, S.Y.; Shama, S.; Zhao, Y.Q.; Sameen, D.E.; He, L.; Liu, Y.W. Investigation of Ultrasonic Treatment on Physicochemical, Structural and Morphological Properties of Sodium Alginate/AgNPs/Apple Polyphenol Films and Its Preservation Effect on Strawberry. *Polymers* **2020**, *12*, 2096. [CrossRef]
33. Dai, H.Y.; Ji, S.J.; Zhou, X.; Wei, B.D.; Cheng, S.C.; Zhang, F.; Wang, S.Y.; Zhou, Q. Postharvest Effects of Sodium Nitroprusside Treatment on Membrane Fatty Acids of Blueberry (Vaccinium Corymbosum, cv. Bluecrop) Fruit. *Sci. Hortic.* **2021**, *288*, 110307. [CrossRef]
34. Han, H.N.; Wang, Q.; Wei, L.; Liang, Y.; Dai, J.L.; Xia, G.M.; Liu, S.W. Small RNA and Degradome Sequencing Used to Elucidate the Basis of Tolerance to Salinity and Alkalinity in Wheat. *Bmc Plant Biol.* **2018**, *18*, 195. [CrossRef]
35. Zhao, S.Q.; Li, J.; Liu, J.P.; Xiao, S.Y.J.; Yang, S.M.; Mei, J.H.; Ren, M.Y.; Wu, S.Z.; Zhang, H.Y.; Yang, X.L. Secondary Metabolites of Alternaria: A Comprehensive Review of Chemical Diversity and Pharmacological Properties. *Front. Microbiol.* **2023**, *13*, 1085666. [CrossRef]
36. Ghuffar, S.; Irshad, G.; Shahid, M.; Naz, F.; Riaz, A.; Khan, M.A.; Mehmood, N.; Sattar, A.; Asadullah, H.M.; Gleason, M.L. First Report of *Alternaria alternata* Causing Fruit Rot of Grapes in Pakistan. *Plant Dis.* **2018**, *102*, 1659. [CrossRef]
37. Li, Z.; Chang, P.P.; Gao, L.L.; Wang, X.P. The Endophytic Fungus Albifimbria verrucaria from Wild Grape as an Antagonist of *Botrytis cinerea* and Other Grape Pathogens. *Phytopathology* **2020**, *110*, 843–850. [CrossRef] [PubMed]
38. Wang, Y.D.; Yang, C.D.; Jin, M.J.; Zhong, J.X.; Mei, D.H.; Wei, X.Y.; Wang, H. Isolation and Identification of Tussilago Farfara Leaf Spot Caused by *Alternaria alternata* in China. *Microb. Pathog.* **2022**, *172*, 105750. [CrossRef] [PubMed]
39. da Silva, B.D.; do Rosário, D.K.A.; Neto, L.T.; Lelis, C.A.; Conte, C.A. Antioxidant, Antibacterial and Antibiofilm Activity of Nanoemulsion-Based Natural Compound Delivery Systems Compared with Non-Nanoemulsified Versions. *Foods* **2023**, *12*, 1901. [CrossRef]
40. Sivakumar, D.; Bautista-Banos, S. A Review on the Use of Essential Oils for Postharvest Decay Control and Maintenance of Fruit Quality during Storage. *Crop Prot.* **2014**, *64*, 27–37. [CrossRef]
41. Nemani, N.; Dong, Z.W.; Daw, C.C.; Madaris, T.R.; Ramachandran, K.; Enslow, B.T.; Rubannelsonkumar, C.S.; Shanmughapriya, S.; Mallireddigari, V.; Maity, S.; et al. Mitochondrial Pyruvate and Fatty Acid Flux Modulate MICU1-Dependent Control of MCU Activity. *Sci. Signal.* **2020**, *13*, eaaz6206. [CrossRef] [PubMed]
42. Kim, J.E.; Lee, J.E.; Huh, M.J.; Lee, S.C.; Seo, S.M.; Kwon, J.H.; Park, I.K. Fumigant Antifungal Activity via Reactive Oxygen Species of Thymus vulgaris and Satureja hortensis Essential Oils and Constituents against Raffaelea quercus-mongolicae and Rhizoctonia solani. *Biomolecules* **2019**, *9*, 561. [CrossRef] [PubMed]
43. Nadi, R.; Golein, B.; Gomez-Cadenas, A.; Arbona, V. Developmental Stage- and Genotype-Dependent Regulation of Specialized Metabolite Accumulation in Fruit Tissues of Different Citrus Varieties. *Int. J. Mol. Sci.* **2019**, *20*, 1245. [CrossRef] [PubMed]
44. Paniagua, C.; Pose, S.; Morris, V.J.; Kirby, A.R.; Quesada, M.A.; Mercado, J.A. Fruit Softening and Pectin Disassembly: An Overview of Nanostructural Pectin Modifications Assessed by Atomic Force Microscopy. *Ann. Bot.* **2014**, *114*, 1375–1383. [CrossRef] [PubMed]
45. Verdeguer, M.; Rosello, J.; Castell, V.; Llorens, J.A.; Santamarina, M.P. Cherry Tomato and Persimmon Kaki Conservation with a Natural and Biodegradable Film. *Curr. Res. Food Sci.* **2020**, *2*, 33–40. [CrossRef]
46. Du, G.A.; Guo, Q.; Qiang, L.Y.; Chang, S.D.; Yan, X.H.; Chen, H.; Yuan, Y.H.; Yue, T.L. Influence of Encapsulated Lactobacillus Plantarum and Eugenol on the Physicochemical Properties and Microbial Community of Fresh-Cut Apples. *Food Chem.* **2023**, *17*, 100563. [CrossRef]
47. Chen, C.Y.; Cai, N.; Wan, C.P.; Huang, Q.; Chen, J.Y. Cell Wall Modification and Lignin Biosynthesis Involved in Disease Resistance against Diaporthe Citri in Harvested Pummelo Fruit Elicited by Carvacrol. *J. Sci. Food Agric.* **2022**, *102*, 3140–3149. [CrossRef]

48. Joradol, A.; Uthaibutra, J.; Lithanatudom, P.; Saengnil, K. Induced Expression of NOX and SOD by Gaseous Sulfur Dioxide and Chlorine Dioxide Enhances Antioxidant Capacity and Maintains Fruit Quality of 'Daw' Longan Fruit during Storage through H_2O_2 Signaling. *Postharvest Biol. Technol.* **2019**, *156*, 110938. [CrossRef]
49. Araniti, F.; Miras-Moreno, B.; Lucini, L.; Landi, M.; Abenavoli, M.R. Metabolomic, Proteomic and Physiological Insights into the Potential Mode of Action of Thymol, a Phytotoxic Natural Monoterpenoid Phenol. *Plant Physiol. Biochem.* **2020**, *153*, 141–153. [CrossRef]

Disclaimer/Publisher's Note: The statements, opinions and data contained in all publications are solely those of the individual author(s) and contributor(s) and not of MDPI and/or the editor(s). MDPI and/or the editor(s) disclaim responsibility for any injury to people or property resulting from any ideas, methods, instructions or products referred to in the content.

foods

MDPI

Article

Alginate/Pectin Film Containing Extracts Isolated from Cranberry Pomace and Grape Seeds for the Preservation of Herring

Gabrielė Urbonavičiūtė [1], Gintarė Dyglė [1], Darius Černauskas [2], Aušra Šipailienė [1], Petras Rimantas Venskutonis [1,*] and Daiva Leskauskaitė [1]

[1] Department of Food Science and Technology, Kaunas University of Technology, Radvilenu pl 19, LT-50254 Kaunas, Lithuania; ausra.sipailiene@ktu.lt (A.Š.)

[2] Food Institute, Kaunas University of Technology, Radvilenu pl 19, LT-50254 Kaunas, Lithuania

* Correspondence: rimas.venskutonis@ktu.lt

Abstract: Alginate/pectin films supplemented with extracts from cranberry pomace (CE) or grape seeds (GE) were developed and applied to herring fillets that were stored for 18 days at 4 °C. Herring coated with films containing GE and CE inhibited the growth of *Listeria monocytogenes* and *Pseudomonas aeruginosa* during the storage period, whereas pure alginate/pectin films did not show an antimicrobial effect against the tested pathogens. The application of alginate/pectin films with CE and GE minimised pH changes and inhibited total volatile basic nitrogen (TVN) and the formation of thiobarbituric acid-reactive substances (TBARS) in the herring fillets. The coating of herring fillets with films with CE or GE resulted in three- and six-fold lower histamine formation and one-and-a-half- and two-fold lower cadaverine formation, respectively, when compared to unwrapped herring samples after 18 days of storage. The incorporation of 5% extracts isolated from cranberry pomace or grape seeds into the alginate/pectin film hindered herring spoilage due to the antimicrobial and antioxidant activity of the extracts.

Keywords: alginate/pectin film; cranberry pomace extract; grape seeds extract; herring; biogenic amines

1. Introduction

Since fish is one of the most perishable foods, the search for a means to preserve it remains a relevant topic among food researchers and technologists. The main reason for the spoilage of fish and fish products is direct microbial propagation or the emergence of bacterial enzymes and their metabolism of fish nutrients such as biogenic amines (BAs) [1]. BAs such as histamine (HI), putrescine (PUT), cadaverine (CAD), spermidine (SPD), spermine (SPM), tyramine (TY), phenylethylamine (PHY) and tryptamine (TR) are formed by the decarboxylation of free amino acids as a result of metabolic processes in microorganisms [2]. Delaying the spoilage of fish involves inactivating pathogens and preventing BA formation by inhibiting bacterial growth or decarboxylase enzyme activity, and several such methods have recently been proposed [3]. Edible films containing natural products with preservative properties are very promising for a variety of reasons. Firstly, they are composed of food-grade biopolymers (proteins or polysaccharides) that are biodegradable [4,5]. Secondly, edible films can act as carriers of functional ingredients (e.g., antimicrobials, antioxidants) from natural sources [6]. Applied directly to fish fillets by forming a thin layer of biopolymer with incorporated functional ingredients, edible films can provide considerable protection against microbial growth and metabolism that can result in the formation of toxic metabolites (e.g., BAs) [7].

The inclusion of various plant extracts in edible films or coatings proved to be an effective strategy for improving fish quality and increasing its shelf life. Products packaged in the alginate edible film containing 15% *Aloe vera* extract showed significantly higher lipid stability and microbial quality [8]. Farsanipour et al. (2020) used chitosan in combination with whey protein isolate and tarragon essential oil for the preparation of edible films and applied them to *Scomberoides commersonnianus* fillets [9]. The authors reported the strong activity of the prepared edible films against bacteria alongside improved antioxidant activity. The preservation of rainbow trout fillets was also improved by coating them with a carboxymethylcellulose film incorporated with *Zataria multiflora* Boiss essential oil, grape seed extract (0.5–1%) and combinations thereof [10]. The inhibition of microbial growth was also observed in a chitosan coating with either grape seed extract or tea polyphenols during storage to preserve red drum fillets [11]. Another study indicated that the edible films of sodium alginate-carboxymethylcellulose matrix incorporated with *Ziziphora clinopodioides* essential oil, apple peel extract or zinc oxide nanoparticles increased the shelf life (microbial, chemical and sensory properties) of silver carp fillet and inhibited the growth of *Listeria monocytogenes* during the refrigerated storage of fish over a period of 2 weeks [12]. Similarly, a film composed of sodium alginate and tea polyphenols applied on bream fillets enhanced the shelf life of fish fillets from 15 to 21 days [13].

In the case of BA formation in fish, the use of various plant extracts with antimicrobial activity has shown promising inhibitory effects [14]. Extracts prepared from safflower (*Carthamus tinctorius*) and bitter melon (*Momordica charantia*) demonstrated bactericidal effects on some fish spoilage (*Acinetobacter lwoffii*, *Pseudomonas oryzihabitans*, *Enterobacter cloacae*, *Shigella* spp., *Morganella psychrotolerans* and *Photobacterium phosphoreum*) and food-borne pathogens (*Staphylococcus aureus*, *Klebsiella pneumoniae* and *Salmonella* Paratyphi A). They also reduced the number of BAs, such as HI, PUT, CAD, SPD, SPM, TY and PHY [15]. Furthermore, the treatment of vacuum-packed sardine fillets with mint (*Mentha spicata*) and sagebrush (*Artemisia campestris*) extracts restricted the production of BAs and reduced the presence of HI, TY and CAD [16].

Recently, several investigations have focused on edible active packaging technologies that can reduce and control BA levels in fish. However, published research results are contradictory. For example, active double-layered furcellaran/gelatin hydrolysate films containing Ala-Tyr peptide were developed and used on Atlantic mackerel stored at −18 °C for 4 months [17]. In general, the use of films did not inhibit the formation of BAs in these mackerel samples. Moreover, in the case of TY and SPD, the use of the film increased their formation. Meanwhile, a study conducted by Hao et al. (2017) showed that in abalone (*Haliotis discus hannai*) treated with a sodium alginate coating and bamboo leaf extract or rosemary extract stored under chilled conditions, PUT and CAD were the main BAs [18]. Moreover, the authors noted that coating abalone with rosemary extract greatly inhibited BA formation. Both the total BAs and specific BAs were far below the recommended limits.

Although many studies have reported the effects of edible films containing plant extracts rich in phenolic compounds on fish quality and the inhibition of microbial growth, few reports have investigated the influence of edible films combined with plant extracts on BA formation. Therefore, our study aimed to incorporate phenol-rich berry extracts into alginate/pectin films to develop edible packaging for the active preservation of fish products. The antimicrobial effectiveness of cranberry pomace extract and grape seed extract incorporated in alginate/pectin films on the growth of food-borne pathogens (*L. monocytogenes* and *P. aeruginosa*) and BA formation was evaluated in herring fillets. Finally, the quality properties (pH, TBARS, volatile nitrogenous base formation) of herring fillets with edible films applied during storage were estimated. To the best of our knowledge, this is the first study on the synergy of cranberry pomace extract or grape seed extract and alginate/pectin films in the preservation of herring.

2. Materials and Methods

2.1. Materials and Chemicals

Skinned herring (*Clupea harengus*) fillets with an average weight of 120 ± 10 g were supplied by a local retailer (Orkos, Kaunas, Lithuania).

Food-grade alginic acid sodium salt from brown algae (low viscosity) (Sigma-Aldrich®, Darmstadt, Germany) and pectin from citrus peel (Sigma-Aldrich®, Darmstadt, Germany) were provided by Labochema (Vilnius, Lithuania). Calcium chloride for the cross-linking reaction and glycerol for improving film plasticity were obtained from Eurochemicals (Vilnius, Lithuania).

Biogenic amines, namely, putrescine (PUT, $C_4H_{12}N_2$), cadaverine (CAD, $C_5H_{14}N_2$), histamine (HI, $C_5H_9N_3$), tyramine (TY, $C_8H_{11}NO$) and spermine (SPD, $C_{10}H_{26}N_4$)) standards, were purchased from Sigma-Aldrich (Darmstadt, Germany).

The frozen cranberries were donated by the Fudo Company (Kaunas, Lithuania). Thawed berries were pressed in a Philips HR1880/01 juicer. The pomace was air dried at 35 °C (final moisture content: 5.83%) and ground in a centrifugal mill (Retsch ZM200, Haan, Germany) using a 0.2 mm sieve. The lipophilic fraction of dried cranberry pomace was isolated in a pilot-scale supercritical CO_2 extractor (Applied Separation, Allentown, PA, USA) as previously described [19]. The defatted residue was extracted with agricultural-origin ethanol (Stumbras, Kaunas, Lithuania) in a pilot-scale Soxhlet-type extractor. After extraction, the ethanol was removed in a rotary vacuum evaporator and stored at 4 °C until used. The extract (CE) contained 111.29 ± 0.24 mg/g of polyphenolics (expressed in gallic acid equivalents (GAE)) and 333.1 ± 7.0 mg/100g of procyanidins (determined spectrophotometrically).

Grape seed extract (GE) was purchased from DRT the Best of Nature (Vielle-Saint-Girons, France). The GE contained 626.32 ± 12.96 mg GAE/g of polyphenolics and 1161.00 ± 66.47 mg/100 g procyanidins (determined spectrophotometrically). Overall, 19.6% of procyanidins were monomers, and 40.8% were dimers and trimers. Consequently, the CE and GE can be considered rich in phenols.

All culture media for microbial analysis were supplied by Biometrija (Kaunas, Lithuania): Plate Count Agar (REF 310040, Liofilchem, Italy), Pseudomonas Agar Base (CM559 Oxoid), CFC Supplement (SR103), Agar Listeria Ottaviani Agosti (REF 4016052 Biolife), ALOA Enrichment Supplement (REF 423501) and ALOA Selective Supplement (REF 423501).

Analytical-grade chemicals were obtained from various suppliers: perchloric acid ($HClO_4$, Chempur, Germany); 1,7-diaminoheptane ($C_7H_{18}N_2$, Sigma-Aldrich, Germany); sodium hydroxide (NaOH, Eurochemicals, Lithuania); sodium bicarbonate ($NaHCO_3$, Lachema, Czech Republic); dansyl chloride (5-dimethylaminonaphtalene-1-sulfonyl chloride, Sigma-Aldrich, Germany); ammonia (25%, NH_3, Chempur, Germany); ammonium acetate (0,1 mol/l, CH_3COONH_4, Reachem, Slovakia); acetonitrile (Carlo Erba, France); HCl (Stanlab, Poland).

2.2. Test Microorganisms

Staphylococcus aureus ATCC 25923, *Listeria monocytogenes* ATCC 19117, *Pseudomonas aeruginosa* ATCC 27853, *Salmonella* Typhimurium ATCC 14028, *Escherichia coli* ATCC 8739 and *Escherichia coli* NCTC 12900 were provided by Biometrija (Kaunas, Lithuania). For the preparation of bacterial suspension, the microorganisms were grown overnight on a slant Plate Count Agar (REF 310040, Liofilchem, Italy) at 37 °C and suspended in sterile saline (0.85% NaCl) to achieve a cell concentration of 10^5–10^6 colony-forming units per millilitre (cfu/ml), corresponding to 0.5 McFarland.

2.3. Preparation of Alginate/Pectin Films with Extracts

First, a mixture was made by mixing alginate and pectin powder in a ratio of 7:3. Then, 3 g of the alginate/pectin mixture was dissolved in 100 mL of 1% (w/w) glycerol. Solubilisation was carried out at 30 ± 1 °C for 1 h with continuous stirring on a magnetic stirrer/hot plate. After adding 5 g of CE or GE to the solution, it was further stirred at 30 ± 1 °C for 30 min. For film formation, 10 g of each film-forming solution was poured into Petri dishes (90 mm diameter), which were placed on flat, level trays and kept at 35 ± 1 °C for 48 h (KB8182 in a cooling incubator, Termaks, Bergen, Norway). The film without extract was prepared and referred to as the control with film without extracts.

The films were cross-linked by spraying with a 5% (*w*/*w*) calcium chloride solution. The cross-linked films were peeled from the Petri dishes and stored in desiccators at 65% RH for further analysis. Films without extract were used as controls.

2.4. Evaluation of Antimicrobial Properties of Film via the Agar Diffusion Method

The antibacterial properties of the film were evaluated by agar diffusion assay [20] with slight modification. First, 1 mL of tested *S. aureus* ATCC2 5923, *L. monocytogenes* ATCC 19117, *S.* Typhimurium ATCC 14028, *E. coli* ATCC 8739 and *E. coli* NTCT 12900 bacterial suspensions were prepared separately for each bacterial strain (as previously described) and mixed with 100 mL of the Plate Count Agar (REF 310040, Liofilchem, Italy). The obtained mixture of media and bacterial suspension was poured into Petri dishes (90 mm diameter) at 12 mL each and allowed to solidify surface.

The pieces of the film (10 × 10 mm) were cut and placed onto the agar surface. For the antimicrobial assessment of extracts, their 5% (*w*/*v*) water solutions were prepared and poured into the agar wells (diameter 8 mm). The Petri dishes were incubated in a KB8182 incubator (Termaks) at 37 °C for 24 h. After incubation, the inhibitory zone around the coating pieces and/or the contact area of the film with the agar surface was measured with a Vernier caliper. Results were expressed (in mm) as average zone areas.

2.5. Preparation of Herring Fillets

Herring fillets were aseptically cut into 10.00 ± 0.05 g samples. Then, the samples were fully coated with prepared film. The coated samples were placed in Petri dishes and stored at 4 ± 1 °C for 18 days. Samples were randomly collected after storage for 1, 2, 4, 6, 12 and 18 days for chemical, microbiological and biogenic amine (BA) analysis. For the determination of tested microbial counts in herring, the samples were spread with 10-µL bacterial suspensions of *P. aeruginosa* ATCC 27853 or *L. monocytogenes* ATCC 19117 (prepared as previously described) to achieve a level of 4.0 lg cfu/g. All samples were analysed after the removal of the film.

2.6. Determination of Tested Microorganism Counts in Herring

Total bacterial counts in coated herring samples were determined after 0, 1, 2, 4, 6, 12 and 18 days of storage. After removal of the film at each sampling interval, 10 g fillet portions were prepared in ten-fold dilutions. Plates inoculated via the streak plate method were incubated aerobically at 37 °C for 48–72 h. *P. aeruginosa* and *L. monocytogenes* bacteria counts were determined on Pseudomonas Agar Base (CM559 Oxoid) with CFC Supplement (SR103) and Agar Listeria Ottaviani Agosti (REF 4016052 Biolife) with ALOA Enrichment Supplement (REF 423501), respectively. The results were expressed as lg cfu/g.

2.7. Determination of Biogenic Amines (BAs)

BAs were quantitatively identified by high-performance liquid chromatography (HPLC). For extraction, 5 g of sample was homogenised with 20 mL of perchloric acid solution (0.4 mol/L) in a 50 mL screw-cap tube, and 250 µL of internal standard 1,7-diaminoheptane stock solution (1 mg/mL) was added to achieve a 1 µg/mL concentration in the injection volume. The mixture was centrifuged (MPW-260RH, MPW Med. Instruments, Poland) at 4000 rpm, and the supernatant was rinsed into a 25 mL bottle through Whatman no. 1 filter paper (180 µm thickness and 11 µm particle retention rating at 98% efficiency). Filtrate was adjusted to 25 mL with a perchloric acid solution (0.4 mol/L).

For the derivatisation, 500 µL of sample extract was made alkaline by adding 100 µL of sodium hydroxide solution (2 mol/L). The sample was then buffered by adding 150 µL of saturated sodium bicarbonate. Then, 1 mL of dansyl chloride (5-dimethylaminonaphtalene-1-sulfonyl chloride) solution (10 mg/mL) was added and mixed thoroughly using a shaker-mixer (IKA mini Shaker TS1, Germany). The reaction mixture was then transferred to a 40 °C incubator for 45 min. After incubation, it was cooled to room temperature for 10 min, and residual dansyl chloride was removed by adding 50 µL of ammonia (25%).

The mixture was then mixed with a shaker-mixer. After 30 min, it was adjusted to 5 mL with an ammonium acetate (0.1 mol/L):acetonitrile mixture (1:1, *v/v*) and mixed well with a shaker-mixer. The mixture was then filtered through a 0.20 μm nylon filter (UptiDisc, Interchim, France), and the solution was injected into an analytical column.

A Shimadzu Prominence LC20AD (Shimadzu, Japan) coupled to a UV detector SPD/20 A chromatographic system was utilized with a LabSolution (Shimadzu, Japan) integrator using a Hydrosphere C18 (5 μm, 12 nm), 150 × 4.6 I.D. column and YMC pre-column ProC18 (10 × 3.0 mml.D., S-3 μm, 12 nm) (YMC Co., Ltd., Japan). LC mobile phase A: ammonium acetate (0.1 mol/L), phase B acetonitrile. Operating conditions: flow rate 0.9 mL/min; injection volume 20 μL; column temperature 40 °C; peaks were detected at 254 nm; gradient 50% B to 90% B in 19 min; run time 20 min; post-run before next run, 50% B, 8 min.

Five BAs were quantified: PUT, CAD, HI, TY and SPD. Stock solutions (1 mg/mL) for each amine were prepared in 0.1 M HCl and stored at 4 ± 1 °C. For amine identification, standard solutions of individual BAs were chromatographed separately and mixed to determine the retention times and responses of each. Standard curves with correlation coefficients for stock solutions were obtained via the external standard method. All results were expressed in mg/kg.

2.8. Determination of the Total Volatile Basic Nitrogen (TVN)

TVN was determined according to Commission Regulation (EC) No. 2074/2005 [21]. Briefly, 10 ± 0.1 g of ground herring sample was mixed with 90.0 mL perchloric acid solution (6 g/100 mL) and filtered through No. 2 Whatman filter paper. Then, 50.0 mL of the obtained extract was subjected to steam distillation in a 2100 Kjeltec Distillation Unit (FOSS Tecator AB) following alkalinisation with 0.2 M NaOH. Then, 100 mL of distillate was collected in 100 mL of 3% (*w/v*) boric acid solution containing Tashiro's indicator. The TVN contained in the distillate solution was determined by titration with 0.01 M HCl, and its concentration was calculated via the following formula:

$$\text{TVN (mg/100 g sample)} = (V_1 - V_0) \times 0.14 \times 2 \times 100/W, \tag{1}$$

where V_0 is the volume of hydrochloric acid used in blank titration; V_1 is the volume of hydrochloric acid used in sample titration; and W is the weight of fish sample in grams.

2.9. pH Measurements

pH was measured directly in fish samples using an XS instrument pH7 digital pH meter (EU, P.R.C.) with a penetrating probe.

2.10. Determination of Thiobarbituric Acid-Reactive Substances (TBARS)

A TBARS assay was performed to evaluate lipid oxidation as described by Tarladgis et al. (1960) [22] with some modification. First, 10 g of homogenised sample were mixed with 98 mL of distilled water. Then, 1.25 mL of 0.4 N HCl were added to 50 mL of the prepared mixture. The mixture was then heated with steam distillation. Subsequently, 5 mL of distillate and 5 mL of thiobarbituric reactive reagent containing 0.02 M TBA in 90% glacial acetic acid were transferred into a glass tube. The tube was then placed into a boiling water bath for 1 h. After cooling, the absorbance of the pink solution was measured at 532 nm using a Biochrom spectrophotometer (Holliston, MA, USA). The constant 7.8 was used to calculate the TBARS number. The TBARS value was expressed as mg malonaldehyde equiv. (MDA)/kg sample.

2.11. Statistical Analysis

Data are expressed as means ± SD. An analysis of variance (ANOVA) followed by Fisher's LSD test was performed using Statistica 8.0 software. A *p*-value of <0.05 was considered statistically significant. All experiments and analyses were run in triplicate.

3. Results and Discussion

3.1. Antimicrobial Properties of Films with CE and GE

To gain insights into the antimicrobial properties of CE, GE and alginate/pectin films with extracts, the inhibition zone diameters against *S. aureus* ATCC25923, *L. monocytogenes* ATCC19117, *S.* Typhimurium ATCC14028, *E. coli* ATCC8739 and *E. coli* NTCT12900 were determined (Figure 1). It should be noted that the amount of incorporated extracts in the 10×10 mm (1 cm^2) piece of film, which was used in this assay, was approximately three-fold higher than that directly added to the well with the extract solution, 7.9 mg vs. 2.6 mg.

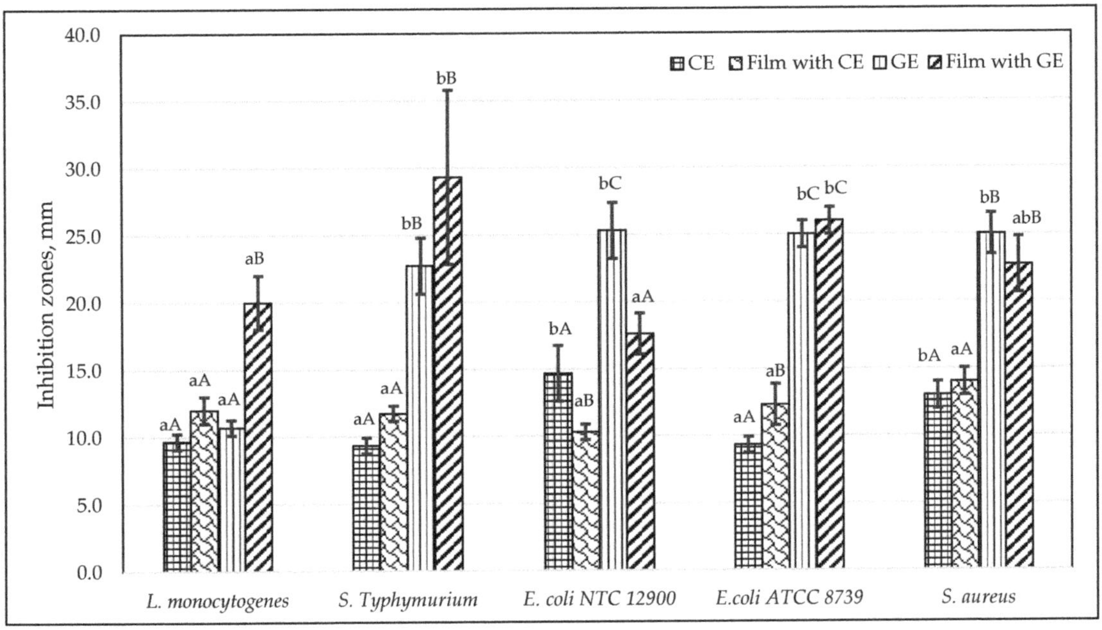

Figure 1. Antimicrobial activities of alginate/pectin films with CE and GE against tested microorganisms. The columns are drawn from the average of at least three independent experiments. Error bars denote ± one standard deviation; small letters indicate significant ($p < 0.05$) differences between bacteria sensitivity to tested alginate/pectin films and extracts; uppercase letters indicate significant ($p < 0.05$) differences between alginate/pectin films and extracts.

This preliminary evaluation indicated that GE both in pure form and incorporated into the film in most cases was a stronger antimicrobial agent than CE, except for GE extract's effect on *L. monocytogenes*, when its activity was not significantly different from the CE. The antimicrobial activity of films with CE was not significantly ($p < 0.05$) different compared with pure extract against the tested bacteria, except for *E. coli* NTC12900, when the film with CE demonstrated lower activity than the extract. The results showed that the films with CE produced inhibition zones in the range of 10.0–14.0 mm. Such zones of inhibition can be considered small. However, the antimicrobial activity of the film with CE was detected against all microorganisms tested, and a wide impact is a definite advantage of the film supplemented with CE. The antimicrobial activity of different cranberry extracts alone against *S. aureus* and *E. coli*, *S. enteritidis* and *L. monocytogenes* has been shown by several studies [23–25]. It was concluded that both a low pH and the concentration of bioactive phenolics are responsible for the antimicrobial properties of cranberries [26]. However, direct comparison of our results with those previously published is very difficult because of different extract preparation procedures and the doses applied. For instance, Côté et al. (2011) tested full cranberry juice and extracts [23], and Laplante et al. (2012) used proanthocyanidin standardized cranberry extracts [24]. The latter study demonstrated that antimicrobial effects were strongly dependent on the proanthocyanidins concentration; the minimal inhibitory concentration (MIC) of the extracts with 210 mg/g in most cases was >30-fold lower than that of the extracts with 8 mg/g [24]. For comparison, the concentration of proanthocyanidins in CE applied in our study was only 3.33 mg/g. Some other studies inoculated whole cranberry extracts in ground beef [25], and cranberry pomace extracts were tested in the commonly used meat fermentation starter cultures [26]. Severo et al. (2021) demonstrated that chitosan films with the incorporated whole CEs inhibited *E. coli* and *S. aureus* biofilm formation, which was explained by the presence of phenolic compounds [27]. Similarly, our previous studies on whey protein-chitosan film incorporating cranberry juice showed their antimicrobial activities against *S.* Typhimurium, *L. sakei*, *L. plantarum*, *S. agona* and *C. jejuni* [28].

Our study showed significantly ($p < 0.05$) higher sensitivity of the tested microorganisms against the films supplemented with GE in comparison to the films with CE. It is interesting to note that the applied amounts of CE and GE in most cases gave similar inhibition zones (the differences were not significant) for the extract solution and the film. This may be explained by the fast impact of inhibiting compounds, which are rapidly diffusing into the bacteria culture after dosing into the well, whereas their release from the cross-linked films proceeds more slowly. From this point of view, approximately three-fold higher amounts of extracts in the films were a quite reasonable selection for this assay. However, the antimicrobial effects were also dependent on the bacteria species. Thus, among tested food-borne pathogens, the smallest inhibition zone was obtained for *E. coli* NTCT, with an average value of 17.7 ± 1.5 mm, whereas for *S.* Typhimurium, the inhibition zone of the film with GE was remarkably larger, 29.3 ± 6.1 mm. Again, the good antimicrobial activity of the film with GE can be attributed to the phenolic compounds present in GE [29]. Our study used GE containing a three-and-a-half-fold higher concentration of proanthocyanidins than CE. For instance, Sogut and Seydim (2018) demonstrated that GE-incorporated chitosan films inhibited *E. coli*, *L. monocytogenes*, *S. aureus* and *P. aeruginosa* more efficiently than chitosan films alone, and the effect of 5% GE (the concentration used in our study) on the count of coliform bacteria was significantly lower that that of 10 and 15% [30].

Our results for the antimicrobial properties of alginate/pectin films supplemented with CE and GE suggest that they can be used in food systems to control food-borne pathogens. Therefore, further antimicrobial studies were performed by applying films on herring that had previously been infected with pathogenic bacteria.

3.2. Herring Preservation by Applying Alginate/Pectin Films with CE or GE

3.2.1. Viability of *L. monocytogenes* and *P. aeruginosa*

In general, the viable cell count method is used to assess the antimicrobial effectiveness of edible films for coated or wrapped fish products. Figure 2 presents the viability results of *L. monocytogenes* and *P. aeruginosa* in herring wrapped in alginate/pectin films supplemented with CE or GE during storage for 18 days. *L. monocytogenes* can be embedded into fish during handling or storage processes, and *P. aeruginosa* is a normal part of the fish microbiota. The tested pathogens showed different sensibilities to films with CE and GE. During storage, the *L. monocytogenes* content in control samples (without film and wrapped in a film without extract) increased from 3.87 to 7.52 lg CFU/g and from 3.75 to 6.69 lg CFU/g, respectively (Figure 2A). Wrapping herring samples in the film with CE resulted in more moderate growth of the *L. monocytogenes* population: during 12 days of storage, it increased from 3.75 to 6.07 lg CFU/g and remained unchanged during subsequent storage. The herring treatment with a film containing GE had a significant ($p < 0.05$) influence on the growth of *L. monocytogenes* during storage: after 18 days, the population of *L. monocytogenes* remained at the same level observed at the beginning of storage. The applied films in the herring samples showed stronger inhibitory effects on gram-positive *L. monocytogenes* bacteria as compared to gram-negative *P. aeruginosa* bacteria.

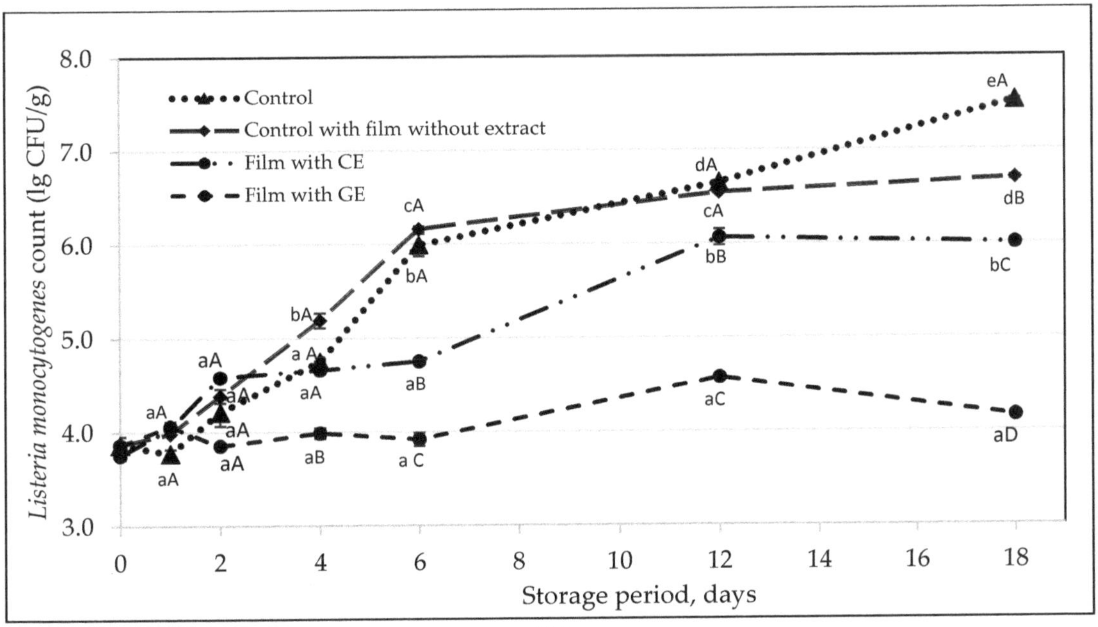

(A)

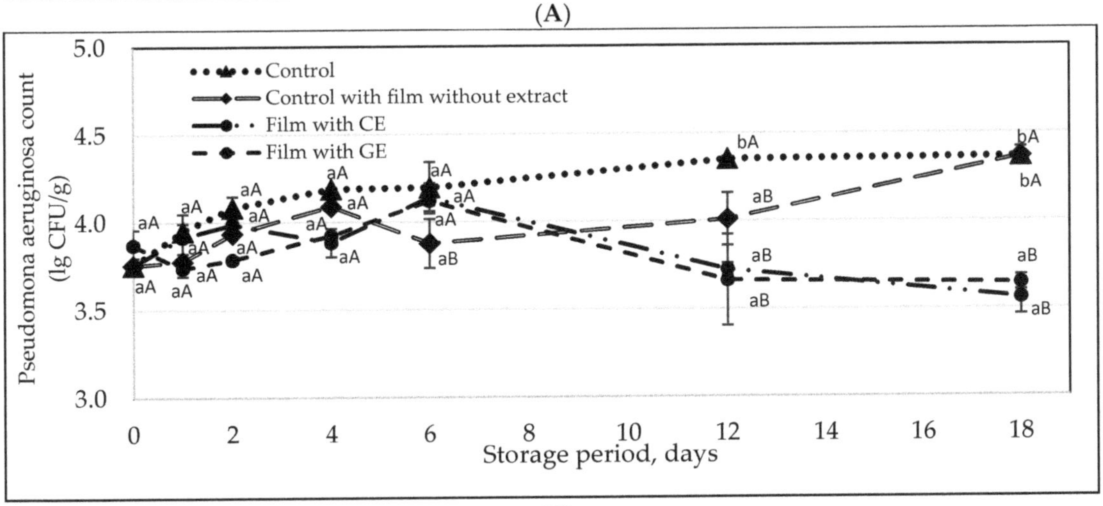

(B)

Figure 2. Effect of alginate/pectin film with CE and GE on the growth of *L. monocytogenes* (**A**) and *P. aeruginosa* (**B**) of herring fillets during storage for 18 days. The curves are drawn from the average of at least three independent experiments. Error bars denote ± one standard deviation. Different characters (a, b, c, d, e) indicate significant ($p < 0.05$) differences between storage duration, and A, B, C, D indicate significant ($p < 0.05$) differences between the films. Note: So far as the CFU measurements were not continuous, the lines cannot be used for estimating the precise number of CFU at any storage period between the points.

Notably, weak growth inhibition during storage was observed for *P. aeruginosa* in herring samples wrapped in films containing CE and GE with 0.31 and 0.23 lg reductions, respectively (Figure 2B). The antibacterial effect of polysaccharide films with incorporated anthocyanins is related to the interaction between them and bacterial membranes. Several mechanisms have been suggested to explain the antibacterial action of anthocyanins, such as differences in cell wall structure, cell physiology, metabolism, destabilisation or permeability of the cytoplasmic membrane [31,32]. In control samples (without film and wrapped in film without extract), the *P. aeruginosa* content increased from 3.75 to 4.36 lg CFU/g after 18 days of storage. Our results are in accordance with Nešić et al. (2017), who indicated that pure alginate/pectin films did not show any antimicrobial effect against tested *S. aureus*, *E. coli* and *C. albicans* pathogens [33]. However, alginate/pectin films were demonstrated to be good vehicles for antimicrobial substances. The major advantage of composite alginate/pectin films with natamycin addition is their low diffusion coefficient [34]. De'Nobili et al. (2015) indicated that alginate and pectin composite film supplemented with ascorbic acid has great potential to be used in antimicrobial packaging to inhibit food spoilage [35]. Undoubtedly, the composition of polyphenols with antimicrobial properties, as well as the amount added to a film, have a significant effect on the antimicrobial properties of films [36]. In our study, GE contained considerably more polyphenols in comparison to CE (626.32 ± 12.96 mg/g and 111.29 ± 0.24 mg/g, respectively). Furthermore, the presence of a high amount of procyanidins in GE—with dimers and trimers predominating (40.8%)—may have also led to better inhibition of pathogenic microorganisms determined for the film with GE. Recent studies on novel food packaging materials containing procyanidins demonstrated outstanding antimicrobial activity [37]. Chitosan film with procyanidins was also applied in salmon muscle perseveration and showed the potential ability to prevent microorganism contamination and texture deterioration for 10 days.

3.2.2. Biogenic Amines

The effect of alginate/pectin films alone—and with CE or GE—on the BA production in herring fillets during storage was investigated. The results of HI, CAD, PUT and TY contents during the storage of unwrapped herring samples and those wrapped in different films are presented in Table 1.

Table 1. Effect of alginate/pectin film with CE and GE on the changes in biogenic amines (BAs)' concentrations (mg/kg) of herring fillets during storage for 18 days.

Sample		Storage Duration, Days					
		1	2	4	6	12	18
HIS	Control	<5 ^{aA}	6.11 ± 0.33 ^{bA}	<5 ^{aA}	<5 ^{aA}	58.86 ± 1.53 ^{cA}	354.8010.32 ^{dA}
	Control with pure film	<5 ^{aA}	<5 ^{aB}	<5 ^{aA}	<5 ^{aA}	104.09 ± 2.70 ^{bB}	99.35 ± 7.57 ^{bB}
	Film with CE	<5 ^{aA}	<5 ^{aB}	<5 ^{aA}	18.67 ± 1.03 ^{bB}	38.42 ± 3.42 ^{cC}	19.52 ± 2.89 ^{bC}
	Film with GE	<5 ^{aA}	<5 ^{aB}	<5 ^{aA}	<5 ^{aA}	9.30 ± 0.42 ^{bD}	31.11 ± 2.96 ^{cD}
CAD	Control	13.02 ± 0.14 ^{aA}	17.04 ± 0.2 ^{bA}	19.54 ± 0.76 ^{bA}	16.30 ± 0.42 ^{bA}	139.03 ± 6.32 ^{cA}	438.99 ± 9.48 ^{dA}
	Control with pure film	<5 ^{aB}	<5 ^{aB}	<5 ^{aB}	15.50 ± 0.71 ^{bA}	136.54 ± 5.77 ^{cA}	291.32 ± 18.07 ^{dB}
	Film with CE	<5 ^{aB}	<5 ^{aB}	<5 ^{aB}	51.20 ± 0.28 ^{bB}	26.46 ± 0.91 ^{cB}	103.87 ± 6.04 ^{dC}
	Film with GE	5.04 ± 0.03 ^{aB}	<5 ^{aB}	6.23 ± 0.33 ^{bC}	<5 ^{aC}	26.40 ± 2.97 ^{cB}	162.26 ± 4.33 ^{dD}
PUT	Control	6.62 ± 0.26 ^{aA}	15.53 ± 0.47 ^{bA}	<5 ^{cA}	14.75 ± 0.64 ^{bA}	54.37 ± 3.35 ^{dA}	76.81 ± 5.81 ^{eA}
	Control with pure film	<5 ^{aB}	<5 ^{aB}	<5 ^{aA}	13.71 ± 1.54 ^{bA}	47.73 ± 3.01 ^{cB}	118.67 ± 3.07 ^{dB}
	Film with CE	<5 ^{aB}	6.53 ± 0.75 ^{bC}	<5 ^{aA}	6.75 ± 0.35 ^{bB}	24.30 ± 0.42 ^{cC}	15.08 ± 1.53 ^{dC}
	Film with GE	<5 ^{aB}	<5 ^{aB}	<5 ^{aA}	8.37 ± 0.35 ^{bC}	19.78 ± 1.70 ^{cD}	56.65 ± 1.77 ^{dD}
TY	Control	<5 ^{aA}	5.89 ± 0.30 ^{bA}	6.00 ± 0.00 ^{bA}	<5 ^{aA}	6.08 ± 0.11 ^{bA}	25.9 ± 4.24 ^{cA}
	Control with pure film	<5 ^{aA}	<5 ^{aB}	<5 ^{aB}	<5 ^{aA}	22.10 ± 1.41 ^{bB}	49.43 ± 5.98 ^{cB}
	Film with CE	<5 ^{aA}	<5 ^{aB}	<5 ^{aB}	5.18 ± 0.26 ^{aA}	17.24 ± 2.89 ^{bC}	48.02 ± 4.24 ^{cB}
	Film with GE	<5 ^{aA}	<5 ^{aB}	<5 ^{aB}	<5 ^{aA}	6.71 ± 0.79 ^{bA}	6.39 ± 0.25 ^{bC}

Different characters (a, b, c, d, e) indicate significant ($p < 0.05$) differences between storage duration, and A, B, C, D indicate significant ($p < 0.05$) differences between the films.

HI remained at low levels in all samples at the early stages of herring storage (6 days). After 18 days of storage, HI accumulation in unwrapped samples was generally higher (358.80 ± 10.32 mg/kg) than in samples coated with films. Moreover, films with CE or GE inhibited the formation of HI more efficiently in comparison to film alone. At the end of the storage period, the HI content in the samples coated with films with CE and GE was 19.52 ± 2.86 mg/kg and 31.11 ± 2.96 mg/kg, respectively. A significant ($p < 0.05$) increase in CAD was observed in all herring fillet samples during storage for 18 days. The unwrapped sample showed a higher CAD level (13.02–19.54 mg/kg) at the beginning of storage, which greatly increased up to 139.03 ± 6.32 mg/kg and 438.99 ± 9.48 mg/kg after 12 and 18 days of storage, respectively. In the film-coated samples, the CAD level exceeded 5 mg/kg only after 6 days of storage. Although the amount of CAD in the film-coated samples also increased during further storage, its amount at the end of storage was approximately two-fold lower when compared to the control sample.

Furthermore, the coating of herring fillets with alginate/pectin films resulted in decreased PUT formation during 12 days of storage. After 18 days of storage, PUT content in the unwrapped samples was lower in comparison to film-coated samples. However, these differences were statistically insignificant and relatively low. Moreover, levels of TY higher than 5 mg/kg were only detected in all samples after 12 days of storage. After 18 days of storage, the TY content in the unwrapped samples was 25.9 ± 4.24 mg/kg. In samples coated with film alone, the TY content was 49.43 ± 5.98 mg/kg. Additionally, in samples coated with film with CE or GE, the TY content was 28.01 ± 2.84 mg/kg and 6.07 ± 0.18 mg/kg, respectively. Thus, only herring fillets with a film coating containing GE inhibited the formation of TY during storage.

In our study, high levels of all analysed BAs in the unwrapped herring fillets were observed at 18 days of storage: 354.80 ± 10.32 mg of HI, 438.99 ± 9.48 mg of CAD, 76.81 ± 5.81 mg of PU and 25.9 ± 4.24 mg of TY in 1 kg of muscle. Our results are similar to those reported by Özogul et al. (2002), who found that the storage of herring fillets for 16 days without ice resulted in the accumulation of HI, CAD, PU and TY levels of 369.4, 329.9, 74.2 and 0 mg/kg, respectively [38]. Coating herring fillets with alginate/pectin film supplemented with CE or GE resulted in approximately three- and six-fold lower HI formation and about one-and-a-half- and two-fold lower CAD formation when compared to unwrapped herring samples after 18 days of storage. The effectiveness of extracts on reducing PUT accumulation in herring fillets was not observed. TY production was only considerably suppressed in the presence of films with GE. Previous research has shown that BA formation by fish spoilage bacteria in the presence of various plant extracts was dependent on the bacterial strain, extract type and dose [39]. Although *Morganella morganii*, *Klebsiella pneumoniae* and *Hafnia alvei* are reported as the strongest producers of BAs [40], Gram-positive bacteria (*S. aureus*, *E. faecalis* and *L. monocytogenes*) were able to decarboxylate more than one amino acid and produce HI, CAD and other amines [41]. Gram-negative bacteria (*E. coli*, *K. pneumoniae*, *A. hydrophila* and *P. aeruginosa*) have also been identified as BA-producing bacteria [15]. In our study, the antimicrobial effect of films containing polyphenol-rich CE and GE against *S. aureus* ATCC25923, *L. monocytogenes* ATCC19117, *S.* Typhimurium ATCC14028, *E. coli* ATCC 8739 and *E. coli* NTCT 12900 was observed. Therefore, the identified negative influence of extracts on the formation of BAs can be related to the recorded antibacterial properties against bacteria that can be BA producers.

Similarly, HI, CAD and PUT accumulation in vacuum-packed sardine fillets was suppressed by sage tea (*Salvia officinalis*) and rosemary (*Rosmarinus officinalis*) extracts [42]. Others have observed the high antimicrobial activity of garlic extract against *Bacillus licheniformis* strains and the greatest inhibitory effect on BA content in Korean salted and fermented anchovy [43]. Additionally, Özyurt et al. (2012) suggested that the icing of sardine (*Sardinella aurita*) along with rosemary extract maintained BA content at low levels, especially HI and PUT [44]. We found no data regarding the effect of CE on BA formation. Notably, GE was proven to decrease the formation of BAs in bacon: TY (30.7%), CAD (37.1%), PUT (29.4%) and HI (73.6%) [45].

3.2.3. Monitoring pH during Storage

The observed pH value changes of herring fillets wrapped in alginate/pectin films with and without CE or GE are presented in Figure 3.

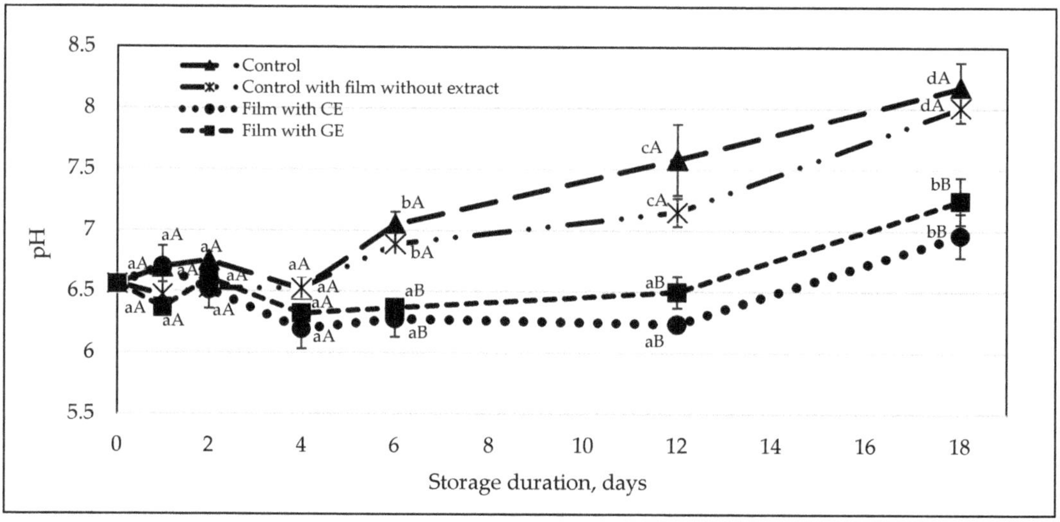

Figure 3. Effect of alginate/pectin film with CE and GE on pH of herring fillets during storage for 18 days. The curves are drawn from the average of at least three independent experiments. Error bars denote ± one standard deviation. Different characters (a, b, c, d) indicate significant ($p < 0.05$) differences between storage duration, and A, B indicate significant ($p < 0.05$) differences between the films.

At the beginning of storage, the pH of all samples was not different (pH = 6.56). During 12 days of storage, the pH values of herring fillets wrapped in films with CE or GE remained constant or slightly decreased. After 18 days of storage, a moderate increase in the pH value (approximately pH 7.0) was recorded for these samples. A different change in pH was observed during the storage of control samples (unwrapped or wrapped in alginate/pectin film without extract). The pH of unwrapped herring fillets continuously increased from 6.56 to 7.05 during 6 days of storage and then drastically increased up to 8.17 after 18 days of storage. Similarly, the pH values of herring fillets coated in film without extract increased to their highest value, 8.06, after 18 days of storage. The increase in pH could be due to autolytic processes such as endogenous enzymes and microbiological enzymatic action causing protein degradation and the accumulation of nitrogen compounds such as primary, secondary and tertiary amines. At the end of storage, these processes were more intensive in the control samples in comparison to the herring samples coated with films with extract, as recorded by TVN value changes (Figure 4). The results indicate that alginate/pectin films with CE and GE significantly ($p < 0.05$) minimised pH changes and hindered herring spoilage due to the antimicrobial and antioxidant activity of polyphenol-rich extracts. Similar results were previously observed during the storage of *Scomberoides commersonnianus* coated with chitosan and chitosan-whey protein films with tarragon essential oil [9] and common carp samples treated with *Carum copticum* and lactic acid [46].

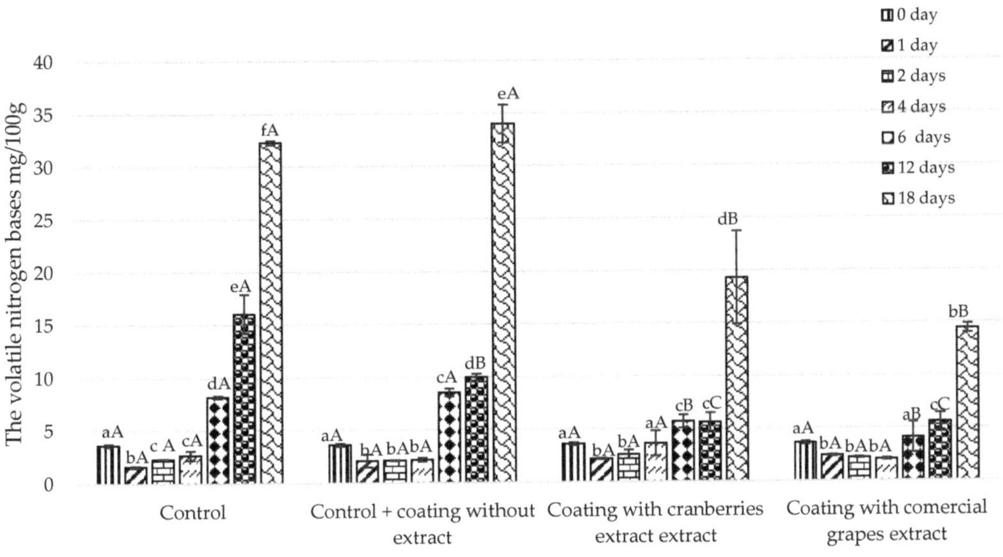

Figure 4. Effect of alginate/pectin film with CE and GE on the total volatile basic nitrogen in herring fillets during storage for 18 days. The columns are drawn from the average of at least three independent experiments. Error bars denote ± one standard deviation. Different characters (a, b, c, d, e, f) indicate significant ($p < 0.05$) differences between storage duration, and A, B, C indicate significant ($p < 0.05$) differences between the films.

3.2.4. Total Volatile Basic Nitrogen (TVN)

At the beginning of storage, the TVN value was 3.6 mg/100 g in all samples (Figure 4). During storage, TVN values increased by approximately 10 times and reached 32.26 mg/100 g and 34.00 mg/100 g in control samples without film and coated with film without extract, respectively. However, in the herring samples wrapped in films with CE and GE, significantly ($p < 0.05$) lower values of TVN were detected at the end of storage when compared to control samples. After 18 days of storage, the TVN values were 19.29 mg/100 g in the samples coated with film with CE and 14.40 mg/100 g in the samples coated with film with GE. Furthermore, in all control samples, the TVN values exceeded 10 mg/100 g—even after 12 days of storage. In the samples coated with films containing extracts at 12 days of storage, the TVN values were between 5.57 and 5.61 mg/100 g. From the obtained data, it can be stated that alginate/pectin film alone does not protect herring from the formation of volatile nitrogen bases, and the incorporation of CE or GE into the films has almost the same effect on reducing the degradation of nitrogenous compounds in herring fillets. In our studies, TVN values correlated well with the pH data.

The TVN parameter identifies primary, secondary and tertiary amines and is recognised as an indicator of muscle tissue deterioration [47]. In light of the recommendations of the European Commission (CEC) (1995) [48] and various scientists [49,50], 25–35 mg of N per 100 g is an upper acceptability limit for spoilage initiation in fresh fish. In our study, only those herring samples that were wrapped in film with CE or GE demonstrated TVN values below this limit of acceptability after 18 days of storage. Similar results were reported by Günlü and Koyun (2013) in sea bass (*Dicentrarchus labrax*) fillets wrapped with chitosan-based edible film during cold storage at 4 °C [51]. According to Farsanipour et al. (2020), chitosan coating in combination with whey protein and tarragon essential oil also has the ability to retard the increase in TVN content in *Scomberoides commersonnianus* muscle during storage [9]. In both studies, TVN values correlated well with the microbiological data, indicating that the TVN parameter is a useful index for fish spoilage.

3.2.5. Thiobarbituric Acid-Reactive Substances (TBARS)

Changes in the TBARS values of wrapped and unwrapped herring fillets during storage are presented in Figure 5. TBARS is also considered an indicator of quality for fish exhibiting secondary lipid oxidation products. At the beginning of storage (0 days), the TBARS value was 2.09 mg MDA/kg. The initial TBARS value in our study was slightly lower than that reported at the onset of the refrigerated storage of herring fillets by Tolstorebov et al. (2014) (2.42 mg MDA/kg) [52].

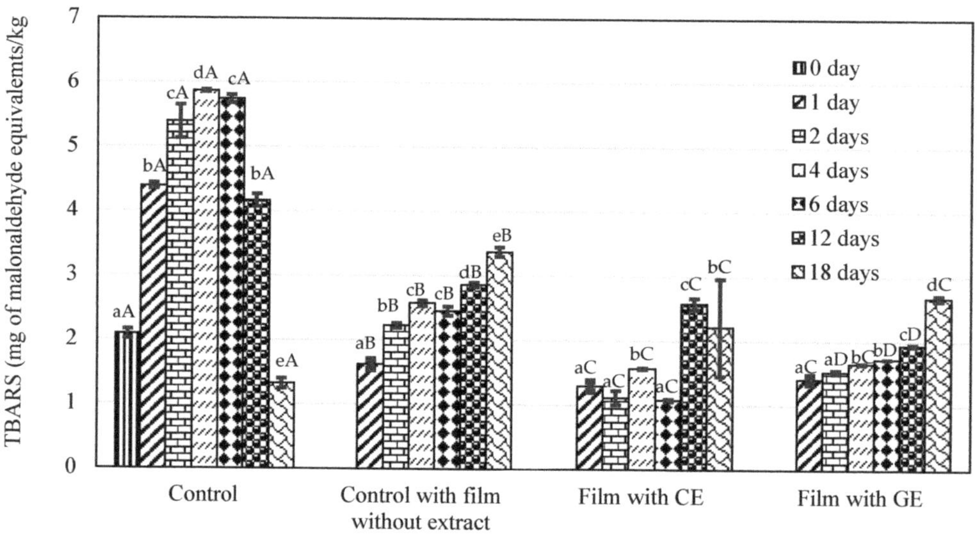

Figure 5. Effect of alginate/pectin film with CE and GE on TBARS in herring fillets during storage for 18 days. The columns are drawn from the average of at least three independent experiments. Error bars denote ± one standard deviation. Different characters (a, b, c, d, e) indicate significant ($p < 0.05$) differences between storage duration, and A, B, C, D indicate significant ($p < 0.05$) differences between the films.

In the unwrapped control sample, secondary lipid oxidation products accumulated, and TBARS values increased to 5.74 mg MDA/kg during 6 days of storage. On the 12th day of storage, a sharp decrease in the TBARS value was recorded, with a further decline during storage. Similar results with increasing and decreasing TBARS values have been recorded in *Scomberoides commersonnianus* during refrigerated storage [9]. Authors have assigned these changes to the partial dehydration of fish and losses of the secondary lipid oxidation products formed during the initial storage period. In our study, this explanation is also reasonable, because the TBARS values only increased and decreased during the storage of the unwrapped herring samples; therefore, they could be dehydrated during storage.

A consistent increase in TBARS during storage was registered for the herring samples wrapped in the different films. However, TBARS values in the herring samples wrapped in composite alginate/pectin film alone were significantly ($p < 0.05$) higher than in those in samples wrapped in films with CE or GE. At the end of the storage period (18 days), TBARS values in the samples coated with film with CE and GE were 2.21 and 2.68 mg MDA/kg, respectively. Higher levels of TBARS were recorded in the samples wrapped in film alone after 18 days of storage (3.37 mg MDA/kg). These results suggest that the supplementation of alginate/pectin films with cranberry or grape seed extracts enhanced the antioxidant properties of the films. The high levels of polyphenols in CE, as well as a large number of procyanidins in GE, act as strong antioxidants that scavenge free radicals and hinder the oxidation chain reactions [53,54]. Cranberry extracts were proven to be suitable for presenting antioxidant (DPPH-scavenging ability) activity in chitosan-based films [27]. Similarly, incorporating grape seed extract into chitosan films improved the oxidative stability of coated red drum fillets [11] and fresh pork [29] under storage with significantly reduced TBARS values.

Notably, all alginate/pectin film-coated samples showed TBARS values of <5 mg MDA/kg throughout the storage period. Meanwhile, the control sample (uncoated) exceeded this value after 4 days of storage. According to Sallam (2007), the maximum TBARS value indicating good fish quality (frozen, chilled or stored with ice) is 5 MDA/kg of tissue [55]. Thus, it can be concluded that according to the values of the lipid oxidation indicator, the quality of the herring remained acceptable during storage when the fish was wrapped in the composite alginate/pectin film supplemented with CE and GE.

4. Conclusions

Composite alginate/pectin films supplemented with polyphenol-rich extracts from defatted cranberry pomace (CE) or grape seeds (GE) and applied on herring fillets showed better preservation properties when compared to pure films. The presence of polyphenol-rich extracts considerably suppressed the viability of pathogenic microorganisms (*L. monocytogenes* and *P. aeruginosa*) on the herring fillets. The results indicated a negative influence of films with extracts on the formation of biogenic amines (BAs), which was related to the recorded antibacterial properties of films against bacteria that can be BA producers (e.g., *S. aureus, L. monocytogenes, S.* Typhimurium and *E. coli*). The application of alginate/pectin films with CE and GE extended the shelf life of herring fillets by minimising pH changes, preventing lipid oxidation and protein degradation and inhibiting microbial growth during storage for 18 days at 4 °C. The results of our study indicate that alginate/pectin films with CE and GE have the potential to be used for the preservation of herring fillets. However, further studies are required to test the influence of these films on the sensory properties of fish.

Author Contributions: Conceptualization, D.L., A.Š. and P.R.V.; methodology, A.Š.; investigation, G.U., G.D. and D.Č.; data curation, A.Š.; writing—original draft preparation, D.L. and P.R.V.; writing—review and editing, D.L.; supervision, D.L. and A.Š.; funding acquisition, P.R.V. All authors have read and agreed to the published version of the manuscript.

Funding: This research was funded by the European Regional Development Fund according to the supported activity "Research Projects Implemented by World-class Researcher Groups" under Measure No. 01.2.2-LMT-K-718.

Data Availability Statement: The data presented in this study are available on request from the corresponding author.

Conflicts of Interest: The authors declare no conflict of interest.

References

1. Huss, H.H. *Quality and Quality Changes in Fresh Fish*; FAO Fisheries Technical Paper No. 348; Food and Agriculture Organization of the United Nations: Rome, Italy, 1995.
2. Taylor, S.L.; Summer, S. Determination of histamine, putrescine and cadaverine. In *Seafood Quality Determination*; Kramer, B.D.T., Liston, J., Eds.; Elsevier Science Publisher: Amsterdam, The Netherlands, 1986; pp. 235–245.
3. Garavand, F.; Rouhi, M.; Razavi, S.H.; Cacciotti, I.; Mohammadi, R. Improving the integrity of natural biopolymer films used in food packaging by crosslinking approach: A review. *Int. J. Biol. Macromol.* **2017**, *104*, 687–707. [CrossRef] [PubMed]
4. Khanzadi, M.; Jafari, S.M.; Mirzaei, H.; Chegini, F.K.; Maghsoudlou, Y.; Dehnad, D. Physical and mechanical properties in biodegradable films of whey protein concentrate–pullulan by application of beeswax. *Carbohydr. Polym.* **2015**, *118*, 24–29. [CrossRef] [PubMed]
5. Tajik, S.; Maghsoudlou, Y.; Khodaiyan, F.; Jafari, S.M.; Ghasemlou, M.; Aalami, M. Soluble soybean polysaccharide: A new carbohydrate to make a biodegradable film for sustainable green packaging. *Carbohydr. Polym.* **2013**, *97*, 817–824. [CrossRef] [PubMed]
6. Ribeiro, A.M.; Estevinko, B.N.; Rocha, F. Preparation and incorporation of functional ingredients in edible films and coatings. *Food Bioprocess Tech.* **2021**, *14*, 209–231. [CrossRef]

7. Gram, L.; Dalgaard, P. Fish spoilage bacteria—Problems and solutions. *Curr. Opin. Biotechnol.* **2002**, *13*, 262–266. [CrossRef] [PubMed]
8. Mahajan, K.; Kumar, S.; Bhat, Z.F.; Naqvi, Z.; Jayawardena, R. Development of bioactive edible film using phytochemicals from Aloe Vera for improved microbial and lipid oxidative stability of frozen dairy products. *Food Bioprocess Tech.* **2021**, *14*, 2120–2133. [CrossRef]
9. Farsanipour, A.; Khodanazary, A.; Hosseini, S.M. Effect of chitosan-whey protein isolated coatings incorporated with tarragon Artemisia dracunculus essential oil on the quality of *Scomberoides commersonnianus* fillets at refrigerated condition. *Int. J. Biol. Macromol.* **2020**, *155*, 766–771. [CrossRef]
10. Raeisi, M.; Tajik, H.; Aliakbarlu, J.; Mirhosseini, S.H.; Hosseini, S.M.H. Effect of carboxymethyl cellulose-based coatings incorporated with *Zataria multiflora* Boiss. essential oil and grape seed extract on the shelf life of rainbow trout fillets. *LWT-Food Sci. Technol.* **2015**, *64*, 898–904. [CrossRef]
11. Li, T.; Li, J.; Hu, W.; Li, X. Quality enhancement in refrigerated red drum (*Sciaenops ocellatus*) fillets using chitosan coatings containing natural preservatives. *Food Chem.* **2013**, *138*, 821–826. [CrossRef]
12. Rezaei, F.; Shahbazi, Y. Shelf-life extension and quality attributes of sauced silver carp fillet: A comparison among direct addition, edible coating and biodegradable film. *LWT-Food Sci. Technol.* **2018**, *87*, 122–133. [CrossRef]
13. Song, Y.; Liu, L.; Shen, H.; You, J.; Luo, Y. Effect of sodium alginate-based edible coating containing different anti-oxidants on quality and shelf life of refrigerated bream (*Megalobrama amblycephala*). *Food Control* **2011**, *22*, 608–615. [CrossRef]
14. Jaguey-Hernández, Y.; Aguilar-Arteaga, K.; Ojeda-Ramirez, D.; Añorve-Morga, J.; González-Olivares, L.G.; Castañeda-Ovando, A. Biogenic amines levels in food processing: Efforts for their control in foodstuffs. *Food Res. Int.* **2021**, *144*, 110341. [CrossRef] [PubMed]
15. Kuley, E.; Özogul, F. Synergistic and antagonistic effect of lactic acid bacteria on tyramine production by food-borne pathogenic bacteria in tyrosine decarboxylase broth. *Food Chem.* **2011**, *127*, 1163–1168. [CrossRef] [PubMed]
16. Houicher, A.; Kuley, E.; Özogul, F.; Bendeddouche, B. Effect of natural extracts (*Mentha spicata* L. and Artemisia campestris) on biogenic amine formation of sardine vacuum-packed and refrigerated (*Sardina pilchardus*) fillets. *J. Food Process. Preserv.* **2015**, *39*, 2393–2403. [CrossRef]
17. Jamróz, E.; Kulawik, P.; Tkaczewska, J.; Guzik, P.; Zając, M.; Juszczak, L.; Krzyściak, P.; Turek, K. The effects of active double-layered furcellaran/gelatin hydrolysate film system with Ala-Tyr peptide on fresh Atlantic mackerel stored at −18 °C. *Food Chem.* **2021**, *338*, 127867. [CrossRef]
18. Hao, R.; Liu, Y.; Sun, L.; Xia, L.; Jia, H.; Li, Q.; Pan, J. Sodium alginate coating with plant extract affected microbial communities, biogenic amine formation and quality properties of abalone (*Haliotis discus hannai Ino*) during chill storage. *LWT-Food Sci. Technol.* **2017**, *81*, 1–9. [CrossRef]
19. Tamkutė, L.; Liepuoniūtė, R.; Pukalskienė, M.; Venskutonis, P.R. Recovery of valuable lipophilic and polyphenolic fractions from cranberry pomace by consecutive supercritical CO_2 and pressurized liquid extraction. *J. Supercrit. Fluids* **2020**, *159*, 104755. [CrossRef]
20. Zaika, L.L. Spices and herbs: Their antimicrobial activity and its determination. *J. Food Saf.* **1988**, *9*, 97–118. [CrossRef]
21. EC 2005, COMMISSION REGULATION (EC) No 2074/2005 of the Commission of the European Communities of 5 December 2005. Official Journal of the European Union. Available online: http://data.europa.eu/eli/reg/2005/2074/oj (accessed on 19 December 2022).
22. Tarladgis, B.G.; Watts, B.M.; Younathan, M.T.; Dugan, L.A. Distillation method for the quantitative determination of malonaldehyde in rancid foods. *J. Am. Oil Chem. Soc.* **1960**, *37*, 44–48. [CrossRef]
23. Côté, J.; Caillet, S.; Doyon, G.; Dussault, D.; Sylvain, J.-F.; Lacroix, M. Antimicrobial effect of cranberry juice and extracts. *Food Control* **2011**, *22*, 1413–1418. [CrossRef]
24. Laplante, K.L.; Sarkisian, S.A.; Woodmansee, S.; Rowley, D.C.; Seeram, N.P. Effects of cranberry extracts on growth and biofilm production of Escherichia coli and Staphylococcus species. *Phytother. Res.* **2012**, *26*, 1371–1374. [CrossRef] [PubMed]
25. Wu, V.C.H.; Qiu, X.; de los Reyes, B.G.; Lin, C.-S.; Pan, Y. Application of cranberry concentrate (*Vaccinium macrocarpon*) to control *Escherichia coli* O157:H7 in ground beef and its antimicrobial mechanism related to the down regulated slp, hdeA and cfa. *Food Microbiol.* **2009**, *26*, 32–38. [CrossRef] [PubMed]
26. Lau, A.T.Y.; Barbut, S.; Ross, K.; Diarra, M.S.; Balamurugan, S. The effect of cranberry pomace ethanol extract on the growth of meat starter cultures, *Escherichia coli* O157:H7, *Salmonella enterica* serovar *Enteritidis* and *Listeria monocytogenes*. *LWT-J. Food Sci.* **2019**, *115*, 108452. [CrossRef]
27. Severo, C.; Anjos, I.; Souza, V.G.L.; Canejo, J.P.; Bronze, M.R.; Fernando, A.L.; Coelhoso, I.; Bettencourt, A.F.; Ribeiro, I.A.C. Development of cranberry extract films for the enhancement of food packaging antimicrobial properties. *Food Packag. Shelf Life* **2021**, *28*, 100646. [CrossRef]
28. Brink, I.; Šipailienė, A.; Leskauskaitė, D. Antimicrobial properties of chitosan and whey protein films applied on fresh cut turkey pieces. *Int. J. Biol. Macromol.* **2019**, *130*, 810–881. [CrossRef]
29. Xiong, Y.; Chen, M.; Warner, R.D.; Fang, Z. Incorporating nisin and grape seed extract in chitosan-gelatine edible coating and its effect on cold storage of fresh pork. *Food Control* **2020**, *110*, 107018. [CrossRef]

30. Sogut, E.; Seydim, A.C. The effects of chitosan and grape seed extract-based edible films on the quality of vacuum packaged chicken breast fillets. *Food Packag. Shelf Life* **2018**, *18*, 13–20. [CrossRef]
31. de Oliveira Filho, J.G.; Braga, A.R.C.; de Oliveira, B.R.; Gomes, F.P.; Moreira, V.L.; Pereira, V.A.C.; Egea, M.B. The potential of anthocyanins in smart, active, and bioactive eco-friendly polymer-based films: A review. *Food Res. Int.* **2021**, *142*, 110202. [CrossRef]
32. Wang, F.; Xie, C.; Tang, H.; Hao, W.; Wu, J.; Sun, Y.; Sun, J.; Liu, Y.; Jiang, L. Development, characterization and application of intelligent/active packaging of chitosan/chitin nanofibers films containing eggplant anthocyanins. *Food Hydrocoll.* **2023**, *139*, 108496. [CrossRef]
33. Nešić, A.; Onjia, A.; Davidović, S.; Dimitrijević, S.; Errico, M.E.; Santagata, G.; Malinconico, M. Design of pectin-sodium alginate based films for potential healthcare application: Study of chemico-physical interactions between the components of films and assessment of their antimicrobial activity. *Carbohydr. Polym.* **2017**, *157*, 981–990. [CrossRef]
34. Bierhalz, A.C.K.; da Silva, M.A.; Kieckbusch, T.G. Natamycin release from alginate/pectin films for food packaging applications. *J. Food Eng.* **2012**, *110*, 18–25. [CrossRef]
35. De'Nobili, M.D.; Rojas, A.M.; Abrami, M.; Lapasin, R.; Grassi, M. Structure characterization by means of rheological and NMR experiments as a first necessary approach to study the l-(+)-ascorbic acid diffusion from pectin and pectin/alginate films to agar hydrogels that mimic food materials. *J. Food Eng.* **2015**, *165*, 82–92. [CrossRef]
36. Zhu, F. Polysaccharide based films and coatings for food packaging: Effect of added polyphenols. *Food Chem.* **2021**, *359*, 129871. [CrossRef] [PubMed]
37. Bi, J.; Tian, C.; Zhang, G.; Hao, H.; Hou, H. Novel procyanidins-loaded chitosan-graft-polyvinyl alcohol film with sustained antibacterial activity for food packaging. *Food Chem.* **2021**, *365*, 130534. [CrossRef] [PubMed]
38. Özogul, F.; Taylor, K.D.A.; Quantick, P.; Özogul, Y. Changes in biogenic amines in herring stored under modified atmosphere and vacuum pack. *J. Food Sci.* **2002**, *67*, 2497–2501. [CrossRef]
39. Houicher, A.; Bensid, A.; Regenstein, J.M.; Özogul, F. Control of biogenic amine production and bacterial growth in fish and seafood products using phytochemicals as biopreservatives: A review. *Food Biosci.* **2021**, *39*, 100807. [CrossRef]
40. Kim, S.H.; Velazquez, J.B.; Gigirey, B.B.; Eun, J.B.; Jun, S.H.; Wei, C.I.; An, H.J. Identification of the main bacteria contributing to histamine formation in seafood to ensure product safety. *Food Sci. Biotechnol.* **2003**, *12*, 451–460.
41. Özogul, F.; Hamed, I.; Gokdogan, S. The impact of natural clinoptilolite on ammonia, cadaverine and other polyamine formation by food-borne pathogen in lysine decarboxylase broth. *LWT-Food Sci. Technol.* **2016**, *65*, 703–710. [CrossRef]
42. Özogul, F.; Kuley, E.; Kenar, M. Effects of rosemary and sage tea extract on biogenic amines formation of sardine (*Sardina pilchardus*) fillets. *Int. J. Food Sci. Technol.* **2011**, *46*, 761–766. [CrossRef]
43. Mah, J.; Kim, Y.J.; Hwang, H. Inhibitory effects of garlic and other spices on biogenic amine production in Myeolchi-jeot, Korean salted and fermented anchovy product. *Food Control* **2009**, *20*, 449–454. [CrossRef]
44. Özyurt, G.; Kuley, E.; Balikçi, E.; Kaçar, Ç.; Gökdogan, S.; Etyemez, M.; Özogul, F. Effect of the icing with rosemary extract on the oxidative stability and biogenic amine formation in sardine (*Sardinella aurita*) during chilled storage. *Food Bioproc. Technol.* **2012**, *5*, 2777–2786. [CrossRef]
45. Wang, X.; Zhang, Y.; Ren, H. Effects of grape seed extract on lipid oxidation, biogenic amine formation and microbiological quality in Chinese traditional smoke-cured bacon during storage. *J. Food Saf.* **2018**, *38*, 12426. [CrossRef]
46. Noori, S.M.A.; Khanzadi, S.; Fazlara, A.; Najafzadehvarzi, H.; Azizzadeh, M. Effect of lactic acid and ajwain (*Carum copticum*) on the biogenic amines and quality of refrigerated common carp (*Cyprinus carpio*). *LWT-Food Sci. Technol.* **2018**, *97*, 434–439. [CrossRef]
47. Fan, W.; Sun, J.; Chen, Y.; Qiu, J.; Zhang, Y.; Chi, Y. Effects of chitosan coating on quality and shelf life of silver carp during frozen storage. *Food Chem.* **2009**, *115*, 66–70. [CrossRef]
48. CEC. Commission of the European Community, Decision 95/149/EC of 8 March 1995 Fixing the Total Volatile Basic Nitrogen (TVB-N) Limit Values for Certain Categories of Fishery Products and Specifying the Analysis Methods to Be Used. Brussels, Belgium. 1995. Available online: https://op.europa.eu/en/publication-detail/-/publication/91dc1ed4-6450-4a3c-8cab-45009644715c (accessed on 19 September 2020).
49. Giménez, B.; Roncalés, P.; Beltrán, J.A. Modified atmosphere packaging of filleted rainbow trout. *J. Sci. Food Agric.* **2002**, *82*, 1154–1159. [CrossRef]
50. Masniyom, P.; Benjakul, S.; Visessanguan, W. Shelf-life extension of refrigerated seabass slices under modified atmosphere packaging. *J. Sci. Food Agric.* **2002**, *82*, 873–880. [CrossRef]
51. Günlü, A.; Koyun, E. Effects of vacuum packaging and wrapping with chitosan-based edible film on the extension of the shelf life of Sea Bass (*Dicentrarchus labrax*) fillets in cold storage (4 °C). *Food Bioprocess Technol.* **2013**, *6*, 1713–1719. [CrossRef]
52. Tolstorebrov, I.; Eikevik, T.M.; Indergård, E. The influence of long-term storage, temperature and type of packaging materials on the lipid oxidation and flesh color of frozen Atlantic herring fillets (*Clupea harengus*). *Int. J. Refrig.* **2014**, *40*, 122–130. [CrossRef]
53. Caillet, S.; Côté, J.; Doyon, G.; Sylvain, J.-F.; Lacroix, M. Antioxidant and antiradical properties of cranberry juice and extracts. *Food Res. Int.* **2011**, *44*, 1408–1413. [CrossRef]

54. Cong, J.; Cui, J.; Zhang, H.; Dzah, C.S.; He, Y.; Duan, Y. Binding affinity, antioxidative capacity and in vitro digestion of complexes of grape seed procyanidins and pork, chicken and fish protein. *Food Res. Int.* **2020**, *136*, 109530. [CrossRef]
55. Sallam, K.I. Antimicrobial and antioxidant effects of sodium acetate, sodium lactate, and sodium citrate in refrigerated sliced salmon. *Food Control* **2007**, *18*, 566–575. [CrossRef] [PubMed]

foods

MDPI

Article

Identification of Antibacterial Components and Modes in the Methanol-Phase Extract from a Herbal Plant *Potentilla kleiniana* Wight et Arn

Yingping Tang [1,2], Pan Yu [1,2] and Lanming Chen [1,2,*]

1 Key Laboratory of Quality and Safety Risk Assessment for Aquatic Products on Storage and Preservation (Shanghai), Ministry of Agriculture and Rural Affairs of the People's Republic of China, Shanghai 201306, China
2 College of Food Science and Technology, Shanghai Ocean University, Shanghai 201306, China
* Correspondence: lmchen@shou.edu.cn

Abstract: The increase in bacterial resistance and the decline in the effectiveness of antimicrobial agents are challenging issues for the control of infectious diseases. Traditional Chinese herbal plants are potential sources of new or alternative medicine. Here, we identified antimicrobial components and action modes of the methanol-phase extract from an edible herb *Potentilla kleiniana* Wight et Arn, which had a 68.18% inhibition rate against 22 species of common pathogenic bacteria. The extract was purified using preparative high-performance liquid chromatography (Prep-HPLC), and three separated fragments (Fragments 1–3) were obtained. Fragment 1 significantly elevated cell surface hydrophobicity and membrane permeability but reduced membrane fluidity, disrupting the cell integrity of the Gram-negative and Gram-positive pathogens tested ($p < 0.05$). Sixty-six compounds in Fragment 1 were identified using Ultra-HPLC and mass spectrometry (UHPLC-MS). The identified oxymorphone (6.29%) and rutin (6.29%) were predominant in Fragment 1. Multiple cellular metabolic pathways were altered by Fragment 1, such as the repressed ABC transporters, protein translation, and energy supply in two representative Gram-negative and Gram-positive strains ($p < 0.05$). Overall, this study demonstrates that Fragment 1 from *P. kleiniana* Wight et Arn is a promising candidate for antibacterial medicine and food preservatives.

Keywords: *Potentilla kleiniana* Wight et Arn; antibacterial component; antibacterial mode; pathogenic bacteria; transcriptome; traditional Chinese herb

1. Introduction

Infectious diseases caused by pathogenic bacteria continue to be a global concern for public health, causing millions of deaths worldwide per year [1]. Since the introduction of sulfonamides in 1933, a large number of antibiotics have been applied in clinics [2]. Nevertheless, in recent decades, the overuse and/or misuse of antibiotics have accelerated the spread of antibiotic-resistant bacteria, leading to ineffective drug treatment [3]. It was estimated that at least 700,000 people worldwide die each year due to antimicrobial resistance [4].

Pharmacophagous plants are recognized as a rich source of phytochemicals with antimicrobial potential [5]. Phytocompounds extracted from such plants are long known for their therapeutic uses, and characterized by safety and low toxicity [6]. The application of herbal products may be a better choice for the extensive and imprudent use of synthetic antibiotics [7]. For example, In China, approximately 34,984 native higher plant species have been recorded [8]. Of these, the herbal plant *Potentilla kleiniana* Wight et Arn was first recorded in the earliest pharmaceutical book "Divine Farmer's Classic of Materia Medica" during the Warring States period (475–221 B.C.) in China. It belongs to the phylum of Angiospermae, the class of Dicotyledoneae, the order of Rosales Bercht. and J. Presl, and the family of Rosaceae Juss. *P. kleiniana* Wight et Arn is widely distributed in China, and many Asian countries such as Japan, India, Malaysia, Indonesia, and North Korea.

Its whole plant has been used as a traditional Chinese medicine to treat fever, arthritis, malaria, insect and snake bites, hepatitis, and traumatic injury [9]. Recently, Zhou et al. identified bioactive components from *P. kleiniana* Wight et Arn with anti-human immunodeficiency virus-1 (HIV-1) protease activity [10]. Liu et al. developed an efficient method for the rapid screening and separation of α-glucosidase inhibitors from *P. kleiniana* Wight et Arn [11]. Li et al. [12] found antihyperglycemic and antioxidant effect of the total flavones of *P. kleiniana* Wight et Arn in streptozotocin induced diabetic rats, which may be helpful in the prevention of diabetic complications associated with oxidative stress [12]. However, to the best of our knowledge, there are few studies so far in the current literature on antibacterial activity of *P. kleiniana* Wight et Arn. Tao et al. [9] reported that total flavonoids from *P. kleiniana* Wight et Arn (TFP) inhibited biofilm formation and virulence factor production in methicillin-resistant *Staphylococcus aureus* (MRSA). The TFP also damaged cell membrane integrity of *Pseudomonas aeruginosa*. These results supported potential application of the TFP as a novel natural bioactive preservative in food processing [13]. Song et al. also reported that bioactive components extracted from *P. kleiniana* Wight et Arn showed antibacterial effects against *S. aureus*, *Candida albicans*, *P. aeruginosa*, and *Escherichia coli*, but not against the mold *Aspergillus niger* [14].

To further exploit bioactive nature products in *P. kleiniana* Wight et Arn, in the present study, we extracted bacteriostatic components in *P. kleiniana* Wight et Arn using the methanol and chloroform method [15,16]. Antimicrobial action modes of the methanol-phase extract were further investigated. The results of this study provide useful data for potential pharmaceutical application of *P. kleiniana* Wight et Arn against the common pathogenic bacteria.

2. Results and Discussion

2.1. Antibacterial Activity of Crude Extracts from P. kleiniana Wight et Arn

Antibacterial substances in the fresh *P. kleiniana* Wight et Arn were extracted using the methanol and chloroform method [15,16]. The water loss rate of the fresh plant sample was 94.12% after freeze-drying treatment of the sample. The extraction rates of the methanol-phase and chloroform-phase crude extracts were 31.13% and 25.43%, respectively. As shown in Table 1, the chloroform-phase extract from *P. kleiniana* Wight et Arn had a 50.00% inhibition rate, which inhibited one species of Gram-positive bacterium *S. aureus*, and 10 species of Gram-negative bacteria, including *Bacillus cereus* A1-1, *B. cereus* A2-2, *Enterobacter cloacae* ATCC13047, *Salmonella typhimurium* ATCC15611, *Shigella dysenteriae* CMCC51252, *Shigella flexneri* CMCC51572, *Shigella sonnei* ATCC25931, *Vibrio cholerae* Q10-54, *Vibrio mimicus* bio-56759, *Vibrio parahemolyticus* ATCC33847, *V. parahemolyticus* B3-13, *V. parahemolyticus* B5-29, *V. parahemolyticus* B9-35, *V. parahemolyticus* A1-1, and *Vibrio vulnificus* ATCC27562 (Table 1).

Of note, the methanol-phase crude extract from *P. kleiniana* Wight et Arn inhibited the growth of 15 bacterial species, including one species of Gram-positive *S. aureus*, and 14 species of Gram-negative bacteria, *P. aeruginosa* ATCC9027, *S. typhimurium* ATCC15611, *S. dysenteriae* CMCC51252, *S. flexneri* CMCC51572, *S. flexneri* CMCC51574, *S. sonnei* ATCC25931, *V. alginolyticus* ATCC17749, *V. cholerae* Q10-54, *V. fluvialis* ATCC33809, *V. mimicus* bio-56759, *V. parahemolyticus* ATCC17802, and *V. vulnificus* ATCC27562, which showed a 68.18% inhibition rate (Table 1, Figure 1).

In this study, the methanol and chloroform extract method exhibited a broader antibacterial spectrum, consistent with our previous reports [15,16]. Previous studies also reported effective extraction of bioactive compounds from *P. kleiniana* Wight et Arn. For example, Tao et al. [13] extracted TFP in *P. kleiniana* Wight et Arn using an ethanol-water solution, and the obtained extract was further partitioned using petroleum ethers, chloroform and ethyl acetate. The extracted TFP inhibited survival and virulence of *P. aeruginosa*, and MRSA. Song et al. [14] extracted bioactive compounds from *P. kleiniana* Wight et Arn using ethanol and ethyl acetate, and the obtained extract showed antibacterial activity against *P. aeruginosa*, *S. aureus*, *C. albicans*, and *E. coli*. The difference in bioactive compounds extracted from *P. kleiniana* Wight et Arn using the different methods may explain the distinct antibacterial profiles between this study and the previous reports [13,14].

Table 1. Antibacterial activity of crude extracts from *P. kleiniana* Wight et Arn.

Strain	Inhibition Zone (Diameter, mm)		MIC (mg/mL)	
	CPE	MPE	CPE	MPE
Aeromonas hydrophila ATCC35654	-	-	-	-
Bacillus cereus A1-1	7.03 ± 0.01	10.54 ± 0.48	50	6.25
Bacillus cereus A2-2	7.11 ± 0.02	10.54 ± 0.75	50	1.56
Enterobacter cloacae ATCC13047	7.00 ± 0.11	7.11 ± 0.26	50	50
Enterobacter cloacae C1-1	-	-	-	-
Escherichia coli ATCC8739	-	7.62 ± 0.37	-	25
Escherichia coli ATCC25922	-	-	-	-
Escherichia coli K12	-	7.51 ± 0.29	-	25
Enterobacter sakazakii CMCC45401	-	-	-	-
Klebsiella pneumoniae 8-2-10-8	-	-	-	-
Klebsiella pneumoniae 8-2-1-14	-	-	-	-
Pseudomonas aeruginosa ATCC9027	-	10.51 ± 0.41	-	6.25
Pseudomonas aeruginosa ATCC27853	-	8.14 ± 0.32	-	25
Salmonella enterica subsp. *enterica* (*ex* Kauffmann and Edwards) ijms-2270933-finalLe Minor and Popoff serovar Choleraesuis ATCC13312	-	-	-	-
Salmonella paratyphi-A CMCC50093	-	-	-	-
Salmonellaenterica subsp. *enterica* (*ex* Kauffmann and Edwards) ijms-2270933-finalLe Minor and Popoff serovar Vellore ATCC15611	7.09 ± 0.09	10.11 ± 0.61	50	6.25
Salmonella E1-1	-	-	-	-
Shigella dysenteriae CMCC51252	7.02 ± 0.11	9.29 ± 0.51	50	12.5
Shigella flexneri CMCC51572	7.82 ± 0.20	10.17 ± 0.20	25	6.25
Shigella flexneri ATCC12022	-	-	-	-
Shigella flexneri CMCC51574	-	9.17 ± 0.21	-	12.5
Shigella sonnei ATCC25931	7.00 ± 0.11	8.19 ± 0.51	50	25
Shigella sonnet CMCC51592	-	-	-	-
Staphylococcus aureus ATCC25923	7.03 ± 0.14	9.41 ± 0.27	50	12.5
Staphylococcus aureus ATCC8095	7.07 ± 0.15	10.15 ± 0.24	50	6.25
Staphylococcus aureus ATCC29213	7.78 ± 0.10	9.21 ± 0.01	25	12.5
Staphylococcus aureus ATCC6538	7.62 ± 0.61	9.55 ± 0.37	25	12.5
Staphylococcus aureus D1-1	7.11 ± 0.25	7.00 ± 0.51	50	50
Vibrio alginolyticus ATCC17749	-	10.11 ± 0.24	-	3.13
Vibrio alginolyticus ATCC33787	-	-	-	-
Vibrio cholerae GIM1.449	-	7.00 ± 0.14	-	50
Vibrio cholerae Q10-54	7.22 ± 0.10	7.02 ± 0.21	50	50
Vibrio fluvialis ATCC33809	-	7.12 ± 0.03	-	50
Vibrio harvey ATCC BAA-1117	-	-	-	-
Vibrio harveyi ATCC33842	-	-	-	-
Vibrio mimicus bio-56759	7.21 ± 0.41	11.00 ± 0.32	25	3.13
Vibrio parahemolyticus ATCC17802	-	10.67 ± 1.21	-	1.56
Vibrio parahemolyticus ATCC33847	8.63 ± 0.24	7.14 ± 0.12	12.5	50
Vibrio parahemolyticus B3-13	7.17 ± 0.29	12.33 ± 0.65	50	3.13
Vibrio parahemolyticus B4-10	-	11.26 ± 0.34	-	6.25
Vibrio parahemolyticus B5-29	7.17 ± 0.04	13.77 ± 0.85	50	3.13
Vibrio parahemolyticus B9-35	7.20 ± 0.09	13.15 ± 0.44	25	3.13
Vibrio parahemolyticus A1-1	7.13 ± 0.15	10.35 ± 0.58	50	3.13
Vibrio vulnificus ATCC27562	7.65 ± 0.44	7.01 ± 0.23	25	50

Note: CPE: chloroform-phase extract. MPE: methanol-phase extract. -: no bacteriostasis activity. Inhibition zone: diameter includes the disk diameter (6 mm). MIC: minimum inhibitory concentration. Values were means ± standard deviation (S.D.) of three parallel measurements.

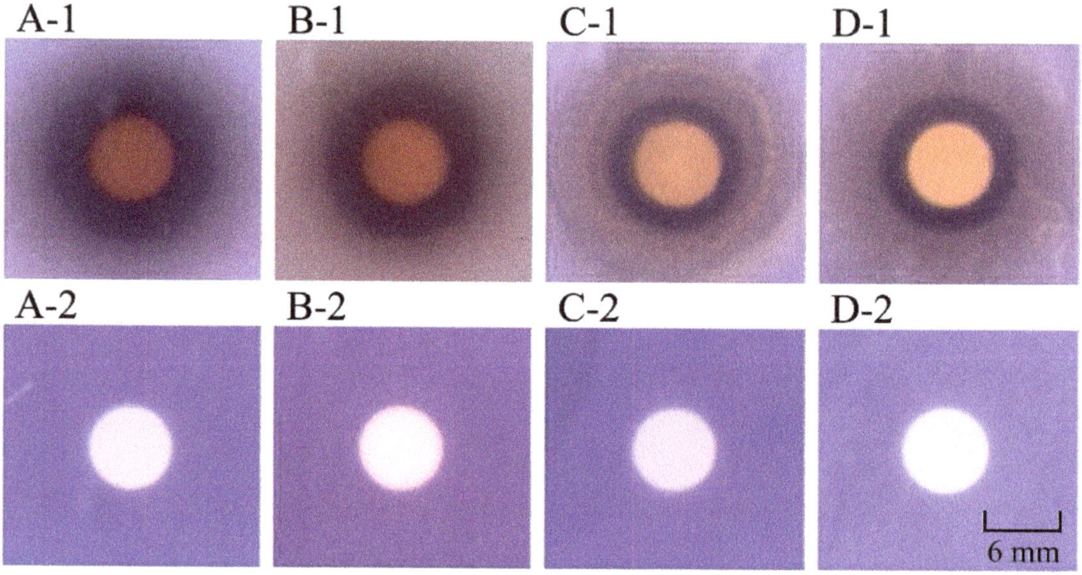

Figure 1. Inhibition activity of the methanol-phase crude extract from *P. kleiniana* Wight et Arn against the four representative bacterial strains. (**A-1–D-1**) *V. parahemolyticus* B5-29, *V. parahemolyticus* ATCC17802, *S. aureus* ATCC25923, and *S. aureus* ATCC8095, respectively. (**A-2–D-2**) corresponding negative controls, respectively.

We further determined minimum inhibitory concentrations (MICs) of the crude extracts from *P. kleiniana* Wight et Arn, and the results are shown in Table 1. The MICs of the chloroform-phase extract ranged from 12.5 mg/mL to 50 mg/mL against the eleven species of the bacteria. Notably, for the methanol-phase extract, the MICs were between 1.56 mg/mL and 50 mg/mL against the fifteen bacterial species. Of these, the growth of *B. cereus* A2-2 and *V. parahemolyticus* ATCC17802 was the most strongly repressed by the methanol-phase extract with the MICs of 1.56 mg/mL, followed by *V. alginolyticus* ATCC17749, *V. mimicus* bio-56759, *V. parahemolyticus* B3-13, *V. parahemolyticus* B5-29, *V. parahemolyticus* B9-35, and *V. parahemolyticus* A1-1 with MICs of 3.13 mg/mL. In addition, the growth of *B. cereus* A1-1, *P. aeruginosa* ATCC9027, *S. typhimurium* ATCC15611, *S. flexneri* CMCC51572, *S. aureus* ATCC8095, and *V. parahemolyticus* B4-10 was also inhibited by the methanol-phase extract with lower MICs (6.25 mg/mL). Of these pathogens, for example, *V. alginolyticus* is a foodborne marine *Vibrio* that can cause gastroenteritis, otitis media, otitis externa, and septicemia in humans [17]. *V. mimicus* can also cause gastroenteritis in humans due to contaminated fish consumption and seafood [18]. *P. aeruginosa* is an opportunistic pathogen and can cause serious infections, especially in patients with compromised immune systems [19].

Recently, Song et al. [14] reported that the ethyl acetate extract of *P. kleiniana* Wight et Arn inhibited *E. coli*, *P. aeruginosa*, and *C. albicans*, with MICs of 5 mg/mL, 2.5 mg/mL, and 5 mg/mL, respectively. Tao et al. reported the MIC value of the TFP against MRSA was 20 μg/mL [9].

These results indicated that the methanol-phase crude extract had a higher inhibition rate (68.18%), showing a more broad inhibitory profile with much lower MICs (1.56–50 mg/mL) against the pathogens tested, as compared to the chloroform-phase crude extract (50.00%; 12.5–50 mg/mL). Thus, the methanol-phase crude extract was chosen for further analysis in this study.

2.2. Purification of the Methanol-Phase Crude Extract from P. kleiniana Wight et Arn

Based on the obtained results, a large amount of the methanol-phase crude from *P. kleiniana* Wight et Arn was prepared and further purified using Prep-HPLC analysis. As shown in Figure S1, three separated fragments (designated Fragments 1–3) were observed via scanning at OD_{211} for 12 min, including Fragment 1 (2.45 min), Fragment 2 (6.75 min), and Fragment 3 (9.83 min). The main peak of the methanol-phase crude was observed to occur at 2.45 min, wherein the absorption peak of Fragment 1 reached its maximum.

The three single fragments were subjected for antibacterial activity analysis. Fragment 1 had strong inhibitory effects on *V. parahemolyticus* ATCC17802, *V. parahemolyticus* B5-29, *V. parahemolyticus* B9-35, *V. parahemolyticus* B3-13, and *V. parahemolyticus* B4-10. In addition, the growth of the other six strains was also effectively repressed, including *B. cereus* A2-2, *V. parahemolyticus* A1-1, *S. flexneri* CMCC51572, *S. aureus* ATCC25923, *S. aureus* ATCC8095, and *S. aureus* ATCC6538 (Table 2). Of these, *V. parahaemolyticus* is a Gram-negative halophilic bacterium that can cause diseases in marine animals, leading to huge economic losses to the aquaculture. *V. parahaemolyticus* can also cause gastrointestinal infections and other health complications in humans [20]. *B. cereus* is a Gram-positive foodborne pathogen that can cause diarrhea and emesis [21]. *S. flexneri* is a Gram-negative intracellular pathogen that invades colonic cells and causes bloody diarrhea in humans [22]. *S. aureus* is a Gram-positive opportunistic pathogen leading to food poisoning as well as human and animal infectious diseases [23,24].

Table 2. Antibacterial activity of Fragment 1 of the methanol-phase extract from *P. kleiniana* Wight et Arn.

Strain	Inhibition Zone (Diameter, mm)	MIC (mg/mL)
B. cereus A2-2	8.03 ± 0.45	6.25
S. flexneri CMCC51572	7.50 ± 0.50	6.25
S. aureus ATCC25923	8.03 ± 0.40	12.5
S. aureus ATCC8095	9.53 ± 0.35	6.25
S. aureus ATCC6538	7.10 ± 0.36	50.0
V. parahemolyticus ATCC17802	10.31 ± 0.62	6.25
V. parahemolyticus A1-1	8.57 ± 0.60	25.0
V. parahemolyticus B3-13	10.37 ± 0.32	6.25
V. parahemolyticus B4-10	10.30 ± 0.50	12.5
V. parahemolyticus B5-29	11.30 ± 0.26	6.25
V. parahemolyticus B9-35	11.27 ± 0.40	12.5

We also determined MICs of Fragment 1 against the four species of pathogenic bacteria (Table 2). The synergistic effect may explain the observed MICs of Fragment 1 (6.25–50 mg/mL), as compared to the methanol-phase extract from *P. kleiniana* Wight et Arn. Among the Gram-negative pathogens, *V. parahemolyticus* ATCC17802 and *V. parahemolyticus* B5-29 were the most sensitive strains to Fragment 1, with MICs of 6.25 mg/mL. For the Gram-positive pathogen, the growth of *S. aureus* ATCC8095 and *S. aureus* ATCC25923 was also effectively repressed, with MICs of 6.25 mg/mL and 12.5 mg/mL, respectively.

Conversely, the other two peaks (Fragments 2 and 3) showed weak or no antibacterial activity. To further investigate possible antibacterial modes of Fragment 1, the two Gram-negative strains *V. parahemolyticus* ATCC17802 and *V. parahemolyticus* B5-29, and two Gram-positive stains *S. aureus* ATCC8095 and *S. aureus* ATCC25923 were chosen for the further analysis in this study.

2.3. Bacterial Cell Surface Hydrophobicity, Membrane Fluidity and Permeability Changes Triggered by Fragment 1 from P. kleiniana Wight et Arn

2.3.1. Cell Surface Hydrophobicity

Cell surface hydrophobicity is an important cellular biophysical parameter that affects cell surface interactions and cell–cell communication [25]. In this study, the hexadecane was used as a probe to assess cell surface hydrophobicity change. The difference between before and after the absorbance value of bacterial fluid can indicate the change of hydrophobicity, and the larger the difference, the more hydrophobicity of the surface [26]. The cell surface hydrophobicity of the four experimental groups (1× MIC of Fragment 1) was significantly increased, as compared to the control groups ($p < 0.05$) (Figure 2A). For instance, after being treated with Fragment 1 for 2 h, bacterial cell surface hydrophobicity was significantly increased, including *V. parahaemolyticus* B5-29 (8.62%, 1.42-fold), *V. parahaemolyticus* ATCC17802 (8.27%, 1.50-fold), *S. aureus* ATCC25923 (10.34%, 1.24-fold), and *S. aureus* ATCC8095 (12.20%, 1.19-fold) ($p < 0.05$). Increasing treatment time, the cell surface hydrophobicity was further increased. After the 4 h treatment, the cell surface hydrophobicity was the most significantly increased (11.97%, 1.97-fold) in the *V. parahaemolyticus* B5-29 treatment group. The highest increase (15.96%, 2.63-fold) was also observed in *V. parahaemolyticus* B5-29, after treatment for 6 h. The results indicated that Fragment 1 from *P. kleiniana* Wight et Arn can significantly increase the cell surface hydrophobicity of both Gram-negative *V. parahemolyticus* and Gram-positive *S. aureus* pathogens.

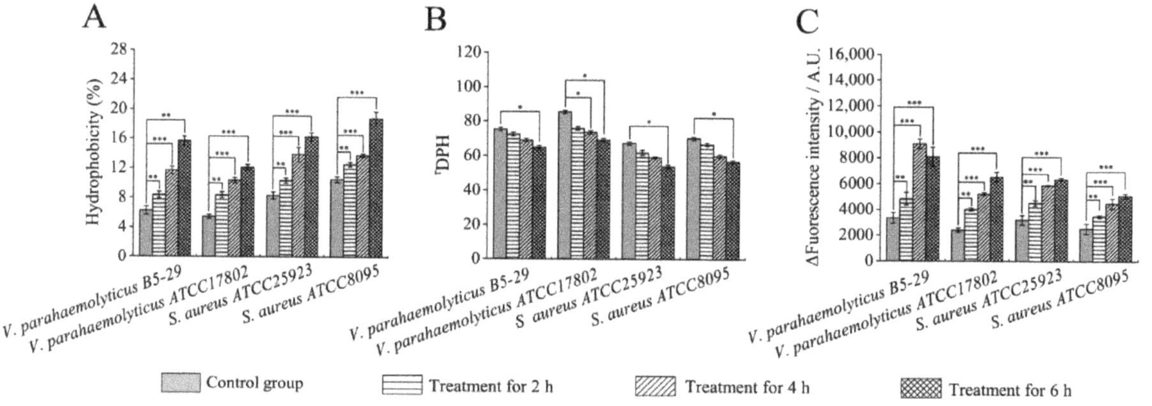

Figure 2. Effects of Fragment 1 (1× MIC) from *P. kleiniana* Wight et Arn on cell surface hydrophobicity, membrane fluidity and outer membrane permeability of the four bacterial strains. (**A–C**) cell surface hydrophobicity, membrane fluidity, and outer membrane permeability, respectively. *: $p < 0.05$; **: $p < 0.01$; and ***: $p < 0.001$.

2.3.2. Cell Membrane Fluidity

Cell membrane is a natural barrier to prevent extracellular substances from freely entering the cell [27]. In this study, as shown in Figure 2B, when compared to the control groups, the membrane fluidity of *V. parahaemolyticus* B5-29, *S. aureus* ATCC25923, and *S. aureus* ATCC8095 did not change significantly after treatment with Fragment 1 (1× MIC) for 2 h and 4 h. However, a significant decrease (1.16-fold, 1.25-fold, and 1.24-fold) was observed in these three treatment groups after treatment for 6 h, respectively ($p < 0.05$). In addition, a significant decrease in cell membrane fluidity was only found in *V. parahaemolyticus* ATCC17802 after treatment for 4 h (1.16-fold) and 6 h (1.24-fold), respectively ($p < 0.05$). These results indicated that Fragment 1 from *P. kleiniana* Wight et Arn can significantly reduce the cell membrane fluidity of both Gram-negative *V. parahemolyticus* and Gram-positive *S. aureus* pathogens.

2.3.3. Cell Membrane Permeability

β-galactosidase is a macromolecular protein naturally found in the interior of cells that can hydrolyze the substrate o-nitrophenyl-β-D-galactopyranosi (ONPG) to galactose and o-nitrophenol in yellow. If the inner membrane of bacterial cells is damaged, ONPG will quickly enter the cell [28]. In this study, the ONPG was used as a probe to assess whether the bacterial inner membrane is damaged. As illustrated in Figure 3D, the inner cell membrane permeability of *S. aureus* ATCC8095 did not change significantly after treatment with

Fragment 1 (1× MIC) from *P. kleiniana* Wight et Arn for 2 h ($p > 0.05$); conversely, significant increases were observed in *V. parahaemolyticus* B5-29, *V. parahaemolyticus* ATCC17802, and *S. aureus* ATCC25923 treatment groups (1.15-fold, 1.18-fold, and 1.04-fold), respectively ($p < 0.05$). After being treated for 4 h, the highest increase was found in *V. parahaemolyticus* B5-29 (1.22-fold). After treatment for 6 h, significant increases were also observed in *V. parahaemolyticus* B5-29, *V. parahaemolyticus* ATCC17802, *S. aureus* ATCC25923, and *S. aureus* ATCC8095 (1.20-fold, 1.17-fold, 1.07-fold, and 1.08-fold), respectively ($p < 0.05$). These results indicated that Fragment 1 from *P. kleiniana* Wight et Arn can significantly increase the inner cell membrane permeability of both Gram-negative *V. parahemolyticus* and Gram-positive *S. aureus* pathogens.

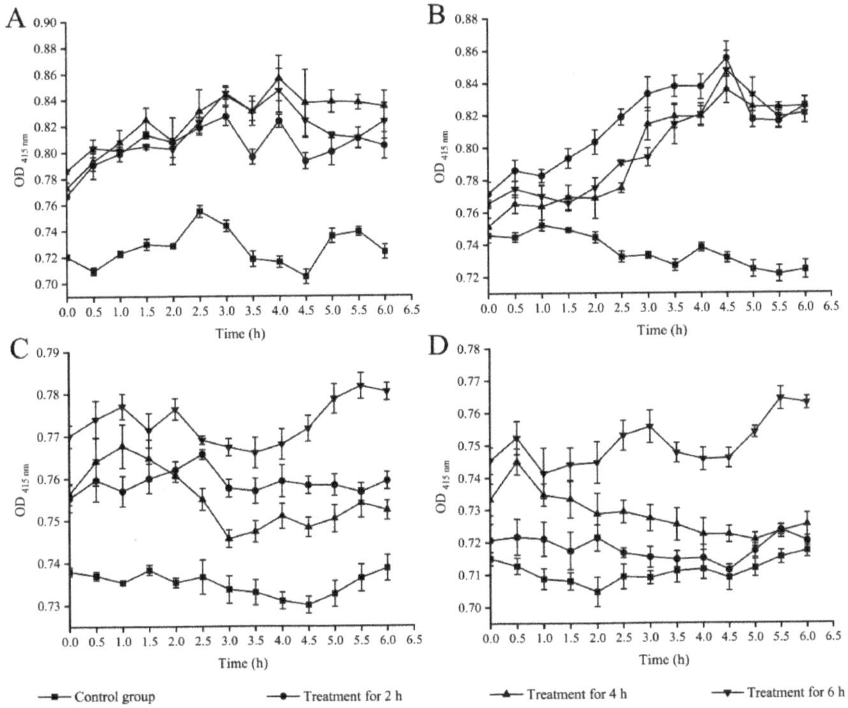

Figure 3. Effects of Fragment 1 (1× MIC) from *P. kleiniana* Wight et Arn on the bacterial inner cell membrane permeability. (**A–D**) *V. parahaemolyticus* B5-29, *V. parahaemolyticus* ATCC17802, *S. aureus* ATCC25923, and *S. aureus* ATCC8095, respectively. The treatment groups were overall significantly different from the control groups ($p < 0.05$), except the *S. aureus* ATCC8095 group treated for 2 h (**D**).

Outer membrane permeability was assessed by measuring the uptake of a hydrophobic fluorescent probe N-phenyl-1-naphthylamine (NPN) [29]. The outer membrane permeability increased significantly in the four treatment groups, after being treated with Fragment 1 for 2 h (1.38-fold to 1.66-fold) ($p < 0.01$), and 4 h (1.77-fold to 2.72-fold), respectively ($p < 0.001$) (Figure 2C). The highest increase was found in *V. parahaemolyticus* ATCC17802 (2.70-fold), after treatment for 6 h. These results indicated that Fragment 1 from *P. kleiniana* Wight et Arn can significantly increase the outer cell membrane permeability of the Gram-negative *V. parahemolyticus* and Gram-positive *S. aureus* pathogens. Recently, Tao et al. also reported that the TFP from *P. kleiniana* Wight et Arn increased cell membrane permeability of MRSA [13].

Taken together, the results of this study demonstrated that Fragment 1 (1× MIC) from *P. kleiniana* Wight et Arn can significantly increase the cell surface hydrophobicity and membrane permeability, but decreases the cell membrane fluidity of both

Gram-negative *V. parahemolyticus* and Gram-positive *S. aureus* pathogens. Antibacterial compounds (e.g., flavonoids) in Fragment 1 from *P. kleiniana* Wight et Arn may have interacted with lipid components of the bacterial cell membrane. The disorder in lipid chains resulted in changed permeability and fluidity of the bacterial cell membrane [30]. The compounds may also have interacted with the bacterial cell surface proteins, leading to the altered nanomechanical properties, which consequently changed cell surface hydrophobicity and fluidity [31]. The two common pathogens *V. parahemolyticus* and *S. aureus* were chosen for further analysis in this study. The former is the leading sea foodborne pathogen worldwide [20], while the latter leads to food poisoning, as well as human and animal infections [23].

2.4. Bacterial Cell Surface Structure Changes Triggered by Fragment 1 from P. kleiniana Wight et Arn

Based on the obtained results in this study, the representative Gram-negative *V. parahaemolyticus* ATCC17802 and Gram-positive *S. aureus* ATCC25923 strains were chosen for further scanning electron microscope (SEM) analysis. As shown in Figure 4, the cells of *V. parahaemolyticus* ATCC17802 were intact in shape with a flat surface, showing a typical rod-like structure, while those of *S. aureus* ATCC25923 were also intact and clear, showing a typical spherical structure. In remarkable contrast to the control groups, the bacterial morphological structures were altered to varying degrees in the treatment groups triggered by Fragment 1 (1× MIC) for different times.

For the Gram-negative *V. parahaemolyticus* ATCC17802, its cell surface was slightly shrunken after being treated with Fragment 1 for 2 h. After 4 h of treatment, the cell surface was more wrinkled and was slightly depressed, the cell membrane was folded and some contents were exuded. After 6 h of the treatment, the cells were severely deformed and crumpled, with a large amount of content leaked.

For the Gram-positive *S. aureus* ATCC25923, its cell surface was rough and slightly wrinkled, but certain cells were depressed, with a small amount of content leaked after the treatment for 2 h. Upon the increased treatment time (4 h), more cells were obviously wrinkled and deformed with the irregularly spherical, and more content leaked out. The cell morphological structure was seriously damaged after being treated for 6 h.

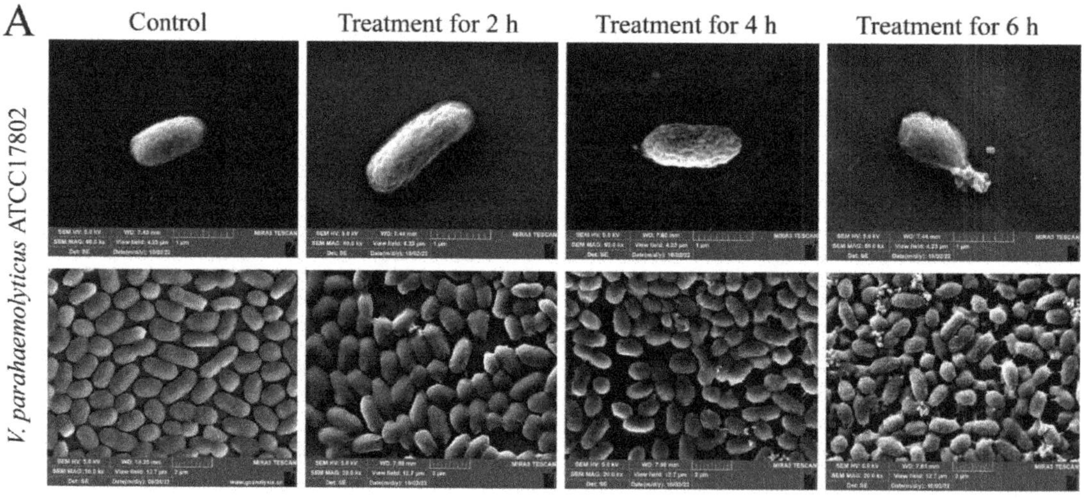

Figure 4. *Cont.*

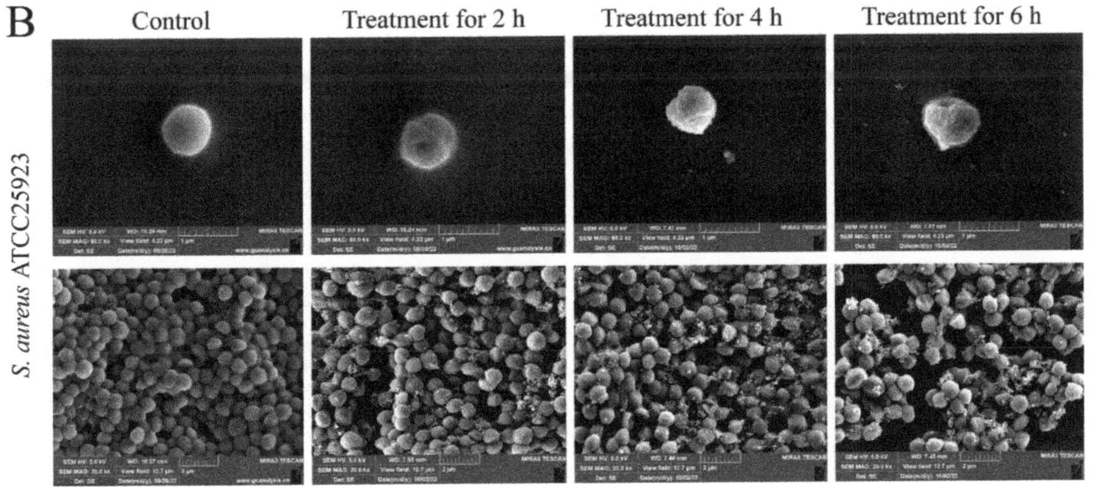

Figure 4. The SEM observation of cell surface structure of the two bacterial strains treated with the 1× MIC of Fragment 1 for different times. (**A**): *V. parahaemolyticus* ATCC17802; (**B**): *S. aureus* ATCC 25923.

These results demonstrated that Fragment 1 (1× MIC) from *P. kleiniana* Wight et Arn can severely damage the cell surface structure of both Gram-negative *V. parahaemolyticus* and Gram-positive *S. aureus* after treatment for 6 h.

2.5. Identification of Potential Antibacterial Compounds in Fragment 1 from P. kleiniana Wight et Arn

Potential antibacterial components in Fragment 1 from *P. kleiniana* Wight et Arn were further identified using UHPLC-MS analysis. As shown in Table 3, a total of 66 different compounds were identified. The highest relative percentage of the compounds was D-maltose (6.77%), followed by oxymorphone (6.29%), rutin (6.29%), D-proline (5.41%), and L-proline (5.41%). In addition, alkaloids, flavonoids, phenols, sesquiterpenoids, fatty acyls, and organic acids were also detected (Table 3).

Highly concentrated sugar solutions, such as the D-maltose identified in this study, are known to be effective antimicrobial agents [32]. Previous research has indicated that the antibacterial activity of phenanthrenes and derivatives, such as the oxymorphone identified in this study, was primarily related to the destruction of the bacterial cell wall structure [33]. Plant extracts contain a large number of bioactive compounds, mainly polyphenols including flavonoids and phenolic compounds. Flavonoids, such as the rutin identified in this study, could exert antibacterial activity via damaging the cytoplasmic membrane, inhibiting energy metabolism and synthesis of nucleic acids [34]. Tao et al. also reported the major compounds of the TFP were 3-O-methylducheside A, naringenin, rutin and quercetin [9,13]. Phenols, such as the p-octopamine identified in this study, are potent antibacterial agents against both Gram-positive and Gram-negative bacteria via the disruption of the bacterial membrane, leading to bacterial lysis and leakage of intracellular contents [35]. Indole alkaloids, such as the indole identified in this study, possess not only intriguing structural features but also biological/pharmacological activities e.g., antimicrobial activity [36]. Additionally, amino acids and its derivatives, such as the D-proline, L-proline, glutamic acid, 5-aminovaleric acid, lysine, pipecolic acid, and L-valine identified in this study, are a kind of antibacterial agent with the advantages of being not easily drug-resistant, and having low toxicity or harmless metabolites [37].

Table 3. Compounds identified in Fragment 1 from *P. kleiniana* Wight et Arn via UHPLC–MS analysis.

Peak No.	Identified Compound	Compound Nature	Rt (min)	Formula	Exact Mass	Peak Area (%)
1	D-Maltose	Carbohydrates	0.76	$C_{12}H_{22}O_{11}$	342.1162	6.77%
2	Oxymorphone	Phenanthrenes and derivatives	11.18	$C_{17}H_{19}NO_4$	301.1314	6.29%
3	Rutin	Flavonoids	12.99	$C_{27}H_{30}O_{16}$	281.0899	6.29%
4	D-Proline	Amino acid and derivatives	0.76	$C_5H_9NO_2$	115.0633	5.41%
5	L-Proline	Amino acid and derivatives	0.73	$C_5H_9NO_2$	115.0633	5.41%
6	L-Glutamic acid	Amino acid and derivatives	0.66	$C_5H_9NO_4$	147.0532	5.20%
7	Sucrose	Carbohydrates	0.89	$C_{12}H_{22}O_{11}$	342.1162	3.62%
8	Cynaroside	Flavonoids	12.98	$C_{21}H_{20}O_{11}$	282.162	3.37%
9	Piperlonguminine	Alkaloids	10.57	$C_{16}H_{19}NO_3$	273.1365	3.21%
10	5-Aminovaleric acid	Amino acid and derivatives	1.11	$C_5H_{11}NO_2$	117.079	3.12%
11	D-Glutamine	Carboxylic acids and derivatives	0.66	$C_5H_{10}N_2O_3$	146.0691	2.99%
12	L-Lysine	Amino acid and derivatives	0.64	$C_6H_{14}N_2O_2$	146.1055	2.99%
13	p-Octopamine	Phenols	3.84	$C_8H_{11}NO_2$	153.079	2.96%
14	Oleic acid	Fatty acyls	13.03	$C_{18}H_{34}O_2$	282.2559	2.91%
15	Isoquercitrin	Flavonoids	10.58	$C_{21}H_{20}O_{12}$	274.1933	2.44%
16	L-Pipecolic acid	Amino acid and derivatives	0.69	$C_6H_{11}NO_2$	129.079	2.31%
17	Moracin C	Phenols	0.67	$C_{19}H_{18}O_4$	129.0426	2.31%
18	Kojibiose	Fatty acyls	0.72	$C_{12}H_{22}O_{11}$	342.1162	2.22%
19	Gluconic acid	Carbohydrates	0.69	$C_6H_{12}O_7$	196.0583	1.97%
20	Betaine	Alkaloids	1.06	$C_5H_{11}NO_2$	117.079	1.51%
21	L-Valine	Amino acid and derivatives	0.93	$C_5H_{11}NO_2$	117.079	1.49%
22	D-alpha-Aminobutyric acid	Carboxylic acids and derivatives	0.65	$C_4H_9NO_2$	103.0633	1.46%
23	cis-Aconitic acid	Organic acids and derivatives	1.46	$C_6H_6O_6$	174.0164	1.34%
24	Lactulose	Organooxygen compounds	0.77	$C_{12}H_{22}O_{11}$	342.1162	1.33%
25	Turanose	Fatty acyls	0.79	$C_{12}H_{22}O_{11}$	342.1162	1.33%
26	L-Pipecolic acid	Amino acid and derivatives	1.47	$C_6H_{11}NO_2$	129.079	1.15%
27	DL-Norvaline	Amino acid and derivatives	1.05	$C_5H_{11}NO_2$	117.079	1.11%
28	L-Asparagine	Amino acid and derivatives	0.64	$C_4H_8N_2O_3$	132.0535	1.11%
29	Malic acid	Hydroxy acids and derivatives	0.8	$C_4H_6O_5$	134.0215	0.90%
30	Trigonelline	Alkaloids	0.82	$C_7H_7NO_2$	137.0477	0.90%
31	Acetamide	Alkaloids	13.95	C_2H_5NO	59.03711	0.88%
32	Beta-D-fructose 2-phosphate	Organooxygen compounds	0.75	$C_6H_{13}O_9P$	260.0297	0.77%
33	22-Dehydroclerosterol	Steroids	12.59	$C_{29}H_{46}O$	410.3549	0.76%
34	Artemisinin	Sesquiterpenoids	13.02	$C_{15}H_{22}O_5$	282.1467	0.72%
35	Kaempferol-3-O-rutinoside	flavonoids	6.29	$C_{27}H_{30}O_{15}$	594.1585	0.54%
36	L-Homoserine	Amino acid and derivatives	0.67	$C_4H_9NO_3$	119.0582	0.52%
37	L-Threonine	Amino acid and derivatives	0.64	$C_4H_9NO_3$	119.0582	0.50%
38	Palmitic acid	Lipids	12.92	$C_{16}H_{32}O_2$	256.2402	0.49%
39	O-Acetylethanolamine	Alkaloids	0.67	$C_4H_9NO_2$	103.0633	0.46%
40	Galactose 1-phosphate	Organooxygen compounds	0.65	$C_6H_{13}O_9P$	260.0297	0.46%
41	Glucose 1-phosphate	Organooxygen compounds	13	$C_6H_{13}O_9P$	260.0297	0.45%
42	Adenosine 5′-monophosphate	Nucleotide and its derivates	1.38	$C_{10}H_{14}N_5O_7P$	347.0631	0.43%
43	L-Arginine	Amino acid and derivatives	0.6	$C_6H_{14}N_4O_2$	174.1117	0.43%
44	Maltotriose	Organooxygen compounds	1.23	$C_{18}H_{32}O_{16}$	504.169	0.40%
45	Indole	Alkaloids	3.82	C_8H_7N	117.0578	0.38%
46	D-Glucose 6-phosphate	Carbohydrates	0.65	$C_6H_{13}O_9P$	260.0297	0.37%
47	D-Aspartic acid	Alkaloids	0.76	$C_4H_7NO_4$	133.0375	0.36%
48	Vitexin rhamnoside	Flavonoids	6.78	$C_{27}H_{30}O_{14}$	578.1636	0.35%
49	L-Aspartic acid	Amino acid and derivatives	0.63	$C_4H_7NO_4$	133.0375	0.33%
50	Maltol	Flavonoids	0.9	$C_6H_6O_3$	126.0317	0.33%
51	Astragalin	Flavonoids	6.52	$C_{21}H_{20}O_{11}$	448.1006	0.32%
52	3-Hydroxy-3-methylpentane-1,5-dioic acid	Amino acid and derivatives	2.32	$C_6H_{10}O_5$	162.0528	0.31%
53	Campesterol	Steroids and steroid derivatives	12.18	$C_{28}H_{48}O$	400.3705	0.30%
54	L-Ornithine	Amino acid and derivatives	0.55	$C_5H_{12}N_2O_2$	132.0899	0.30%
55	Adenosine	Nucleotide and its derivates	2.58	$C_{10}H_{13}N_5O_4$	267.0968	0.29%
56	Vidarabine	Purine nucleosides	2.28	$C_{10}H_{13}N_5O_4$	267.0968	0.27%
57	Nicotinic acid	Nicotinic acid derivatives	0.73	$C_6H_5NO_2$	123.032	0.27%
58	Pelargonidin-3-O-glucoside	Flavonoids	1.11	$C_{21}H_{20}O_10$	100.0524	0.26%
59	L-Citruline	Amino acid and derivatives	0.66	$C_6H_{13}N_3O_3$	175.0957	0.26%
60	Diallyl disulfide	Miscellaneous	0.68	$C_6H_{10}S_2$	146.0224	0.26%
61	Sarracine	Alkaloids	13.14	$C_{18}H_{27}NO_5$	337.1889	0.22%
62	N-Acetylputrescine	Phenolamides	1.79	$C_6H_{14}N_2O$	130.1106	0.22%
63	Salicylic acid	Organic acid	7.06	$C_7H_6O_3$	138.0317	0.22%
64	5-Methylcytosine	Nucleotide and its derivates	2.26	$C_5H_7N_3O$	125.0589	0.21%
65	Ellagic acid	Phenols	6.12	$C_{14}H_6O_8$	302.0063	0.21%
66	Isodiospyrin	Quinones	11.28	$C_{22}H_{14}O_6$	374.079	0.21%

2.6. Differential Transcriptomes Triggered by Fragment 1 from P. kleiniana Wight et Arn

To obtain the genome-wide gene expression changes triggered by Fragment 1 from *P. kleiniana* Wight et Arn, we determined transcriptomes of the Gram-negative *V. parahaemolyticus* ATCC17802 and the Gram-positive *S. aureus* ATCC25923 pathogens treated with Fragment 1 (1× MIC) for 6 h using the Illumina RNA sequencing technology. A complete list of differently expressed genes (DEGs) in the two strains are available in the National Center for Biotechnology Information (NCBI) SRA database under the accession number PRJNA906658.

2.6.1. The Major Changed Metabolic Pathways in V. parahaemolyticus ATCC17802

Approximately 13.07% (580 of 4436 genes) of *V. parahaemolyticus* ATCC17802 genes were differentially expressed in the treatment group, as compared to the control group. Of these, 238 DEGs showed higher transcriptional levels (fold change $\geq$ 2.0), whereas 342 DEGs were significantly down-regulated (fold change $\leq$ 0.5) ($p < 0.05$). Sixteen significantly altered metabolic pathways were identified in *V. parahaemolyticus* ATCC 17802, including the citrate cycle; glyoxylate and dicarboxylate metabolism; fatty acid degradation; glycine, serine, and threonine metabolism; oxidative phosphorylation; pyruvate metabolism; propanoate metabolism; beta-Lactam resistance; ABC transporters; two-component system; alanine, aspartate, and glutamate metabolism; phosphotransferase system (PTS); butanoate metabolism; lysine degradation; quorum sensing (QS); and nitrogen metabolism (Figure 5, Table 4).

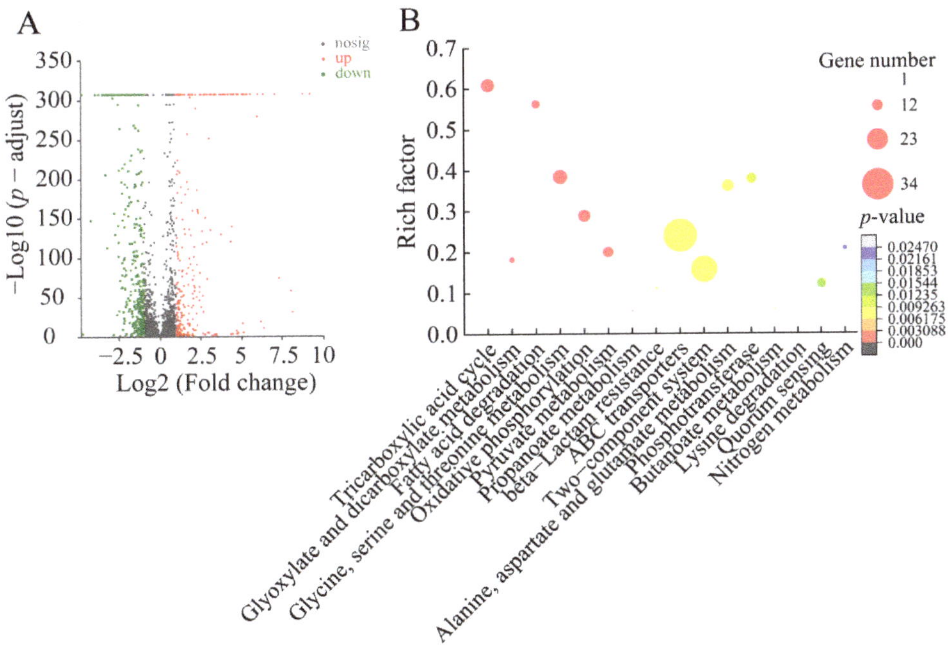

Figure 5. The major changed metabolic pathways in *V. parahaemolyticus* ATCC 17802 mediated by Fragment 1 from *P. kleiniana* Wight et Arn. (**A**) The Volcano plot of the DEGs. (**B**) The significantly altered metabolic pathways in the bacterium. Different colors represented significant levels of the enriched genes.

In the citrate cycle, all the DEGs (n = 14) were significantly repressed (0.146-fold to 0.35-fold) ($p < 0.05$) in *V. parahaemolyticus* ATCC17802 after treatment by Fragment 1 from *P. kleiniana* Wight et Arn. For instance, the DEGs (*sucABCD*, *WU75_19785* and

WU75_19790, *WU75_19795*, and *WU75_19800*), encoding a 2-oxoglutarate dehydrogenase, a dihydrolipoamide succinyltransferase, and succinyl-CoA synthetase subunits alpha and beta, respectively, were highly inhibited (0.146-fold, 0.133-fold, 0.134-fold, and 0.16-fold) ($p < 0.05$). Moreover, the DEGs (*sdhABCD*, *WU75_19775*, *WU75_19780*, *WU75_19765*, and *WU75_19770*) encoding a succinate dehydrogenase were also highly repressed (0.144-fold to 0.199-fold) ($p < 0.05$), which links two essential energy-producing processes, the citrate cycle and oxidative phosphorylation [38]. The inhibited key enzymes in the citrate cycle highlighted inactive energy production in *V. parahaemolyticus* ATCC17802 triggered by Fragment 1.

Table 4. The major altered metabolic pathways in *V. parahaemolyticus* ATCC17802.

Metabolic Pathway	Gene ID	Gene Name	Fold Change	Gene Description
Citrate cycle	*WU75_19785*	*sucA*	0.146	2-oxoglutarate dehydrogenase
	WU75_07425	*pckA*	0.465	Phosphoenolpyruvate carboxykinase
	WU75_19790	*sucB*	0.133	Dihydrolipoamide succinyltransferase
	WU75_11550	*acnB*	0.143	Bifunctional aconitate hydratase 2/2-methylisocitrate dehydratase
	WU75_19795	*sucC*	0.134	Succinyl-CoA synthetase subunit beta
	WU75_19800	*sucD*	0.16	Succinyl-CoA synthetase subunit alpha
	WU75_19770	*sdhD*	0.199	Succinate dehydrogenase
	WU75_19780	*sdhB*	0.157	Succinate dehydrogenase
	WU75_19765	*sdhC*	0.182	Succinate dehydrogenase
	WU75_13785	*fumA*	0.497	Fumarate hydratase
	WU75_09605	*icd*	0.179	Isocitrate dehydrogenase
	WU75_19775	*sdhA*	0.144	Succinate dehydrogenase
	WU75_06430	*mdh*	0.177	Malate dehydrogenase
	WU75_16530	*lpd*	0.35	Dihydrolipoamide dehydrogenase
Glyoxylate and dicarboxylate metabolism	*WU75_19760*	*gltA*	0.129	Type II citrate synthase
	WU75_19150	*aceA*	0.37	Isocitrate lyase
	WU75_19145	*aceB*	0.352	Malate synthase
	WU75_00290	*aceB*	0.315	Malate synthase
	WU75_10840	*phbB*	0.277	3-ketoacyl-ACP reductase
	WU75_03265	*katE*	2.389	Catalase
Fatty acid degradation	*WU75_22235*	*fadB*	0.151	Multifunctional fatty acid oxidation complex subunit alpha
	WU75_08655	*fadE*	0.184	Acyl-CoA dehydrogenase
	WU75_20175	*fadJ*	0.204	Multifunctional fatty acid oxidation complex subunit alpha
	WU75_22230	*fadA*	0.208	3-ketoacyl-CoA thiolase
	WU75_20180	*fadA*	0.305	3-ketoacyl-CoA thiolase
	WU75_10835	*atoB*	0.433	Acetyl-CoA acetyltransferase
	WU75_10445	*atoB*	0.445	Acetyl-CoA acetyltransferase
	WU75_12560	*fadE*	0.452	Acyl-CoA dehydrogenase
	WU75_19885	*fadD*	0.493	Long-chain fatty acid—CoA ligase
Glycine, serine and threonine metabolism	*WU75_14910*	*gcvP*	0.113	Glycine dehydrogenase
	WU75_14915	*gcvH*	0.127	Glycine cleavage system protein H
	WU75_10395	*betA*	0.162	Choline dehydrogenase
	WU75_14930	*gcvT*	0.184	Glycine cleavage system protein T
	WU75_16130	*lysC*	0.187	Aspartate kinase
	WU75_14920	*glyA*	0.203	Serine hydroxymethyltransferase
	WU75_16140	*ectB*	0.222	Diaminobutyrate-2-oxoglutarate aminotransferase
	WU75_16145	*ectA*	0.246	L-2,4-diaminobutyric acid acetyltransferase
	WU75_10400	*betB*	0.259	Betaine-aldehyde dehydrogenase
	WU75_00565	*sdaA*	0.264	Serine dehydratase
	WU75_16135	*ectC*	0.27	Ectoine synthase
	WU75_02030	*trpB*	0.397	Tryptophan synthase subunit beta
	WU75_05755	*thrC*	0.429	Threonine synthase
	WU75_05760	*thrB*	0.47	Serine kinase
	WU75_05330	*glxK*	0.495	Glycerate kinase
Oxidative phosphorylation	*WU75_06010*	*petC*	0.195	Cytochrome C
	WU75_06015	*petB*	0.209	Cytochrome B
	WU75_14570	*ccoO*	0.228	Peptidase S41
	WU75_14575	*ccoN*	0.272	Cbb3-type cytochrome c oxidase subunit I
	WU75_14560	*ccoP*	0.301	Cytochrome Cbb3
	WU75_06485	*ppa*	0.339	Inorganic pyrophosphatase
	WU75_06020	*petA*	0.442	Ubiquinol-cytochrome C reductase

Table 4. *Cont.*

Metabolic Pathway	Gene ID	Gene Name	Fold Change	Gene Description
	WU75_14565	*ccoQ*	0.475	Cytochrome C oxidase
	WU75_02240	*cyoC*	0.478	Cytochrome o ubiquinol oxidase subunit III
	WU75_19125	*ppk2*	2.159	Polyphosphate kinase
	WU75_09420	*cydA*	3.637	Cytochrome d terminal oxidase subunit 1
	WU75_09415	*cydB*	4.11	Cytochrome d ubiquinol oxidase subunit 2
	WU75_09410	*cydX*	5.362	Membrane protein
Pyruvate metabolism	*WU75_01940*	*yiaY*	0.171	Alcohol dehydrogenase
	WU75_03655	*lldD*	0.276	Lactate dehydrogenase
	WU75_22155	*dld*	0.322	Lactate dehydrogenase
	WU75_16665	*oadA*	0.324	Oxaloacetate decarboxylase
	WU75_16060	*aldB*	0.397	Aldehyde dehydrogenase
	WU75_20855	*gloA*	2.451	Lactoylglutathione lyase
	WU75_12805	*pta*	8.464	Phosphate acetyltransferase
	WU75_02150	*ackA*	8.851	Acetate kinase
	WU75_12810	*ackA*	10.365	Acetate kinase
	WU75_09685	*pflD*	12.853	Pyruvate formate-lyase
	WU75_00810	*gloA*	13.536	Glyoxalase
Propanoate metabolism	*WU75_15760*	*prpF*	0.402	3-methylitaconate isomerase
	WU75_15770	*prpC*	0.435	Methylcitrate synthase
beta-Lactam resistance	*WU75_09315*	*acrA*	6.699	Hemolysin D
	WU75_09310	*acrB*	8.911	Multidrug transporter
	WU75_09925	*acrA*	40.366	Hemolysin D
ABC transporters	*WU75_10385*	*proW*	0.106	ABC transporter permease
	WU75_16175	*proX*	0.116	Glycine/betaine ABC transporter substrate-binding protein
	WU75_10390	*proX*	0.122	Glycine/betaine ABC transporter substrate-binding protein
	WU75_12775	*oppC*	0.133	Peptide ABC transporter permease
	WU75_10380	*proV*	0.138	ABC transporter ATP-binding protein
	WU75_09655	*aotM*	0.143	Amino acid ABC transporter permease
	WU75_09665	*aotJ*	0.144	Nickel transporter
	WU75_13090	*yejA*	0.151	Diguanylate cyclase
	WU75_12770	*oppB*	0.164	Oligopeptide transporter permease
	WU75_12780	*oppD*	0.172	Oligopeptide transporter ATP-binding component
	WU75_09660	*aotQ*	0.176	ABC transporter
	WU75_16170	*proW*	0.199	Glycine/betaine ABC transporter permease
	WU75_08085	*oppA*	0.201	Peptide ABC transporter substrate-binding protein
	WU75_07210	*yejA*	0.204	Diguanylate cyclase
	WU75_12765	*oppA*	0.214	Peptide ABC transporter substrate-binding protein
	WU75_07220	*yejB*	0.22	Hypothetical protein
	WU75_07215	*yejE*	0.221	Peptide ABC transporter permease
	WU75_09670	*aotP*	0.228	Amino acid transporter
	WU75_12785	*oppF*	0.228	Peptide ABC transporter ATP-binding protein
	WU75_04720	*oppA*	0.341	Peptide ABC transporter substrate-binding protein
	WU75_16165	*proV*	0.343	Glycine/betaine ABC transporter ATP-binding protein
	WU75_14765	*aapQ*	0.377	Amino acid ABC transporter permease
	WU75_03180	*malE*	0.4	Sugar ABC transporter substrate-binding protein
	WU75_14775	*aapP*	0.405	ABC transporter ATP-binding protein
	WU75_04605	*vcaM*	0.406	Multidrug ABC transporter ATP-binding protein
	WU75_14055	*mdlB*	0.411	Multidrug ABC transporter ATP-binding protein
	WU75_10275	*rbsD*	0.438	D-ribose pyranase
	WU75_05845	*btuF*	0.487	Vitamin B12-binding protein
	WU75_14760	*aapJ*	0.491	Amino acid ABC transporter substrate-binding protein
	WU75_03185	*malK*	2.175	Maltose/maltodextrin transporter ATP-binding protein
	WU75_19815	*znuA*	2.204	Zinc ABC transporter substrate-binding protein
	WU75_19810	*znuC*	2.491	Zinc ABC transporter ATPase
	WU75_02265	*artP*	2.617	Arginine ABC transporter ATP-binding protein
	WU75_19805	*znuB*	2.666	Membrane protein
	WU75_00425	*macB*	14.353	Macrolide transporter
Two-component system	*WU75_07480*	*glnG*	0.186	Nitrogen regulation protein NR(I)
	WU75_13735	*mcp*	0.218	Chemotaxis protein
	WU75_15795	*tctB*	0.237	TctB
	WU75_21750	*dctD*	0.288	C4-dicarboxylate ABC transporter
	WU75_13155	*ttrB*	0.31	4Fe-4S ferredoxin
	WU75_21770	*dctP*	0.31	C4-dicarboxylate ABC transporter
	WU75_01920	*mcp*	0.32	Chemotaxis protein
	WU75_21745	*dctB*	0.352	ATPase
	WU75_10200	*phoA*	0.353	Alkaline phosphatase
	WU75_21765	*dctQ*	0.368	C4-dicarboxylate ABC transporter permease
	WU75_00210	*dctD*	0.406	C4-dicarboxylate ABC transporter

Table 4. *Cont.*

Metabolic Pathway	Gene ID	Gene Name	Fold Change	Gene Description
	WU75_16210	*qseC*	0.423	Histidine kinase
	WU75_23015	*fliC*	0.435	Flagellin
	WU75_07100	*mcp*	0.453	Chemotaxis protein
	WU75_13380	*crp*	0.457	Transcriptional regulator
	WU75_09825	*mcp*	0.471	Chemotaxis protein
	WU75_16525	*hapR*	0.477	LuxR family transcriptional regulator
	WU75_15800	*tctA*	0.485	Tripartite tricarboxylate transporter TctA
	WU75_14800	*mcp*	0.491	Chemotaxis protein
	WU75_06085	*tolC*	2.068	Outer membrane channel protein
	WU75_15630	*dcuB*	2.125	C4-dicarboxylate transporter
	WU75_06045	*degP*	2.148	Serine endoprotease DegQ
	WU75_04355	*mcp*	2.163	Chemotaxis protein
	WU75_10915	*luxQ*	3.377	ATPase
	WU75_22175	*mcp*	4.001	Chemotaxis protein
	WU75_02450	*pfeR*	4.828	Transcriptional regulator
	WU75_18570	*cpxA*	10.981	Two-component sensor protein
	WU75_18575	*cpxR*	26.5	Transcriptional regulator
Alanine, aspartate and glutamate metabolism	WU75_06265	*glmS*	0.037	Glucosamine-fructose-6-phosphate Aminotransferase
	WU75_07465	*glnA*	0.123	Glutamine synthetase
	WU75_04655	*putA*	0.145	Pyrroline-5-carboxylate dehydrogenase
	WU75_14680	-	0.286	NAD-glutamate dehydrogenase
	WU75_05875	*carB*	0.343	Carbamoyl phosphate synthase large subunit
	WU75_05820	*gltB*	0.414	Glutamate synthase
	WU75_05825	*gltD*	0.44	Glutamate synthase
	WU75_05880	*carA*	0.46	Carbamoyl phosphate synthase small subunit
	WU75_18095	*pyrI*	0.462	Aspartate carbamoyltransferase regulatory subunit
	WU75_18090	*pyrB*	0.466	Aspartate carbamoyltransferase catalytic subunit
	WU75_20915	*ansA*	2.141	Cytoplasmic asparaginase I
	WU75_01110	*ansB*	2.718	L-asparaginase II
	WU75_18550	*aspA*	7.015	Aspartate ammonia-lyase
PTS	WU75_03285	*ptsN*	0.462	PTS fructose transporter subunit IIA
	WU75_12990	*ptsG*	0.5	PTS glucose transporter subunit IIBC
	WU75_17910	*celC*	2.36	Molecular chaperone TorD
	WU75_14970	*fruB*	2.451	Bifunctional PTS system fructose-Specific transporter subunit IIA/HPr protein
	WU75_19555	*ptsH*	3.973	PTS sugar transporter
	WU75_00455	*ulaB*	3.977	PTS ascorbate transporter subunit IIB
	WU75_19550	*ptsI*	4.075	Phosphoenolpyruvate-protein Phosphotransferase
	WU75_00460	*cmtB*	4.118	PTS system mannitol-specific Transporter subunit IIA
	WU75_01640	*cmtB*	4.539	PTS mannitol transporter subunit IIA
	WU75_14960	*fruA*	5.096	PTS fructose transporter subunit IIBC
	WU75_00450	*ulaA*	6.946	PTS beta-glucoside transporter subunit IIBC
Butanoate metabolism	WU75_01985	*acsA*	0.334	Acetoacetyl-CoA synthetase
	WU75_10825	*phaC*	0.336	Poly(3-hydroxyalkanoate) synthetase
Lysine degradation	WU75_21960	*ldcC*	7.207	Lysine decarboxylase LdcC
QS	WU75_07805	-	0.109	Cytochrome C
	WU75_07800	-	0.181	ABC transporter permease
	WU75_07795	-	0.202	ABC transporter permease
	WU75_07810	*ddpD*	0.216	ABC transporter ATP-binding protein
	WU75_11620	-	0.218	Peptide ABC transporter permease
	WU75_11630	-	0.233	Peptide ABC transporter substrate-binding protein
	WU75_11625	-	0.261	Peptide ABC transporter permease
	WU75_11610	*ddpF*	0.358	Chemotaxis protein
	WU75_11615	*ddpD*	0.484	Sugar ABC transporter ATP-binding protein
	WU75_21410	*aphA*	2.288	Transcriptional regulator
Nitrogen metabolism	WU75_00760	*ncd2*	0.276	2-nitropropane dioxygenase
	WU75_10810	*napA*	2.286	Nitrate reductase
	WU75_15655	*nirD*	3.934	Nitrite reductase
	WU75_10815	*napB*	6.27	Nitrate reductase
	WU75_08850	*hcp*	63.107	Hydroxylamine reductase

In the propanoate metabolism, all the DEGs (n = 2) were significantly inhibited (0.402-fold to 0.435-fold) in the *V. parahaemolyticus* ATCC17802 treatment group ($p < 0.05$). For example, the DEG (*prpC*, *WU75_15770*) encoding a 2-methylcitrate synthase was significantly inhibited (0.435-fold) ($p < 0.05$). It has been reported that the strategic inhibition of organic acid catabolism in *P. aeruginosa* through inhibition of PrpC activity may be a potent mechanism to halt the growth of this pathogen [39].

In the glyoxylate and dicarboxylate metabolism, five of the six DEGs were significantly repressed (0.129-fold to 0.277-fold) ($p < 0.05$). For instance, the DEGs (*aceAB*, *WU75_19150*, *WU75_19145*, and *WU75_00290*), encoding an isocitrate lyase and a malate synthase of the glyoxylate shunt (GS) carbon cycle, were significantly inhibited (0.315-fold to 0.370-fold) ($p < 0.05$). The GS could avoid unnecessary reactive oxygen species (ROS) generation by bypassing nicotinamide adenine dinucleotide (NADH) production, and respiration, eventually helping cells to survive in harsh conditions [40,41].

In the glycine, serine, and threonine metabolism, all the DEGs ($n = 15$) were significantly inhibited (0.113-fold to 0.495-fold) in *V. parahaemolyticus* ATCC17802 ($p < 0.05$). For example, the DEGs (*ectBAC*, *WU75_16140*, *WU75_16145*, and *WU75_16135*), encoding a diaminobutyrate-2-oxoglutarate aminotransferase, a 2% 2C4-diaminobutyric acid acetyltransferase, and an ectoine synthase, which are involved in the synthesis of ectoine that is commonly found in halophilic and halotolerant microorganisms to maintain cell osmotic balance [42]. Additionally, in the alanine, aspartate, and glutamate metabolism, ten of the thirteen DEGs were significantly down-regulated (0.037-fold to 0.466-fold) in *V. parahaemolyticus* ATCC17802 as well ($p < 0.05$). Conversely, the DEGs (*ansAB*, *WU75_20915*, and *WU75_01110*) were up-regulated (2.141-fold and 2.718-fold) ($p < 0.05$), which encoded a cytoplasmic asparaginase I and a L-asparaginase II. The asparaginase I is required for bacterial growth on asparagine as the sole nitrogen source [43], while asparaginases are important in maintaining nitrogen balance and the levels of amino acids within cells [43]. These results indicated that the amino acid synthesis was inhibited in *V. parahaemolyticus* ATCC17802 mediated by Fragment 1.

For the ABC transporters, 29 of the 35 DEGs were significantly down-regulated (0.106-fold to 0.491-fold) in *V. parahaemolyticus* ATCC17802 ($p < 0.05$). Of these, the DEGs (*proVXW*, *WU75_10380*, *WU75_10390*, and *WU75_10385*), encoding a choline ABC transporter ATP-binding protein, a choline ABC transporter substrate-binding protein, and a choline ABC transporter permease subunit that are responsible for the choline transport, were all significantly repressed (0.106-fold to 0.138-fold). The DEGs (*oppABCDF*, *WU75_12765*, *WU75_12770*, *WU75_12775*, *WU75_12780*, and *WU75_12785*) encoding a peptide ABC transporter substrate-binding protein, an oligopeptide transporter permease, a peptide ABC transporter permease, an oligopeptide transporter ATP-binding component, and a peptide ABC transporter ATP-binding protein, respectively, were all highly repressed (0.172-fold and 0.214-fold). Additionally, the DEGs (*yejABE*, *WU75_13090*, *WU75_07210*, *WU75_07220*, and *WU75_07215*) encoding a diguanylate cyclase, an ABC transporter permease subunit, and a peptide ABC transporter permease, respectively, were highly repressed as well (0.151-fold and 0.220-fold). The ABC transporter YejABEF is required for resistance to antimicrobial peptides and virulence of *Brucella melitensis* [44]. These results indicated that the inhibited ABC transporters likely led to the repressed substance transport and harmful substances discharged in *V. parahaemolyticus* ATCC17802.

In the oxidative phosphorylation, nine of the thirteen DEGs were significantly down-regulated in *V. parahaemolyticus* ATCC17802 (0.195-fold to 0.478-fold) ($p < 0.05$). Oxidative phosphorylation is a major metabolic pathway to obtain energy required for cell growth and proliferation [45] (Huang et al., 2019). For instance, the DEGs (*ccoNOQ*, *WU75_14575*, *WU75_14570*, and *WU75_14565*) were significantly inhibited (0.228-fold to 0.475-fold) ($p < 0.05$), which regulated the bacterial adhesion in environmental stresses in *V. alginolyticus* [45].

In the QS, most DEGs ($n = 9$) were significantly inhibited (0.109-fold to 0.484-fold) ($p < 0.05$), e.g., cytochrome c (*WU75_06010*), cytochrome B (*WU75_06015*), and peptidase S41 (*WU75_14570*). For instance, the cytochrome c mediates electron-transfer in the respiratory chain and acts as a detoxifying agent to dispose of reactive oxygen species (ROS) [46].

In contrast, in the PTS, nine of the eleven DEGs were significantly up-regulated (2.36-fold to 6.946-fold) in the *V. parahaemolyticus* ATCC17802 treatment group ($p < 0.05$).

Of these, the DEGs (*fruA, WU75_14960; ulaA, WU75_00450*), encoding a PTS fructose transporter subunit IIBC and a PTS beta-glucoside transporter subunit IIBC, respectively, were highly up-regulated (5.096-fold and 6.946-fold) ($p < 0.05$).

In the nitrogen metabolism, most of the DEGs ($n = 4$) were significantly up-regulated (2.286-fold to 63.107-fold) ($p < 0.05$). Remarkably, the DEG (*hcp, WU75_08850*) encoding a hydroxylamine reductase was strongly up-regulated (63.107-fold) ($p < 0.05$), and is involved in the processes of scavenging hydroxylamine with NO detoxification [47].

In the two-component system, 19 DEGs were significantly inhibited (0.186-fold to 0.491-fold), whereas 9 DEGs were significantly enhanced (2.068-fold to 26.5-fold) ($p < 0.05$). The two-component system is one of the primary pathways by which bacteria adapt to environmental stresses [48]. For instance, the DEGs (*cpxAR, WU75_18570,* and *WU75_18575*) encoding a two-component sensor protein and a transcriptional regulator were strongly up-regulated (10.981-fold and 26.500-fold) ($p < 0.05$). The CpxAR is a key modulator of capsule export that facilitates *Actinobacillus pleuropneumoniae* survival in the host [49]. It also regulates cell membrane permeability and efflux pump activity and induces multidrug resistance (MDR) in *Salmonella enteritidis* [50].

Additionally, in the beta-lactam resistance, all the DEGs (*acrAB, WU75_09925, WU75_09315,* and *WU75_09310*) were strongly up-regulated (6.699-fold to 40.366-fold) in the *V. parahaemolyticus* ATCC17802 treatment group ($p < 0.05$), which encoded a multidrug efflux resistance nodulation division (RND) transporter periplasmic adaptor subunit and a multidrug transporter. The RND family efflux pumps, including the major pump AcrAB-TolC, are important mediators of intrinsic and evolved antibiotic resistance [51].

Taken together, these results indicated that Fragment 1 from *P. kleiniana* Wight et Arn can significantly change sixteen metabolic pathways in the Gram-negative *V. parahaemolyticus* ATCC17802, which consequently led to repressed substance transporting, energy production, and protein translation, but enhanced stringent response, and harmful substance discharging, and thereby cell death.

2.6.2. The Major Changed Metabolic Pathways in *S. aureus* ATCC25923

Approximately 7.3% (196 of 2672 genes) of *S. aureus* ATCC25923 genes were differentially expressed in the treatment group, as compared to the control group. Of these, 156 DEGs showed higher transcriptional levels (fold changes ≥ 2.0), whereas 40 DEGs were significantly down-regulated (fold changes ≤ 0.5) ($p < 0.05$). Based on the comparative transcriptomic analysis, seven significantly altered metabolic pathways were identified in *S. aureus* ATCC25923, including the two-component system; nitrogen metabolism; riboflavin metabolism; arginine and proline metabolism; atrazine degradation; alanine, aspartate and glutamate metabolism; and pyrimidine metabolism (Figure 6, Table 5).

In the arginine and proline metabolism, all the DEGs ($n = 4$) were significantly down-regulated at the transcription levels (0.109-fold to 0.461-fold) in *S. aureus* ATCC25923 ($p < 0.05$). The arginine metabolism converts L-arginine to urea and L-ornithine, which are further metabolized into proline and polyamides that drive collagen synthesis and bioenergetic pathways critical for cell proliferation, respectively [52]. For instance, the DEG (*rocF, KQ76_11235*) encoding an arginase was significantly down-regulated (0.461-fold) ($p < 0.05$), and was associated with the ability of *Helicobacter pylori* to establish chronic infections [53].

All the DEGs ($n = 4$) in the riboflavin metabolism were also significantly inhibited (*ribBADEH*, 0.3734-fold to 0.480-fold) ($p < 0.05$). In this pathway, the redox cofactors flavin mononucleotide and flavin adenine dinucleotide and their precursor riboflavin play important roles in many cellular processes, such as respiration, DNA repair, biosyntheses of heme groups, cofactors and nucleotides, fatty acid beta-oxidation, and bioluminescence [54].

Bacteria use two-component signal transduction systems to elicit adaptive responses to environmental changes [55]. In this study, seven DEGs in the two-component system were significantly up-regulated (2.117-fold to 28.924-fold) in *S. aureus* ATCC25923 ($p < 0.05$). For instance, the DEGs (*agrB, KQ76_10520*; and *graS, KQ76_03245*) encoding histidine kinases were significantly up-regulated by 2.565-fold and 2.989-fold, respectively ($p < 0.05$). The accessory gene regulator (agr) quorum-sensing system contributes to its pathogenicity of *S. aureus* [56]. GraS, the sensor histidine kinase of the GraXRS system, has been suggested to directly activate the response regulator ArlR [53]. Loss of the ArlR alone impairs the ability of *S. aureus* to respond to host-imposed manganese starvation and glucose limitation [57].

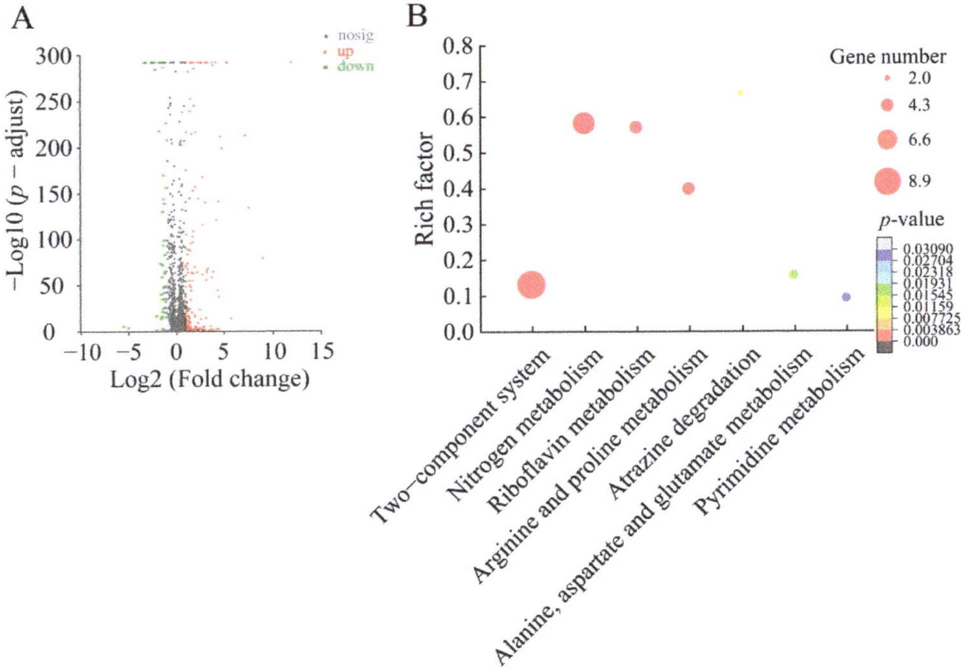

Figure 6. The major changed metabolic pathways in *S. aureus* ATCC25923 triggered by Fragment 1 from *P. kleiniana Wight et Arn*. (**A**) The Volcano plot of the DGEs. (**B**) The significantly altered metabolic pathways in the bacterium.

Interestingly, expression of all the DEGs ($n = 7$) in the nitrogen metabolism was significantly increased at the transcription level (3.529-fold to 10.404-fold) in *S. aureus* ATCC25923 ($p < 0.05$). The seven DEGs (*nirBD*, *narHIJZT*) were all involved in nitrate reduction [58–60]. Of these, the NirD (*KQ76_12515*) was a small subunit of cytoplasmic NADH-dependent nitrite reductase complex NirBD [61,62]. Over-expression of *nirD* limits RelA-dependent accumulation of guanosine 5′-triphosphate 3′-diphosphate ((p)ppGpp) in vivo and can prevent activation of the stringent response during amino acid starvation in *E. coli* [62].

In the alanine, aspartate, and glutamate metabolism, two DEGs (*carBA*, *KQ76_05770* and *KQ76_05765*) encoding carbamoyl phosphate synthase were significantly up-regulated (2.154-fold and 3.084-fold) in *S. aureus* ATCC25923 ($p < 0.05$). The interface residues located near the CarB region of carboxy phosphate synthetic domain plays a key role in carbamoyl phosphate synthetase, aspartate transcarbamoylase, and dihydroorotase (CAD) complex regulation in the pyrimidine biosynthesis [63]. Correspondingly, in the pyrimidine metabolism, four DEGs (*pyrBCR*, *KQ76_05755*, *KQ76_05760*, and *KQ76_05745*) were also significantly up-regulated (2.968-fold to 3.213-fold) ($p < 0.05$), and encoded an aspartate carbamoyltransferase, a dihydroorotase, and a phosphoribosyl transferase, respectively. The pyrimidines are involved in the synthesis of DNA, RNA, lipids, and carbohydrates. The pyrimidine metabolism is involved in the synthesis, degradation, salvage, interconversion, and transport of these compounds [64].

Table 5. The major altered metabolic pathways in *S. aureus* ATCC25923.

Metabolic Pathway	Gene ID	Gene Name	Fold Change	Gene Description
Two-component system	*KQ76_00500*	-	0.373	Capsular biosynthesis protein
	KQ76_00560	*wecC*	0.490	UDP-N-acetyl-D-mannosamine dehydrogenase
	KQ76_12475	*nreC*	2.117	Nitrate respiration regulation response regulator NreC
	KQ76_12480	*nreB*	2.276	Nitrate respiration regulation sensor histidine kinase NreB
	KQ76_12485	*nreA*	2.433	Nitrate respiration regulation accessory nitrate sensor NreA
	KQ76_10520	*agrB*	2.565	Histidine kinase
	KQ76_03245	*graS*	2.989	Histidine kinase
	KQ76_10785	*kdpF*	5.371	ATPase
	KQ76_04230	*dltC*	28.924	Alanine-phosphoribitol ligase
Nitrogen metabolism	*KQ76_12490*	*narI*	3.529	Nitrate reductase
	KQ76_12515	*nirD*	4.199	Nitrite reductase
	KQ76_12520	*nirB*	5.060	Nitrite reductase
	KQ76_12460	*narT*	6.376	Nitrate transporter NarT
	KQ76_12500	*narH*	5.799	Nitrate reductase
	KQ76_12505	*narZ*	8.442	Nitrate reductase
	KQ76_12495	*narJ*	10.404	Nitrate reductase
Riboflavin metabolism	*KQ76_09200*	*ribE*	0.373	Riboflavin synthase subunit alpha
	KQ76_09195	*ribBA*	0.413	GTP cyclohydrolase
	KQ76_09205	*ribD*	0.430	Diaminohydroxyphosphoribosylaminopyrimidine deaminase
	KQ76_09190	*ribH*	0.480	6,7-dimethyl-8-ribityllumazine synthase
Arginine and proline metabolism	*KQ76_09185*	*fadM*	0.109	Proline dehydrogenase
	KQ76_00580	-	0.218	Aldehyde dehydrogenase
	KQ76_13360	-	0.303	1-pyrroline-5-carboxylate dehydrogenase
	KQ76_11235	*rocF*	0.461	Arginase
Atrazine degradation	*KQ76_11915*	*ureC*	0.406	Urease subunit alpha
	KQ76_11910	*ureB*	0.412	Urease subunit beta
Alanine, aspartate and glutamate metabolism	*KQ76_13360*	-	0.303	1-pyrroline-5-carboxylate dehydrogenase
	KQ76_05770	*carB*	2.158	Carbamoyl phosphate synthase large subunit
	KQ76_05765	*carA*	3.084	Carbamoyl phosphate synthase small subunit
Pyrimidine metabolism	*KQ76_05745*	*pyrR*	2.968	Phosphoribosyl transferase
	KQ76_05760	*pyrC*	3.115	Dihydroorotase
	KQ76_05755	*pyrB*	3.213	Aspartate carbamoyltransferase

Taken together, these results indicate that Fragment 1 from *P. kleiniana* Wight et Arn can significantly influence seven metabolic pathways in the Gran-positive *S. aureus* ATCC25923. Of these, the two-component system, alanine, aspartate and glutamate metabolism, and nitrogen metabolism were also changed in the Gram-negative *V. parahaemolyticus* ATCC17802, which led to the enhanced regulation of stringent response in the two pathogens. On the other hand, we also found distinct transcriptomic profiles between the Gram-positive and Gram-negative pathogens triggered by Fragment 1. For example, consistent with the results obtained from the cell structure analysis, *V. parahaemolyticus* ATCC17802 was more sensitive to Fragment 1 treatment, as more metabolic pathways were altered, such as the citrate cycle, glyoxylate and dicarboxylate metabolism, fatty acid degradation, glycine, serine and threonine metabolism, oxidative phosphorylation, pyruvate metabolism, propanoate metabolism, beta-lactam resistance, ABC transporters, PTS, butanoate metabolism, lysine degradation, and QS, which resulted in cell destruction and even death.

In addition, to validate the transcriptome data, we tested 16 representative DEGs (Table S1) via reverse transcription real time-quantitative PCR (RT-qPCR) analysis, and the resulting data were generally correlated with those yielded from the transcriptome analysis (Table S2).

3. Materials and Methods

3.1. Bacterial Strains and Culture Conditions

The bacterial strains and culture media used in this study are listed in Table S3. *Vibrio* strains and non-*Vibrio* strains were incubated as described in our recent studies [15,16,65].

3.2. Extraction of Bioactive Substances from P. kleiniana Wight et Arn

Fresh *P. kleiniana* Wight et Arn was purchased from the Qian Shan Zhen Pin shop in Guiyang City (26°36′5.01″ N, 106°41′19.90″ E), Guizhou Province, China, in October of 2021. Bioactive substances were extracted from the samples using the methanol and chloroform method described in our recent reports [15,16,66]. Briefly, aliquot of a 500 g of the whole plant sample was lyophilized, pulverised, powded, sonicated, and then filtered and collected for the secondary extraction. The methanol and chloroform phases were separated and then concentrated using the Rotary Evaporator (IKA, Staufen, Germany) [15,16].

3.3. Antimicrobial Susceptibility Assay

The susceptibility of the bacterial strains (Table S3) to the extracts from *P. kleiniana* Wight et Arn were determined according to the standard method issued by the Clinical and Laboratory Standards Institute, USA (CLSI, M100-S23, 2018). The antibacterial activity was defined as described previously [15,16]. Broth dilution testing (microdilution) (CLSI, M100-S18, 2018) was used to determine MICs of the extracts. The MIC was defined as described previously [15,16].

3.4. Prep-HPLC Analysis

Aliquots of the extracted samples (10 mg/mL) were resolved, centrifuged, filtered, and subjected for the Prep-HPLC Analysis, using Waters 2707 (Waters, Milford, MA, USA) linked with UPLC Sunfire C18 column (5 μm, 10 × 250 mm) (Waters, Milford, MA, USA) with the same parameters and elution conditions described in our recent reports [15,16].

3.5. UHPLC–MS Analysis

The UHPLC–MS analysis was conducted using the EXIONLC System (Sciex, Framingham, MA, USA) by Shanghai Hoogen Biotech, Shanghai, China [67].

3.6. Bacterial Cell Surface Hydrophobicity and Membrane Fluidity Assays

The cell surface hydrophobicity was measured according to the method of Cui et al. [68]. The cell membrane fluidity was measured according to the method of Kuhry et al. [69], using the 1,6-diphenyl-1,3,5-hexatriene (DPH, Sangon, Shanghai, China).

3.7. Cell Membrane Permeability Analysis

Cell outer membrane permeability was measured according to the method of Wang et al. [70], with the NPN solution (Sangon, Shanghai, China). The inner membrane permeability was measured according to the method of Huang et al. [71], with the ONPG solution (Sangon, Shanghai, China).

3.8. Scanning Electron Microscope (SEM) Assay

The preparation of the samples for the SEM analysis was performed using the method described in our recent reports [15,16,72]. The samples were observed using the Scanning Electron Microscope (Tescan Mira 3 XH, Tescan, Brno, Czech Republic, 5.0 kV, 30,000×).

3.9. Illumina RNA Sequencing

The bacterial cell culture at the mid-LGP was treated with Fragment 1 (1× MIC) from *P. kleiniana* Wight et Arn for 6 h, and then collected via centrifugation for the total RNA preparation [15,16,72]. Three independently prepared RNA samples for each strain were subjected for the Illumina RNA sequencing analysis, using Illumina HiSeq 2500 platform (Illumina, Santiago, CA, USA) [72].

3.10. RT-qPCR Assay

The RT-qPCR assay was performed according to the method described in our recent reports [15,16,72]. The oligonucleotide primers were designed (Table S1), and synthesized via Sangon (Shanghai, China).

3.11. Data Analysis

The DEGs were analyzed as described in our recent reports [15,16,72]. All tests were carried out in triplicate. The data were analyzed using the SPSS statistical analysis software version 17.0 (SPSS Inc., Armonk, NY, USA). One-way analysis of variance (ANOVA) was performed using the least-significant difference (LSD) method and homogeneity of variance test. There was no significant difference between the control and the treatment groups if the generalized *p*-values were more than 0.05; conversely, there was significant difference if *p*-values were less than 0.05.

4. Conclusions

In this study, the methanol-phase extract from *P. kleiniana* Wight et Arn showed an inhibition rate of 68.18% against 22 species of common pathogenic bacteria. The methanol-phase extraction inhibited the growth of one species of Gram-positive *S. aureus*, and 14 species of Gram-negative bacteria, including *B. cereus*, *E. cloacae*, *E. coli*, *P. aeruginosa*, *S. typhimurium* 1, *S. dysenteriae*, *S. flexneri*, *S. flexneri*, *S. sonnei*, *V. alginolyticus*, *V. cholerae*, *V. fluvialis*, *V. mimicus*, *V. parahemolyticus*, and *V. vulnificus* strains. This extract was further purified using the Prep-HPLC, and three separated fragments were obtained. Fragment 1 significantly increased bacterial cell surface hydrophobicity and membrane permeability and decreased membrane fluidity, disrupting the cell integrity of the Gram-positive and Gram-negative bacteria such as *S. aureus* ATCC25923, *S. aureus* ATCC8095, *V. parahaemolyticus* ATCC17802, and *V. parahaemolyticus* B5-29. The MIC values of Fragment 1 ranged from 6.25 mg/mL to 50 mg/mL. A total of 66 different compounds in Fragment 1 were identified. The highest relative percentage of the compounds was D-maltose (6.77%), followed by oxymorphone (6.29%), rutin (6.29%), D-proline (5.41%), and L-proline (5.41%). Highly concentrated sugar solutions, such as the D-maltose identified in Fragment 1, are known to be effective antimicrobial agents. The identified oxymorphone and rutin could exert antibacterial activity via damaging the bacterial cell wall and cytoplasmic membrane, respectively. Multiple cellular metabolic pathways altered by Fragment 1 in the representative Gram-negative *V. parahaemolyticus* ATCC17802 and Gram-positive *S. aureus* ATCC25923 pathogens after treatment with Fragment 1 (1× MIC) for 6 h ($p < 0.05$). These results indicated that the energy supply and protein translation of the tested strains was inhibited, the signal transduction was blocked, and the ability to pump foreign harmful substances was reduced, leading to cell death. Overall, the results of this study demonstrate that Fragment 1 from *P. kleiniana* Wight et Arn is a promising candidate for antibacterial medicine and food preservatives.

Supplementary Materials: The following supporting information can be downloaded at: https://www.mdpi.com/article/10.3390/foods12081640/s1, Table S1: The oligonucleotide primers designed and used in the RT-qPCR assay; Table S2: The relative expression of representative DEGs by the RT-qPCR assay; Table S3: The bacterial strains and media used in this study; Figure S1: The Prep−HPLC diagram of purifying the methanol-phase crude extract from *P. kleiniana* Wight et Arn.

Author Contributions: Y.T.: major experiments, data curation, and writing—original draft; P.Y.: writing—review and editing; L.C.: funding acquisition, conceptualization, and writing—review and editing. All authors have read and agreed to the published version of the manuscript.

Funding: This work was supported by Shanghai Municipal Science and Technology Commission, grant number 17050502200, and National Natural Science Foundation of China, grant number 31671946.

Institutional Review Board Statement: Not applicable.

Informed Consent Statement: Not applicable.

Data Availability Statement: Data is contained within the article or Supplementary Materials. The complete lists of DEGs in the two strains are available in the NCBI SRA database (https://submit.ncbi.nlm.nih.gov/subs/bioproject/, accessed on 29 November 2022) under the accession number PRJNA906658.

Conflicts of Interest: The authors declare no conflict of interest.

References

1. Bueno, E.; Pinedo, V.; Cava, F. Adaptation of *Vibrio cholerae* to hypoxic environments. *Front. Microbiol.* **2020**, *11*, 739. [CrossRef]
2. Stocco, G.; Lucafò, M.; Decorti, G. Pharmacogenomics of antibiotics. *Int. J. Mol. Sci.* **2020**, *21*, 5975. [CrossRef]
3. Bombaywala, S.; Mandpe, A.; Paliya, S.; Kumar, S. Antibiotic resistance in the environment: A critical insight on its occurrence, fate, and eco-toxicity. *Environ. Sci. Pollut. Res. Int.* **2021**, *28*, 24889–24916. [CrossRef]
4. Mancuso, G.; Midiri, A.; Gerace, E.; Biondo, C. Bacterial antibiotic resistance: The most critical pathogens. *Pathogens* **2021**, *10*, 1310. [CrossRef] [PubMed]
5. Gomes, F.; Rodrigues, M.E.; Martins, N.; Ferreira, I.C.F.R.; Henriques, M. Phenolic plant extracts versus penicillin G: In vitro susceptibility of *Staphylococcus aureus* isolated from bovine mastitis. *Pharmaceuticals* **2019**, *12*, 128. [CrossRef]
6. Thomford, N.E.; Senthebane, D.A.; Rowe, A.; Munro, D.; Seele, P.; Maroyi, A.; Dzobo, K. Natural products for drug discovery in the 21st century: Innovations for novel drug discovery. *Int. J. Mol. Sci.* **2018**, *19*, 1578. [CrossRef]
7. Chandra, G.; Mukherjee, D.; Ray, A.S.; Chatterjee, S.; Bhattacharjee, I. Phytoextracts as antibacterials: A review. *Curr. Drug Discov. Technol.* **2020**, *17*, 523–533. [CrossRef] [PubMed]
8. Volis, S. Securing a future for China's plant biodiversity through an integrated conservation approach. *Plant Divers.* **2018**, *40*, 91–105. [CrossRef] [PubMed]
9. Tao, J.; Yan, S.; Zhou, C.; Liu, Q.; Zhu, H.; Wen, Z. Total flavonoids from *Potentilla kleiniana* Wight et Arn inhibits biofilm formation and virulence factors production in methicillin-resistant *Staphylococcus aureus* (MRSA). *J. Ethnopharmacol.* **2021**, *279*, 114383. [CrossRef]
10. Zhou, Y.Q.; Li, S.M.; Wei, X.; Yang, X.; Xiao, J.W.; Pan, B.W.; Xie, S.X.; Zhou, Y.; Yang, J.; Wei, Y. Identification and quantitative analysis of bioactive components from *Potentilla kleiniana* Wight et Arn with anti HIV-1 proteases activity. *Nat. Prod. Res.* **2022**, 1–4. [CrossRef] [PubMed]
11. Liu, M.; Huang, X.; Liu, Q.; Li, X.; Chen, M.; Zhu, Y.; Chen, X. Separation of α-glucosidase inhibitors from *Potentilla kleiniana* Wight et Arn using solvent and flow-rate gradient high-speed counter-current chromatography target-guided by ultrafiltration HPLC-MS screening. *Phytochem. Anal.* **2019**, *30*, 661–668. [CrossRef]
12. Li, S.; Tan, J.; Zeng, J.; Wu, X.W.X.; Zhang, J. Antihyperglycemic and antioxidant effect of the total flavones of *Potentilla kleiniana* Wight et Arn. in streptozotocin induced diabetic rats. *Pak. J. Pharm. Sci.* **2017**, *30*, 171–178.
13. Tao, J.; Yan, S.; Wang, H.; Zhao, L.; Zhu, H.; Wen, Z. Antimicrobial and antibiofilm effects of total flavonoids from *Potentilla kleiniana* Wight et Arn on *Pseudomonas aeruginosa* and its potential application to stainless steel surfaces. *LWT-Food Sci. Technol.* **2022**, *154*, 112631. [CrossRef]
14. Xuan, S.H.; Hong, I.K.; Lee, Y.J.; Kim, J.W.; Park, S.N. Biological activities and chemical components of *Potentilla kleiniana* Wight & Arn. *Nat. Prod. Res.* **2020**, *34*, 3262–3266. [CrossRef] [PubMed]
15. Liu, Y.; Yang, L.; Liu, P.; Jin, Y.; Qin, S.; Chen, L. Identification of antibacterial components in the methanol-phase extract from edible herbaceous plant *Rumex madaio* Makino and their antibacterial action modes. *Molecules* **2022**, *27*, 660. [CrossRef]
16. Fu, J.; Wang, Y.; Sun, M.; Xu, Y.; Chen, L. Antibacterial activity and components of the methanol-phase extract from rhizomes of pharmacophagous plant *Alpinia officinarum* Hance. *Molecules* **2022**, *27*, 4308. [CrossRef]
17. Wang, J.; Ding, Q.; Yang, Q.; Fan, H.; Yu, G.; Liu, F.; Bello, B.K.; Zhang, X.; Zhang, T.; Dong, J.; et al. *Vibrio alginolyticus* triggers inflammatory response in mouse peritoneal macrophages via activation of NLRP3 inflammasome. *Front. Cell. Infect. Microbiol.* **2021**, *11*, 769777. [CrossRef] [PubMed]
18. Hernández-Robles, M.F.; Natividad-Bonifacio, I.; Álvarez-Contreras, A.K.; Tercero-Alburo, J.J.; Quiñones-Ramírez, E.I.; Vázquez-Salinas, C. Characterization of potential virulence factors of *Vibrio mimicus* isolated from fishery products and water. *Int. J. Microbiol.* **2021**, *2021*, 8397930. [CrossRef]
19. Dey, R.; Rieger, A.M.; Stephens, C.; Ashbolt, N.J. Interactions of *Pseudomonas aeruginosa* with *Acanthamoeba polyphaga* observed by imaging flow cytometry. *Cytom. Part A* **2019**, *95*, 555–564. [CrossRef]
20. Liu, J.; Qin, K.; Wu, C.; Fu, K.; Yu, X.; Zhou, L. *De Novo* sequencing provides insights into the pathogenicity of foodborne *Vibrio parahaemolyticus*. *Front. Cell. Infect. Microbiol.* **2021**, *11*, 652957. [CrossRef] [PubMed]
21. Huang, Y.; Flint, S.H.; Palmer, J.S. *Bacillus cereus* spores and toxins—The potential role of biofilms. *Food Microbiol.* **2020**, *90*, 103493. [CrossRef]
22. Ojha, R.; Dittmar, A.A.; Severin, G.B.; Koestler, B.J. *Shigella flexneri* diguanylate cyclases regulate virulence. *J. Bacteriol.* **2021**, *203*, e0024221. [CrossRef] [PubMed]
23. Li, H.; Tang, T.; Stegger, M.; Dalsgaard, A.; Liu, T.; Leisner, J.J. Characterization of antimicrobial-resistant Staphylococcus aureus from retail foods in Beijing, China. *Food Microbiol.* **2021**, *93*, 103603. [CrossRef] [PubMed]
24. Guo, Y.; Song, G.; Sun, M.; Wang, J.; Wang, Y. Prevalence and therapies of antibiotic-resistance in *Staphylococcus aureus*. *Front. Cell. Infect. Microbiol.* **2020**, *10*, 107. [CrossRef]
25. Danchik, C.; Casadevall, A. Role of cell surface hydrophobicity in the pathogenesis of medically-significant fungi. *Front. Cell. Infect. Microbiol.* **2020**, *10*, 594973. [CrossRef] [PubMed]

26. Soundharrajan, I.; Kim, D.; Kuppusamy, P.; Muthusamy, K.; Lee, H.J.; Choi, K.C. Probiotic and triticale silage fermentation potential of *Pediococcus pentosaceus* and *Lactobacillus brevis* and their impacts on pathogenic bacteria. *Microorganisms* **2019**, *7*, 318. [CrossRef]
27. Yu, X.; Sha, L.; Liu, Q.; Zhao, Y.; Fang, H.; Cao, Y.; Zhao, J. Recent advances in cell membrane camouflage-based biosensing application. *Biosens. Bioelectron.* **2021**, *194*, 113623. [CrossRef]
28. Zhang, M.; Yu, Y.; Lian, L.; Li, W.; Ren, J.; Liang, Y.; Xue, F.; Tang, F.; Zhu, X.; Ling, J.; et al. Functional mechanism of antimicrobial peptide bomidin and its safety for *Macrobrachium rosenbergii*. *Probiotics Antimicrob. Proteins* **2022**, *14*, 169–179. [CrossRef]
29. Bojkovic, J.; Richie, D.L.; Six, D.A.; Rath, C.M.; Sawyer, W.S.; Hu, Q.; Dean, C.R. Characterization of an *acinetobacter baumannii* lptD deletion strain: Permeability defects and response to inhibition of lipopolysaccharide and fatty acid biosynthesis. *J. Bacteriol.* **2015**, *198*, 731–741. [CrossRef]
30. Veiko, A.G.; Olchowik-Grabarek, E.; Sekowski, S.; Roszkowska, A.; Lapshina, E.A.; Dobrzynska, I.; Zamaraeva, M.; Zavodnik, I.B. Antimicrobial activity of quercetin, naringenin and catechin: Flavonoids inhibit *Staphylococcus aureus*-induced hemolysis and modify membranes of bacteria and erythrocytes. *Molecules* **2023**, *28*, 1252. [CrossRef]
31. Zdybicka-Barabas, A.; Stączek, S.; Pawlikowska-Pawlęga, B.; Mak, P.; Luchowski, R.; Skrzypiec, K.; Mendyk, E.; Wydrych, J.; Gruszecki, W.I.; Cytryńska, M. Studies on the interactions of neutral *Galleria mellonella* cecropin D with living bacterial cells. *Amino Acids* **2019**, *51*, 175–191. [CrossRef]
32. Mizzi, L.; Maniscalco, D.; Gaspari, S.; Chatzitzika, C.; Gatt, R.; Valdramidis, V.P. Assessing the individual microbial inhibitory capacity of different sugars against pathogens commonly found in food systems. *Lett. Appl. Microbiol.* **2020**, *71*, 251–258. [CrossRef]
33. Fan, X.; Kong, D.; He, S.; Chen, J.; Jiang, Y.; Ma, Z.; Feng, J.; Yan, H. Phenanthrene derivatives from *asarum heterotropoides* showed excellent antibacterial activity against phytopathogenic bacteria. *J. Agric. Food Chem.* **2021**, *69*, 14520–14529. [CrossRef]
34. Tan, Z.; Deng, J.; Ye, Q.; Zhang, Z. The antibacterial activity of natural-derived flavonoids. *Curr. Top. Med. Chem.* **2022**, *22*, 1009–1019. [CrossRef]
35. Kachur, K.; Suntres, Z. The antibacterial properties of phenolic isomers, carvacrol and thymol. *Crit. Rev. Food Sci. Nutr.* **2020**, *60*, 3042–3053. [CrossRef]
36. Wibowo, J.T.; Ahmadi, P.; Rahmawati, S.I.; Bayu, A.; Putra, M.Y.; Kijjoa, A. marine-derived indole alkaloids and their biological and pharmacological activities. *Mar. Drugs* **2021**, *20*, 3. [CrossRef]
37. Li, H.; Li, Y.; Wang, Y.; Liu, L.; Dong, H.; Satoh, T. Membrane-active amino acid-coupled polyetheramine derivatives with high selectivity and broad-spectrum antibacterial activity. *Acta Biomater.* **2022**, *142*, 136–148. [CrossRef] [PubMed]
38. Sharma, P.; Maklashina, E.; Cecchini, G.; Iverson, T.M. The roles of SDHAF2 and dicarboxylate in covalent flavinylation of SDHA, the human complex II flavoprotein. *Proc. Natl. Acad. Sci. USA* **2020**, *117*, 23548–23556. [CrossRef] [PubMed]
39. Dolan, S.K.; Wijaya, A.; Kohlstedt, M.; Gläser, L.; Brear, P.; Silva-Rocha, R.; Wittmann, C.; Welch, M. Systems-wide dissection of organic acid assimilation in *Pseudomonas aeruginosa* Reveals a Novel Path to Underground Metabolism. *Mbio* **2022**, *13*, e0254122. [CrossRef]
40. Park, C.; Shin, B.; Park, W. Alternative fate of glyoxylate during acetate and hexadecane metabolism in *Acinetobacter oleivorans* DR1. *Sci. Rep.* **2019**, *9*, 14402. [CrossRef] [PubMed]
41. Durall, C.; Kukil, K.; Hawkes, J.A.; Albergati, A.; Lindblad, P.; Lindberg, P. Production of succinate by engineered strains of *Synechocystis* PCC 6803 overexpressing phosphoenolpyruvate carboxylase and a glyoxylate shunt. *Microb. Cell Factories* **2021**, *20*, 39. [CrossRef]
42. Zhang, H.; Liang, Z.; Zhao, M.; Ma, Y.; Luo, Z.; Li, S.; Xu, H. Metabolic engineering of *Escherichia coli* for ectoine production with a fermentation strategy of supplementing the amino donor. *Front. Bioeng. Biotechnol.* **2022**, *10*, 824859. [CrossRef] [PubMed]
43. Yun, M.K.; Nourse, A.; White, S.W.; Rock, C.O.; Heath, R.J. Crystal structure and allosteric regulation of the cytoplasmic *Escherichia coli* L-asparaginase I. *J. Mol. Biol.* **2007**, *369*, 794–811. [CrossRef]
44. Wang, Z.; Bie, P.; Cheng, J.; Lu, L.; Cui, B.; Wu, Q. The ABC transporter YejABEF is required for resistance to antimicrobial peptides and the virulence of *Brucella melitensis*. *Sci. Rep.* **2016**, *6*, 31876. [CrossRef] [PubMed]
45. Huang, L.; Huang, L.; Zhao, L.; Qin, Y.; Su, Y.; Yan, Q. The regulation of oxidative phosphorylation pathway on *Vibrio alginolyticus* adhesion under adversities. *Microbiologyopen* **2019**, *8*, e00805. [CrossRef]
46. Santucci, R.; Sinibaldi, F.; Cozza, P.; Polticelli, F.; Fiorucci, L. Cytochrome c: An extreme multifunctional protein with a key role in cell fate. *Int. J. Biol. Macromol.* **2019**, *136*, 1237–1246. [CrossRef]
47. Maza-Márquez, P.; Lee, M.D.; Detweiler, A.M.; Bebout, B.M. Millimeter-scale vertical partitioning of nitrogen cycling in hypersaline mats reveals prominence of genes encoding multi-heme and prismane proteins. *ISME J.* **2022**, *16*, 1119–1129. [CrossRef] [PubMed]
48. Tierney, A.R.; Rather, P.N. Roles of two-component regulatory systems in antibiotic resistance. *Future Microbiol.* **2019**, *14*, 533–552. [CrossRef]
49. Liu, F.; Yao, Q.; Huang, J.; Wan, J.; Xie, T.; Gao, X.; Sun, D.; Zhang, F.; Bei, W.; Lei, L. The two-component system CpxA/CpxR is critical for full virulence in *Actinobacillus pleuropneumoniae*. *Front. Microbiol.* **2022**, *13*, 1029426. [CrossRef]
50. Hu, M.; Huang, X.; Xu, X.; Zhang, Z.; He, S.; Zhu, J.; Liu, H.; Shi, X. Characterization of the role of two-component systems in antibiotic resistance formation in *Salmonella enterica* Serovar Enteritidis. *mSphere* **2022**, *7*, e0038322. [CrossRef]

51. Blair, J.M.A.; Siasat, P.; McNeil, H.E.; Colclough, A.; Ricci, V.; Lawler, A.J.; Abdalaal, H.; Buckner, M.M.C.; Baylay, A.; Busby, S.J.; et al. EnvR is a potent repressor of acrAB transcription in *Salmonella*. *J. Antimicrob. Chemother.* **2022**, *78*, 133–140. [CrossRef]
52. Oberlies, J.; Watzl, C.; Giese, T.; Luckner, C.; Kropf, P.; Müller, I. Regulation of NK cell function by human granulocyte arginase. *J. Immunol.* **2009**, *182*, 5259–5267. [CrossRef]
53. Kim, S.H.; Sierra, R.A.; McGee, D.J.; Zabaleta, J. Transcriptional profiling of gastric epithelial cells infected with wild type or arginase-deficient *Helicobacter pylori*. *BMC Microbiol.* **2012**, *12*, 175. [CrossRef]
54. Yurgel, S.N.; Johnson, S.A.; Rice, J.; Sa, N.; Bailes, C.; Baumgartner, J.; Pitzer, J.E.; Roop, R.M., II; Roje, S. A novel formamidase is required for riboflavin biosynthesis in invasive bacteria. *J. Biol. Chem.* **2022**, *298*, 102377. [CrossRef]
55. Lai, R.Z.; Parkinson, J.S. Monitoring two-component sensor kinases with a chemotaxis signal readout. *Methods Mol. Biol.* **2018**, *1729*, 127–135. [CrossRef]
56. Tan, L.; Huang, Y.; Shang, W.; Yang, Y.; Peng, H.; Hu, Z.; Wang, Y.; Rao, Y.; Hu, Q.; Rao, X.; et al. Accessory gene regulator (*agr*) allelic variants in cognate *Staphylococcus aureus* strain display similar phenotypes. *Front. Microbiol.* **2022**, *13*, 700894. [CrossRef]
57. Párraga Solórzano, P.K.; Shupe, A.C.; Kehl-Fie, T.E. The sensor histidine kinase ArlS is necessary for *Staphylococcus aureus* to activate ArlR in response to nutrient availability. *J. Bacteriol.* **2021**, *203*, e0042221. [CrossRef]
58. Loiseau, L.; Vergnes, A.; Ezraty, B. Methionine oxidation under anaerobic conditions in *Escherichia coli*. *Mol. Microbiol.* **2022**, *118*, 387–402. [CrossRef]
59. Fang, J.; Yan, L.; Tan, M.; Li, G.; Liang, Y.; Li, K. Nitrogen removal characteristics of a marine denitrifying *Pseudomonas stutzeri* BBW831 and a simplified strategy for improving the denitrification performance under stressful conditions. *Mar. Biotechnol.* **2023**, *25*, 109–122. [CrossRef]
60. Alvarez, L.; Sanchez-Hevia, D.; Sánchez, M.; Berenguer, J. A new family of nitrate/nitrite transporters involved in denitrification. *Int. Microbiol.* **2019**, *22*, 19–28. [CrossRef]
61. Harborne, N.R.; Griffiths, L.; Busby, S.J.; Cole, J.A. Transcriptional control, translation and function of the products of the five open reading frames of the *Escherichia coli* nir operon. *Mol. Microbiol.* **1992**, *6*, 2805–2813. [CrossRef]
62. Léger, L.; Byrne, D.; Guiraud, P.; Germain, E.; Maisonneuve, E. NirD curtails the stringent response by inhibiting RelA activity in *Escherichia coli*. *Elife* **2021**, *10*, e64092. [CrossRef] [PubMed]
63. Kanagarajan, S.; Dhamodharan, P.; Mutharasappan, N.; Choubey, S.K.; Jayaprakash, P.; Biswal, J.; Jeyaraman, J. Structural insights on binding mechanism of CAD complexes (CPSase, ATCase and DHOase). *J. Biomol. Struct. Dyn.* **2021**, *39*, 3144–3157. [CrossRef] [PubMed]
64. Garavito, M.F.; Narváez-Ortiz, H.Y.; Zimmermann, B.H. Pyrimidine metabolism: Dynamic and versatile pathways in pathogens and cellular development. *J. Genet. Genom.* **2015**, *42*, 195–205. [CrossRef] [PubMed]
65. Xu, M.; Fu, H.; Chen, D.; Shao, Z.; Zhu, J.; Alali, W.Q.; Chen, L. Simple visualized detection method of virulence-associated genes of *Vibrio cholerae* by loop-mediated isothermal amplification. *Front. Microbiol.* **2019**, *10*, 2899. [CrossRef]
66. Wang, Y.; Chen, L.; Pandak, W.M.; Heuman, D.; Hylemon, P.B.; Ren, S. High Glucose Induces Lipid Accumulation via 25-Hydroxycholesterol DNA-CpG Methylation. *iScience* **2020**, *23*, 101102. [CrossRef]
67. Shan, X.; Fu, J.; Li, X.; Peng, X.; Chen, L. Comparative proteomics and secretomics revealed virulence, and coresistance-related factors in non O1/O139 *Vibrio cholerae* recovered from 16 species of consumable aquatic animals. *J. Proteom.* **2022**, *251*, 104408. [CrossRef]
68. Cui, J.; Hölzl, G.; Karmainski, T.; Tiso, T.; Kubicki, S.; Thies, S.; Blank, L.M.; Jaeger, K.E.; Dörmann, P. The glycine-glucolipid of *Alcanivorax borkumensis* is resident to the bacterial cell wall. *Appl. Environ. Microbiol.* **2022**, *88*, e0112622. [CrossRef]
69. Kuhry, J.G.; Duportail, G.; Bronner, C.; Laustriat, G. Plasma membrane fluidity measurements on whole living cells by fluorescence anisotropy of trimethylammoniumdiphenylhexatriene. *Biochim. Biophys. Acta.* **1985**, *845*, 60–67. [CrossRef]
70. Wang, Z.; Qin, Q.; Zheng, Y.; Li, F.; Zhao, Y.; Chen, G.Q. Engineering the permeability of *Halomonas bluephagenesis* enhanced its chassis properties. *Metab. Eng.* **2021**, *67*, 53–66. [CrossRef]
71. Huang, B.; Liu, X.; Li, Z.; Zheng, Y.; Wai Kwok Yeung, K.; Cui, Z.; Liang, Y.; Zhu, S.; Wu, S. Rapid bacteria capturing and killing by AgNPs/N-CD@ZnO hybrids strengthened photo-responsive xerogel for rapid healing of bacteria-infected wounds. *Chem. Eng. J.* **2021**, *414*, 128805. [CrossRef]
72. Yang, L.; Wang, Y.; Yu, P.; Ren, S.; Zhu, Z.; Jin, Y.; Yan, J.; Peng, X.; Chen, L. Prophage-related gene *VpaChn25_0724* contributes to cell membrane integrity and growth of *Vibrio parahaemolyticus* CHN25. *Front. Cell. Infect. Microbiol.* **2020**, *10*, 595709. [CrossRef]

MDPI AG
Grosspeteranlage 5
4052 Basel
Switzerland
Tel.: +41 61 683 77 34

Foods Editorial Office
E-mail: foods@mdpi.com
www.mdpi.com/journal/foods

www.ingramcontent.com/pod-product-compliance
Lightning Source LLC
Chambersburg PA
CBHW061551210926
51850CB00009B/2074
* 9 7 8 3 7 2 5 8 7 7 6 7 6 *